T0297683

Open Radio Access Network (O-RAN) Systems Architecture and Design

Open Radio Access Network (O-RAN) Systems Architecture and Design

Wim Rouwet

NXP Semiconductors, Austin, TX, United States

Academic Press is an imprint of Elsevier
125 London Wall, London EC2Y 5AS, United Kingdom
525 B Street, Suite 1650, San Diego, CA 92101, United States
50 Hampshire Street, 5th Floor, Cambridge, MA 02139, United States
The Boulevard, Langford Lane, Kidlington, Oxford OX5 1GB, United Kingdom

British Library Cataloguing-in-Publication Data
A catalogue record for this book is available from the British Library

Library of Congress Cataloging-in-Publication Data
A catalog record for this book is available from the Library of Congress

ISBN: 978-0-323-91923-4

For Information on all Academic Press publications
visit our website at https://www.elsevier.com/books-and-journals

Publisher: Mara Conner
Acquisitions Editor: Tim Pitts
Editorial Project Manager: John Leonard
Production Project Manager: Prasanna Kalyanaraman
Cover Designer: Christian Bilbow

Typeset by MPS Limited, Chennai, India

Contents

About the authors

Wim Rouwet is a distinguished member of technical staff at NXP Semiconductors. He has an MSc in electrical engineering from the Eindhoven University of Technology in the Netherlands. He had spent more than 15 years in Motorola, Freescale, and NXP in networking and network processing, wireless algorithm development, and system and modem architecture roles. His focus is on 3GPP LTE and 5G as well as 802.11 processing stacks and their implementation. In his job, he has been responsible for 4G and 5G stack development, small cells, and CRAN implementations associated with many wireless infrastructure projects. He has led key next-generation R&D projects including multistandard modem architecture, virtualization, 5G macro and small cell, and client-side products.

Contributing Authors

David Spencer is a product line manager for Timing Solutions at Skyworks Inc., formerly Silicon Labs I&A division. He has a degree in physics and an over 30-year career in technology spanning technical writing, hardware and software design, applications engineering, and product marketing. For the past 15 years, David has focused on synchronization solutions and is considered an expert in both physical layer sync technologies and IEEE 1588 packet-based timing, having presented papers at multiple international conferences such as WSTS and ITSF and represented various companies in the ITU-T standards body. In his current role, David is responsible for driving the definition and implementation of hardware and software enabling Synchronous Ethernet and IEEE 1588 sync in a wide range of systems including 5G RAN networks.

Vishwapathi Rao Tadinada works as a director of test at NXP Semiconductors where he has to ensure the quality of networking applications and customer solution product lines. He has over 20 years of extensive experience leading product testing in 4G, 5G, Security, Networking, Cloud, SDN, NFV, and Embedded Industry. He has several presentations at international conferences and papers on IoT, security, and virtualization. He has BSc from Kanpur University, India.

Preface

Across the IoT, Autonomous Driving (ADAS), edge compute, and service provider ecosystem, 5G is understood to not "just" be about enabling a faster data connection to the end-user device. 5G evolves both existing technologies (radio access and home networks) while laying the foundation for new applications across consumer and industrial automation with the self-driving car being a prime example.

Even though 5G is very much hyped since 2016, infrastructure rollout has only started in 2019, with subscriber devices having become available in volume starting 2020. At the same time, 4G/LTE standards will continue to evolve and add more subscribers.

Million Mobile Connections (4G, 5G)

3000
2500
2000
1500
1000
500
0
2016 2018 2020 2022 2024 2026

4G and 5G (projected) subscriber growth (multiple public sources, indicated with data points).

During 2G, 3G, and 4G network rollouts, wireless service providers and their system vendors built networks that, although providing predictable and high performance, are often characterized by their complexity, inflexibility, and associated high barrier to entrance.

Like the wireline network transformation when migrating from circuit-switched to packet-switched (PS) networks in the early 2000s, the wireless network is migrating from this proprietary nature to a true open-standards approach that allows "out-of-the-box" thinking and disruptive solutions in physical implementation and deployment. Consider, for example, the cost and execution requirements imposed by new operators like Reliance Jio in India, DISH in the United States, and Rakuten in Japan, needing to support fast ramp-up (quick provisioning), or the Facebook Terragraph project, bringing high-speed connectivity at a fraction of fiber deployment cost through the use of an unlicensed 60 GHz spectrum.

The rapid evolution of 5G standards combined with aggressive deployment schedules and cost targets drives a push for Radio Access Network (RAN) implementations on standard hardware and software platforms. This trend started with 4G networks where upper layer stacks are implemented on general purpose (GPP) compute platforms and the Linux operating system (OS) and is increasing with 5G deployments. Examples include the trend to Ethernet for fronthaul replacing quasistandards such as CPRI as well as rigorous standardization efforts in 3GPP to define standard APIs between different components.

Because of this trend, system vendors are moving increasingly aggressively toward Software Defined Radio (SDR) platforms where most, if not all, of the RAN MAC and PHY layers are implemented in a true software environment on the Linux OS. The expectation is that although the implementation efficiency may not match that of custom hardware and firmware, this is offset by cost and flexibility provided by SDR platforms through the use of GPP multicore devices and broad software enablement. On-chip integration of high-speed (25–100 Gbps) Ethernet hardware reduces the need for custom I/O solutions and allows for reuse of data center switch/routing solutions for fronthaul.

The trend for SDR even extends to the remote radio head/radio unit (RU). Very large compute requirements make it impossible for programmable cores (DSPs) to implement digital front end (DFE) functions as opposed to ASIC implementation. Optimized vector DSPs allow for a software-friendly and flexible deployment, while new chip-level partitioning, with on-chip data converter integration, modern memory interfaces, multi-chip integration, and other innovative techniques, enable the cost reduction required for massive MIMO deployment.

AT&T, China Mobile, Deutsche Telekom, NTT DOCOMO, and Orange jointly announced the creation of the Open-RAN (O-RAN) Alliance in February 2018. Press releases say the O-RAN Alliance is a worldwide, carrier-led effort to drive new levels of openness in the RAN of next-generation wireless systems. But what does this mean? It means that three wireless networking industry trends are coming together into a "perfect storm" moment. First, networking products in the wireline space (switches, routers, and similar equipment) have moved from proprietary, hardware-centric implementations to a software-centric implementation that executes on general-purpose, Arm or x86 hardware. Second, the wireless infrastructure vendor landscape has consolidated dramatically. Consider Nokia, which has been consolidating Alcatel, Lucent, Nokia, Nortel, Siemens, and more companies. Wireless operators, who are the customers of these infrastructure vendors, see a potential for an oligopoly that implies reduced innovation and increased pricing and vendor dependency.

Lastly, there is a geopolitical aspect that worsens the competitive situation. There has been broad coverage on Huawei being banned as an infrastructure vendor, leading Huawei customers to reach out to find alternative paths to acquire equipment.

Given the promise of O-RAN to open up a whole new ecosystem of hardware and software vendors, there is obvious excitement in the industry around the concept of a software-centric ecosystem of vendors that can be mix-and-matched to implement a wireless network, especially if this allows the wireless network to be composed out of systems that are optimized to the task at hand—say rural, dense urban, industrial, or private networks.

At the same time, there is a lot of catching up to do. The institutionalized knowledge from the big wireless systems vendors takes a long time to catch up to, in systems, hardware, and software design. This book is a jump-start to engineers who are developing O-RAN hardware and software systems and provides a top-down approach to O-RAN systems design. We cover wireless and systems history as an introduction into why wireless systems look the way they do today, before introducing relevant O-RAN and 3GPP standards. The remainder of the book discusses hardware and software aspects of O-RAN system design, including dimensioning and performance targets. We include real-life examples of relevant elements of detailed hardware and software design where needed as a guide for developers. Lastly, we show a few practical examples of where O-RAN designs play in the market and how those examples map to hardware and software architectures.

The target audience includes hardware and software engineers as well as product managers and consumers of O-RAN products who are looking to understand better what O-RAN is.

This book is organized as follows:

Chapter 1, *Open Radio Access Network Overview*, gives an overview of the O-RAN standard and its origins and covers related topics like 5G and spectral availability for 5G deployment. This provides a background for a discussion on deployment options, which defines the hardware and software architecture.

Chapter 2, *System Components, Requirements, and Interfaces*, outlines the requirements and architecture of key system components such as the central unit (CU), distributed unit (DU) and RU.

Chapter 3, *Hardware System Dimensioning*, establishes key performance metrics such as front/backhaul throughputs, memory requirements, interconnect performance, as well as latency and other metrics.

Chapter 4, *Hardware Architecture Choices*, talks about the different hardware implementation options, from server-based designs to more embedded implementation options for CU/DU as well as options for RU system design. We discuss the pros and cons of each architecture approach.

Chapter 5, *System Software*, is like the hardware architecture choice but more software centric in nature. We cover OS components such as Linux and DPDK/bbdev as well as required software drivers.

Chapter 6, *User-Plane Application Components*, covers relevant details of the Physical and Layer 2 application stacks. We outline functions implemented by these stacks. This includes 3GPP defined functions such as PDCP, RLC/MAC, and PHY and implementation-specific parts such as Air Interface Scheduler, O-RAN eCPRI fronthaul, and DFE.

Chapter 7, *Wireless Scheduling and Quality of Service Optimization Techniques*, delves deeper into the most complex and differentiating piece of the User-Plane stack: the wireless scheduler algorithm. We explain commonly used algorithms for time and frequency domain scheduling and outline the potential structure of a scheduling algorithmic framework.

Chapter 8, *Synchronization in Open Radio Access Networks*, covers requirements as well as implementation on CU/DU and RU side for time and frequency synchronization including relevant standards and implementation methods.

Chapter 9, *Software Performance,* discusses key performance metrics of the O-RAN systems, such as L1/L2 algorithmic performance as well as user performance in terms such as latency, throughput, and capacity.

Chapter 10, *Interoperability and Test,* establishes standards software techniques for integration and test before covering 3GPP/O-RAN specific system integration and test aspects.

Chapter 11, *Differentiation by Use Case,* shows a few "off the normal" O-RAN use-cases/implementations to give the reader insight into what kinds of products can be established with an O-RAN systems architecture. These systems are discussed as practical implementations that showcase O-RAN strengths of ecosystem-based development and software centricity.

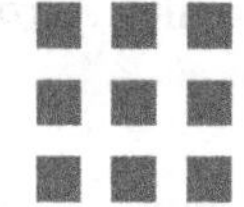

Acronyms

3GPP	Third-Generation Partnership Project
5GC	5G Core
5QI	5G QoS Identifier
ACIA	Alliance for Connected Industries and Automation
ACL	Access Control Lists
ACLR	Adjacent Channel Leakage Ratio
ACPR	Adjacent Channel Power Ratio
ADAS	Advanced Driver-Assistance Systems
ADC	Analog to Digital Converter
AGC	Automatic Gain Control
AM	Acknowledged Mode
AMF	Access and Mobility Management Function
APD	Analog Pre Distortion
API	Application Programming Interface
ARP	Allocation and Retention Priority
ARPU	Average Revenue Per Unit
ARQ	Automatic Repeat Request
ASIC	Application Specific Integrated Circuit
ASN1	Abstract Syntax Notation 1
ATM	Asynchronous Transfer Mode
AWGN	Additive White Gaussian Noise
AxC	Antenna Container
BCCH	Broadcast Control Channel
BCH	Broadcast Channel
BE	Best Effort
BER	Bit Error Rate
BFWA	Broadband Fixed Wireless Access
BLER	Block Error Rate
BPSK	Binary Phase-Shift Keying
BD	Buffer Descriptor
BE	Best Effort
BS	Base Station
BWP	Bandwidth Partitioning
CB	Code Block
CCH	Common Control Channel
CF	Crest Factor
CFR	Crest Factor Reduction
C-ITS	Cooperative Intelligent Transport Systems
CLI	Command Line Interface
CoMP	Cooperative Multipoint
COTS	Commercial Off-The-Shelf
CP	Control Plane
CPI	Certified Professional Installer

CPU	Central Processing Unit
CPRI	Common Public Radio Interface
CRC	Cyclic Redundancy Check
CSR	Cell Site Router
CU	Central Unit
D2D	Device to Device
DAC	Digital to Analog Converter
DAS	Distributed Antenna System
dB	Decibel
DC	Dual Connectivity
DCCH	Dedicated Control Channel
DCI	Downlink Control Information
DDC	Digital Down Conversion
DDR	Double Data Rate
DFE	Digital Front End
DFT	Discrete Fourier Transform
DL	Downlink
DLSCH	Downlink Shared Channel
DMRS	Demodulation Reference Signal
DPD	Digital Pre Distortion
DPDK	Data Plane Development Kit
DRB	Data Radio Bearer
DRX	Discontinuous Reception
DSP	Digital Signal Processing
DSS	Dynamic Spectrum Sharing
DTCH	Dedicated Transport Channel
DTX	Discontinuous Transmission
DU	Distributed Unit
DUC	Digital Up Conversion
DUT	Device Under Test
DVFS	Dynamic Voltage and Frequency Scaling
EBI	EPS Bearer ID
eCPRI	Enhanced Common Public Radio Interface
EIRP	Effective Isotropic Radiated Power
EMC	Electromagnetic Compatibility
eNB	Evolved NodeB or 4G Base Station
EPC	Enhanced Packet Core
eSIM	Electronic Subscriber Identity Module
EMBB	Enhanced Mobile Broadband
ESMC	Ethernet Synchronization Messaging Channel
FAPI	Femto API (Application Programming Interface)
FCC	Federal Communications Commission
FEM	Front End Module
FDD	Frequency Division Duplexing
FFT	Fast Fourier Transform
FGW	Femto Gateway
FHGW	Fronthaul Gateway
FIFO	First In–First Out
FPGA	Field Programmable Gate Array

FR	Frequency Range
FWA	Fixed Wireless Access
GBR	Guaranteed Bit Rate
GFBR	Guaranteed Flow Bit Rate
GNSS	Global Navigation by Satellite Systems
GPP	General-Purpose Processor
GPRS	General Packet Radio Services
GPS	Global Positioning System
GSM	Global System for Mobile communication
GTP	GPRS Tunneling Protocol
GTP-C	GPRS Tunneling Protocol – Control
GTP-U	GPRS Tunneling Protocol – User
GTPS	Giga Transactions Per Second
GW	Gateway
gNB	gNodeB or Next Generation NodeB or 5G Base Station
HAAT	Height Above Average Terrain
HARQ	Hybrid Automatic Repeat Request
HBM	High Bandwidth Memory
HLS	Higher Level Split
HNB	Home NodeB
HPF	High Priority Fronthaul
IBW	Instantaneous BandWidth
IDFT	Inverse Discrete Fourier Transform
IPC	Inter Process Communication
IOT	Internet of Things
IEEE	Institute of Electrical and Electronics Engineers
IETF	Internet Engineering Task Force
IFFT	Inverse Fast Fourier Transform
IP	Internet Protocol or Ingress Protection
IPSec	Internet Protocol Security
ISA	Instruction Set Architecture
ISC	Integrated Small Cell
ITU	International Telecommunications Union
LBRM	Limited Buffer Rate Matchin
LLR	Log Likelihood Radio
LLS	Lower Level Split
LPF	Low Priority Fronthaul
LTE	Long Term Evolution
MAC	Medium Access Control
MEC	Metro Edge Compute
MeNB	Master eNB
MFBR	Maximum Flow Bit Rate
MIB	Master Information Block
MIMO	Multiple Input Multiple Output
MMU	Memory Management Unit
MP	Management Plane
MPF	Medium Priority Fronthaul
MPLS	Multi Protocol Label Switching
MPS	Maximum Payload Size

NAS	Non Access Stratum
NEF	Network Exposure Function
NF	Noise Figure
NFV	Network Function Virtualization
NFVI	Network Function Virtualization Infrastructure
NIC	Network Interface Card
NPF	Network Processor Forum
NPU	Network Processor Unit
NR	New Radio
NRF	NF Repository Function
NSÀ	Non Standalone
NSSF	Network Slice Selection Function
NTP	Network Time Protocol
OAM	Operation and Maintenance
OBSAI	Open Base Station Architecture Initiative
OBW	Occupied Bandwidth
OFDM	Orthogonal Frequency Division Multiplexing
OFDMA	Orthogonal Frequency Division Multiple Access
ONF	Open Networking Foundation
OSS	Operations Support Systems
PA	Power Amplifier
PAPR	Peak to Average Power Ratio
PCCH	Paging Control Channel
PCF	Policy Control Function
PDCCH	Physical Downlink Control Channel
PDCP	Packet Data Convergence Protocol
PDSCH	Physical Downlink Shared Channel
PH	Power Headroom
PI	Preemption Indicator
PLL	Phase Locked Loop
PoE	Power over Ethernet
PRACH	Physical Random Access Channel
PRB	Physical Resource Block
PSS	Primary Synchronization Signal
PTP	Precision Time Protocol
PUCCH	Physical Uplink Control Channel
PUSCH	Physical Uplink Shared Channel
QAM	Quadrature Amplitude Modulation
QFI	QoS Flow Id
QoS	Quality of Service
QPSK	Quadrature Phase Shift Keying
RACH	Random Access Channel
RAN	Radio Access Network
RANAP	Radio Access Network Application Part
RAT	Radio Access Technology
RB	Resource Block
RE	Resource Element
RED	Random Early Discard
RIBS	Radio Interface Based Synchronization

RIC	RAN Intelligent Controller
RF	Radio Frequency
RFC	Request For Comments
RLC	Radio Link Control
RMS	Root Mean Square
RNTI	Radio Network Temporary Identifier
RoHC	Robust Header Compression
RQA	Reflective QoS Attribute
RRC	Radio Resource Control
ROM	Read Only Memory
RRM	Radio Resource Management
RSU	Roadside Unit
RT	Realtime
RTOS	Realtime Operating System
rtPS	real-time Polling Service
RTT	Round Trip Time
RU	Radio Unit
SA	Standalone
SAS	Spectrum Allocation System
SCS	Subcarrier Spacing
SDAP	Service Data Adaptation Protocol
SDN	Software Defined Networking
SDoC	Supplier Declaration of Conformity
SDU	Service Data Unit
SeNB	Slave eNB
SFN	Single Frequency Network
SIB	Secondary Information Block
SIM	Subscribed Identity Module
SIMD	Single Instruction Multiple Data
SLA	Service Level Agreement
SMF	Session Management Function
SN	Sequence Number
SNR	Signal to Noise Ratio
SPS	Semi Persistent Scheduling
SRB	Signaling Radio Bearer
SRS	Sounding Reference Signal
SSB	Synchronization Signal Block
SSE	Streaming Signaling Extensions
SSS	Secondary Synchronization Signal
SUL	Supplementary Uplink
SW	Software
T-BC	Telecom Boundary Clock
T-GM	Telecom Grand Master
T-TSC	Telecom Time Slave Clock
TAI	International Atomic Time
TB	Transport Block
TCM	Three Color Marker
TCP	Transmission Control Protocol
TCXO	Temperature Controlled Oscillator

TIP Telecom Infrastructure Project
TDD Time Division Duplex
TDM Time Division Multiplexing
TEID Tunnel Endpoint Identifier
TM Transparent Mode
ToR Top of Rack
TPC Transmit Power Control
TRP Transmit Receive Point
TSN Time Sensitive Networking
TTI Transmit Time Interval
UAV Unmanned Arial Vehicle
UCI Uplink Control Information
UE User Equipment
UL Uplink
ULSCH Uplink Shared Channel
UM Unacknowledged Mode
UMA Unlicensed Mobile Access
UMTS Universal Mobile Telecommunication System
UP User Plane
UPF User Plane Function
URLLC Ultra Reliable Low Latency Communication
UTC Coordinated Universal Time
V2I Vehicle to Infrastructure
V2P Vehicle to Pedestrian
V2X Vehicle to Anything
VCXO Voltage Controlled Oscillator
WCDMA Wideband Code Division Multiple Access
WG Working Group
WiFi Wireless Fidelity
WiMAX Wireless Interoperability for Microwave Access
WRED Weighted Random Early Discard

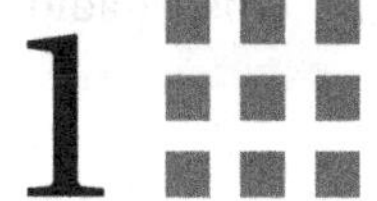

Open radio access network overview

This chapter provides an introduction to the Open Radio Access Networks (O-RAN) (with a dash!) Alliance and its goals and main deliveries. We then discuss related standards bodies as well as 5G spectral aspects and a short history of 3GPP standards and their implementation in terms of systems architecture. Note that although O-RAN standards cover both Long-Term Evolution (LTE) and 5G/NR networks, we are using 5G terminology throughout for simplicity and ease of reading.

The Open Radio Access Networks Alliance

Founding members of the O-RAN Alliance are AT&T, China Mobile, Deutsche Telekom AG, NTT DOCOMO Inc. and Orange but by now membership has increased after including the "Who's Who" in the wireless industry. The O-RAN public website can be found at Ref[1]. O-RAN is defined from the onset to be *operator-driven* rather than system vendor-driven, to ensure that the O-RAN goals (see next) are not interfered with.

The stated target of O-RAN is to break the closed nature of current radio access network (RAN) implementations. O-RAN explicitly aims specifically at 3GPP networks, as opposed to 802.11 (Wi-Fi) and other wireless standards. By opening the 3GPP implementation, O-RAN means to decouple hardware and software implementations allowing vendors (hardware, software, and systems) to focus on providing components rather than complete solution. Aim is to follow what happened in wireline software-defined networking (SDN) (think of switches, routers, and firewalls), which have moved from proprietary, hardware-centric implementations to a software-centric implementation that executes on general-purpose, Arm, RISC-V, or x86 hardware.

O-RAN is organized into working groups that own specific hardware, software, and system components. These working groups are shown in Fig. 1–1.

Note that the contributions of these working groups are "live" and continuously updated. Descriptions and ownership as we are describing below are therefore per definition in flux. Check out the O-RAN member website (where parts of these descriptions are taken from) for the latest updates.

Working Group 1: Use Cases and Overall Architecture Workgroup

WG1 focuses on use cases and system-level requirements as well as organizing proof of concepts to showcase O-RAN products to the wider market. This working group is operator-led (AT&T, CMCC).

Working Group 2: Non-Real-Time RIC and A1 Interface Workgroup

Open Radio Access Network (O-RAN) Systems Architecture and Design. DOI: https://doi.org/10.1016/B978-0-323-91923-4.00013-6

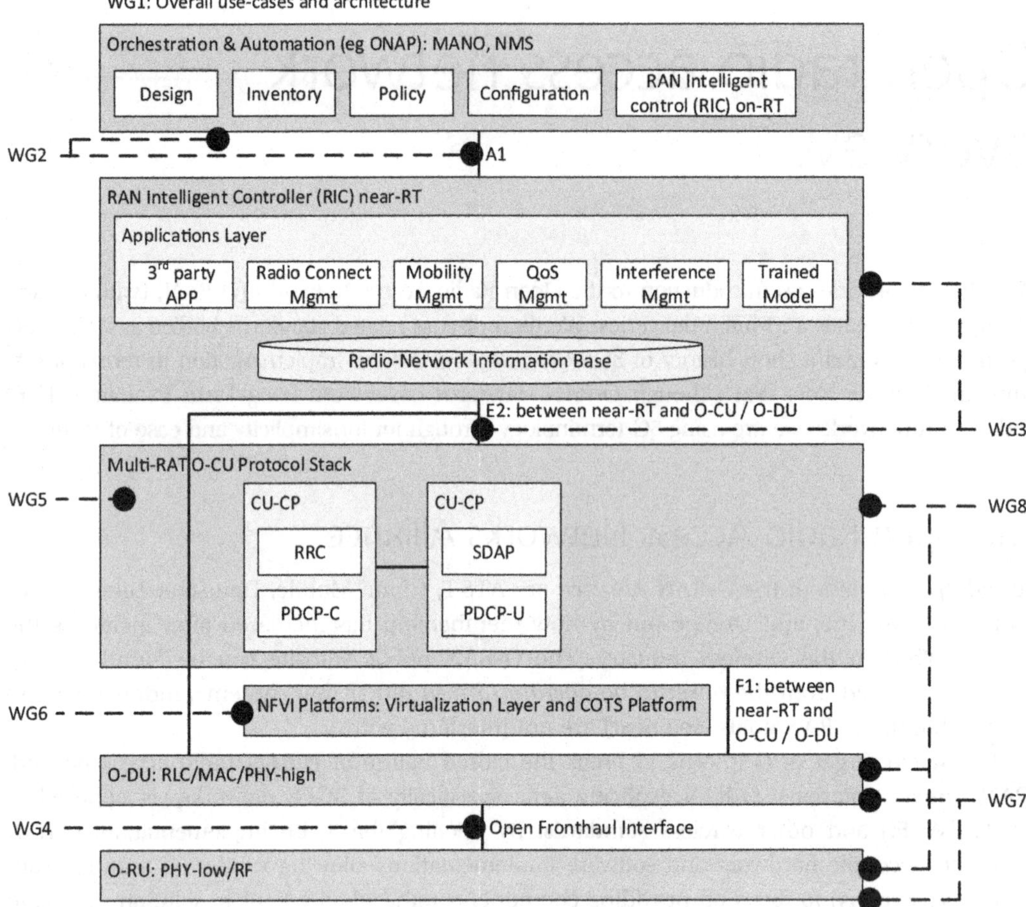

FIGURE 1–1 O-RAN Working Groups. *O-RAN*, Open Radio Access Networks.

WG2 owns the definition of the non-Real-Time (RT) RAN Intelligent Controller (RIC) and the A1 interface. The non-RT RIC controls radio resource management (RRM), higher layer procedure optimization, and RAN policy optimization, including Artificial Intelligence (AI)/ Machine Learning model application. Communication between the near-RT RIC and the non-RT RIC is defined by the A1 interface—functionally, this interface carries policy-based guidance of near-RT RIC functions/use cases and appropriate feedback/input data in the return path.

At its lowest level, non-RT RIC converts system goals (RAN intent) and observed parameters and counters to policies that guide the RT RIC toward fulfilling the system goals. Deliverables include:

- A1 interface specification

Working Group 3: Near-Real-time RIC and E2 Interface Workgroup

WG3 owns the near-RT RIC architecture and functionalities. Deliverables include:

- E2 interface specification. Note that the E2 interface is 3GPP-defined. WG3 provides a framework with a defined subset of 3GPP messages.

Working Group 4: Open Fronthaul Interfaces Workgroup

WG4 owns the definition of fronthaul interfaces. Deliverables include:

- Management plane (MP) specification. A NETCONF/YANG based M-Plane is used for supporting the management features including "start-up" installation, software management, configuration management, performance management, fault management, and file management toward the O-RU.
- Control, user and synchronization plane (CP, UP, SP) specification. This specifies the control plane, user plane, and synchronization plane protocols used to link the distributed unit (DU) with radio unit (RU) assuming a functional split 7, covering both LTE and 5G/NR. Note that the control plane refers specifically to real-time control between O-DU and O-RU and not the 3GPP Control Plane.
- Fronthaul interoperability test specification. This specifies test scenarios to be implemented to confirm interoperability between DU and RU implementations of different vendors. This specification includes several standardized Interoperability and Test (IoT) profiles. The specification covers M-Plane and CU-Planes, including topics such as beamforming, compression, and latency.

Working Group 5: Open F1/W1/E1/X2/Xn Interface Workgroup

WG5 owns the definition of mid/backhaul interfaces, like how WG4 owns the fronthaul interface. Deliverables include:

- O1 interface specification. The O1 interface links the DU with central unit (CU) and Service Management and Orchestration. This interface defines initialization, configuration, and management of the DU, including "start-up" installation, software management, configuration management, performance management, fault management, and file management toward the O-RU. Like the WG4 MP specification for RU, a YANG module is used for DU definition.
- Mid/backhaul IoT specification.

Working Group 6: Cloudification and Orchestration Workgroup

WG6 addresses Cloudification and Orchestration in O-RAN. WG6 identifies use cases that demonstrate the benefits from hardware/software decoupling (cloudification), including RIC, DU, CU, and RU. It also defines deployment scenarios, requirements, and reference designs for the cloud platform, including the Network Function Virtualization Infrastructure (infrastructure), Virtualized Infrastructure Manager (VIM) for container/VM orchestration, and Accelerator Abstraction Layers (AAL).

Accelerator Abstraction Layers allow the definition of hardware accelerated components (e.g., fronthaul, timing, GPRS Tunneling Protocol (GTP)/transport, high-PHY) with a common software interface to the (virtualized) host software.

Examples of other deliverables from this working group cloud deployment scenarios include use-case descriptions and specifications for key requirements, from hardware setups to Linux Kernel PREEMPT_RT latency metrics and container management requirements.

Working Group 7: White Box Hardware Workgroup

WG7 focuses on the architecture of DU and RU reference hardware design development. Contributing companies are mainly device vendors, such as ADI, Intel, NXP, Qualcomm, Xilinx, and others. Various contributions show how DU and RU equipment can physically be put together using (mainly) these companies' components. The aim is to reduce the effort of hardware architecture and design by standardizing pieces of the solution, thus enabling quicker time-to-market. This working group is operator-led (AT&T, CMCC).

Working Group 8: Stack Reference Design Workgroup

Just like WG7 architects reference hardware, WG8 focuses on architecture of reference software.

Working Group 9: Open X-haul Transport Workgroup

WG9 defines an architecture for fronthaul (DU↔RU), mid-haul (CU↔DU), and backhaul (CU↔core) transport networks. It defines the following:

- Transport requirements documents.
- Packet transport network architecture, including timing/synchronization architecture, QoS/queueing models, and target protocol stacks. These documents also include nonideal network architectures, such as DOCSIS (cable modem) and microwave backhaul. Packet transport is defined for both Multi-Protocol Label Switching and Segment Routing over IPv6 dataplane (SRv6) network architectures.

Open Radio Access Networks members

As we said earlier, O-RAN is an operator-driven standards body to make sure that operator goals such as openness and software centricity are met. This does not mean that membership is limited to operators only. Members include white box hardware, software, and systems vendors as well as other companies in the wireless supply chain such as test equipment and component/silicon vendors and research institutes.

The breakdown in Fig. 1–2 shows that the whole O-RAN supply chain is represented in an ever-growing list of members.

Why now?

None of the concepts that are discussed in O-RAN are groundbreaking in nature. This raises the question on why now "O-RAN" is such a buzzword. There are a couple of reasons to point out:

- Historical evidence. Much of what O-RAN is trying to achieve has already been shown effective in the wireline networking world. SDN defines a set of technologies (OpenFlow

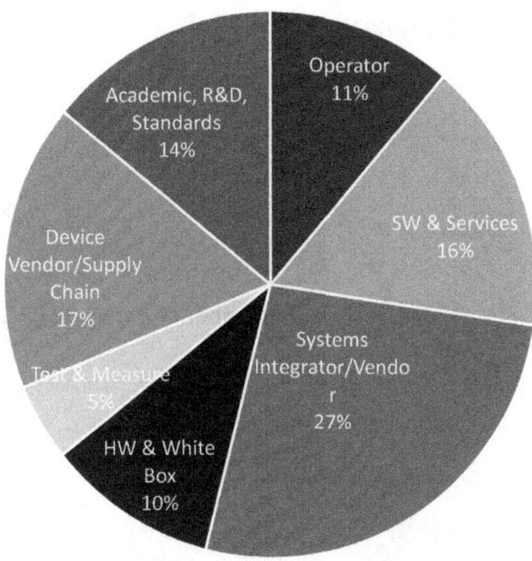

FIGURE 1–2 O-RAN membership breakdown (start of 2021). *O-RAN*, Open Radio Access Networks.

and others) that separate the wireline networking control plane from the user plane, allowing the user plane to be commoditized into white box hardware and the control plane to become fully software-defined. Early work on SDN was done by both the University of Stanford and the University of Berkeley, but by now, SDN is supported by all major networking vendors (Cisco, Juniper, and others) as well as a plethora of open-source options. An IDC estimate says that the SDN market will be worth more than $12 billion in 2022, with a CAGR of 18.5% during the 2017–22 period. This precedent is hard to ignore—why would the same not happen to wireless networking?

- Moore's law. As we will be showing next, the concepts on software-defined radio (SDR) were explored since the 2000s, but O-RAN is only getting serious consideration in the ~2020s. Why is this? One aspect to consider is the advancement in silicon technology that has made a move from proprietary application-specific integrated circuit (ASIC)-based implementations to software-defined implementations possible.
- Standards flexibility. The traditional macro base station architecture has been the mainstay for 1G, 2G, 3G, and 4G networks and will be the cornerstone for initial (macro) 5G network deployments. However, 5G is different from previous standards, in that the standard is design ground-up to support flexibility. Consider support for features such as flexible numerology to support operation in <1 GHz bands up to mmWave bands, bandwidth parts/partitioning (BWP), ultrareliable low-latency communication (URLLC), sidelink communication, and many more. With flexibility in standards comes flexibility in deployment options and this is contrary to an implementation strategy that has been honed over decades to support a single (macro) base station deployment. Flexibility is enabled by software and drives

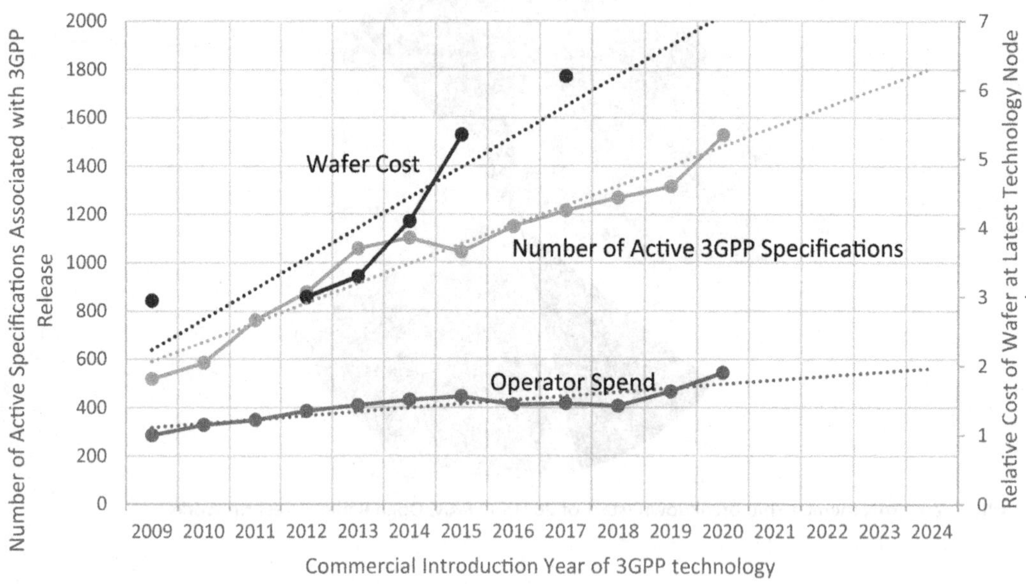

FIGURE 1–3 3GPP standards complexity versus operator equipment spent and component cost.

software-centric implementations from a wide ecosystem of nontraditional vendors. This is exactly what O-RAN enables.

- Vendor lock-in. As we are showing later in this chapter, the telecommunication industry went through a range of mergers and acquisitions the beginning of the century. It is this move toward a vendor oligopoly, combined with geopolitical considerations that is driving operators to invest in alternative solutions that are supported by a wider set of vendors and reduced lock-in by any single supplier.

Perhaps the best way to show what we are talking about is by looking at some numbers. See Fig. 1–3, a graph that combines:

- the (relative) cost of a silicon wafer as a proxy for chip cost,
- the number of 3GPP active specifications as a proxy for standards implementation complexity and flexibility, and
- the operator financial spent on wireless infrastructure over time.

The chart shows how the first two (device cost and requirement on flexibility) grow aggressively over time, while the amount of available money for systems development only increases marginally. Clearly, something has to change, and O-RAN is one of the potential solutions to this challenge.

On C-RAN, Open vRAN, OpenRAN, xRAN, and Telecommunications Infrastructure Project

By now, O-RAN is established as the de facto standard for open, disaggregated, and software-defined 3GPP network implementations. But the O-RAN standards body is not the only (or first) to aim for this goal.

A lot of the initial R&D and concept development on O-RAN topics has been done by China Mobile Research Institute (CMRI) under the *C-RAN* name, where the C stands for Cloud (or, sometimes, Centralized). An initial MOU on C-RAN collaboration was signed in April 2010 between the China Mobile Research Institute (CMRI) and ZTE, IBM, Huawei, and Intel. Six more companies joined in December 2010: Orange, Chuanhua Telecom, Alcatel-Lucent, Datang Mobile, Ericsson, and Nokia Siemens Networks. CMRI hosted the first C-RAN international workshop in 2010.

C-RAN targets cloud computing on general-purpose servers in a centralized location for (at the time) 2G, 3G, and 4G communication standards. According to CMRI, colocation of radio equipment in a centralized location allows for resource sharing, leading to significant cost reduction (OPEX: − 50%, CAPEX: − 15%). Resource sharing is achieved across standards (use the same hardware to support 2G, 3G, 4G, and now 5G) as well as across radio sites and time (residential district busy-hours are in the weekends, business districts busy-hours are in weekdays).

CMRI projected cost savings through lower maintenance and (feature) development through unified development environment and use of off-the-shelf hardware and software components. In addition, capacity increase can be achieved through advanced algorithm support, such as PHY layer interference cancellation through network multiple-input, multiple-output (MIMO) and MAC layer interference avoidance and joint scheduling (cooperativeRRM, CoMP).

The C-RAN concept imposed many early engineering challenges:

- Bandwidth and timing requirements for antenna sample transport are high/strict.
- The concept of a C-RAN virtualized environment assumes a virtualized SDR environment which leads to high-performance requirements.
  - Choice of a single processing environment (such as a general-purpose processor) means a likely choice for suboptimal environments for portions of the processing chain: general-purpose processors are simply not as efficient at signal processing as an optimized digital signal processing (DSP).
  - Limited or no use of fixed-function acceleration.

These engineering challenges triggered the alternative interpretation of the "C" in C-RAN: Centralized. The Centralized RAN architecture does use the resource pooling concept and provides many of the C-RAN advantages (capacity sharing, advanced algorithms) without imposing the engineering challenge of implementing DSP algorithms on (inefficient) general-purpose processors. Keep in mind that O-RAN does not dictate the hardware implementation of the system either but rather defines the interfaces between hardware components, thus allowing for proprietary implementations of those components. Having said that, there is an implied assumption that O-RAN solutions indeed will be software-centric and using general-purpose processors.

The *xRAN Forum* was established in 2016, releasing its first specification in 2018 for fronthaul communication between (what is now called) the DU and the RU. xRAN membership included AT&T, Deutsche Telekom, KDDI, NTT Docomo, SK Telecom, Telstra, and Verizon which made it more operator-centric compared to C-RAN which was more vendor-centric. The group's focus has been on developing standards for interfaces between wireless network components targeting interoperability between those components—much like O-RAN does today. The xRAN Forum defined the use of NETCONF/YANG standards for configuring and managing RAN architecture as well as the first versions of what now are the O-RAN CUS plane standards. At Mobile World Congress (MWC) 2018, xRAN and the O-RAN Alliance announced their merger ensuring adoption of the xRAN standards into O-RAN.

Open vRAN, announced by Cisco in MWC 2018, was a virtualized RAN architecture using general-purpose compute for implementation of wireless standards, initially only supported by a single operator (Reliance Jio). The Open vRAN ecosystem has developed multiple successful demos and trials on 3GPP and O-RAN standards, trying to accelerate the adoption of these solutions and integration with remainder (wireline) network architecture.

The Telecommunications Infrastructure Project (TIP) was founded by Facebook in 2016, supported by Deutsche Telekom, Intel, Nokia, and SK Telecom. TIP now has more than 500 members. When Vodafone joined TIP in 2017, it contributed its concept of a software-defined RAN and started the *OpenRAN* project. Target of OpenRAN is to disaggregate hardware and software and resulting from this to run wireless stacks on general-purpose hardware. TIP is less focused on standards development and more on use-case definition and hardware/software implementations. TIP and the O-RAN Alliance announced a liaison agreement at MWC 2020 to ensure their alignment in developing interoperable O-RAN solutions. This ensures sharing of information, referencing specifications, and conducting joint testing and integration efforts.

Spectrum: enabling 5G

Ultimately, spectrum is the enabler for wireless deployment, and 5G deployments are no exception to this. We differentiate between:

- licensed spectrum that is owned by large operators deploying (typically) nationwide networks,
- licensed spectrum that is owned by private companies, and
- unlicensed spectrum that is available to any anybody adhering to spectral usage models.

Spectral allocations change over time, Fig. 1–4 showing key global 5G spectrum allocations around 2020:

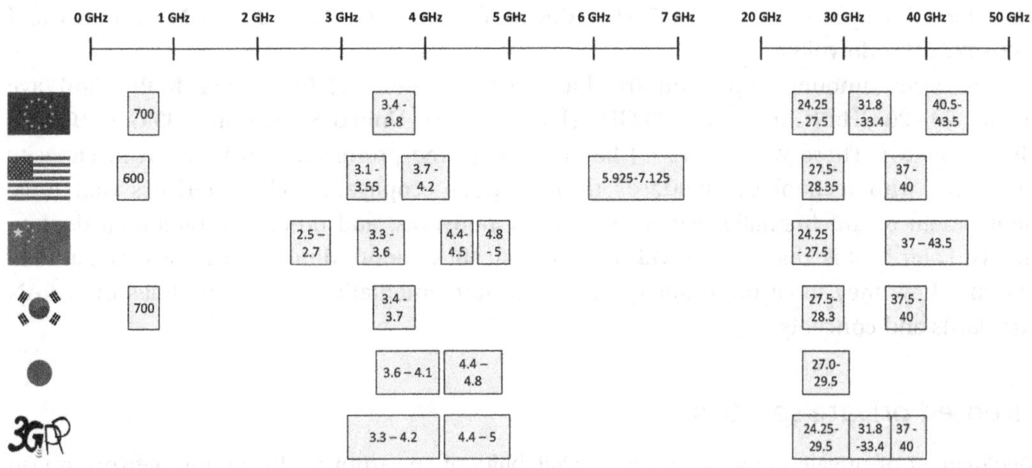

FIGURE 1–4 Global 5G spectrum allocations.

Note that, licensed or unlicensed, mmWave spectrum is unmistakably linked to 5G, given that wide spectrum is the only way to provide the 10× promised throughput that 5G promises over 4G/LTE deployments. *LTE*, Long-Term Evolution

Licensed operator spectrum

As we will discuss later, the 5G waveform is not significantly more efficient (~10%) compared to the 4G waveform. If the spectrum is not being made more broadly available by industry regulators (United States: FCC), the chances for successful deployment of 5G are limited.

The first and most obvious path to making more spectrum available is freeing up previously un- or under-utilized spectrum. Consider the 3.x GHz band that has traditionally been used for everything from fixed wireless access to satellite services. Different slices of this spectrum have been made available across the globe for licensing by wireless operators. This spectrum has relatively large bandwidth (typically up to 100 MHz) and reasonable propagation characteristics. This spectrum is considered the cornerstone of 5G deployment.

In addition, the existing 2G, 3G, and 4G spectrum is slowly being repurposed for 5G use. The 2G (GSM) bands may still have legacy use (IoT and other purposes) but 3G spectrum is underutilized and can be repurposed for 5G deployment. In addition, technologies such as Dynamic Spectrum Sharing enable 4G and 5G networks to share RF bandwidth so to gradually transition spectrum usage as consumer equipment upgrades take place.

Low-band spectrum dedicated to IoT use cases is also under consideration, reusing traditional television bands for 3GPP operation, given that television services are moving to digital/IP-based. Examples include 3GPP-approved frequency bands in 400 MHz frequency range, driving harmonization of chipsets/devices and networking equipment: Band 87,

410–415 MHz uplink and 420–425 MHz downlink, and band 88, 412–417 MHz uplink and 422–427 MHz downlink.

The largest amount of spectrum freed for licensed (operator) 3GPP use is in the mmWave bands, 24–26 GHz (China, EU); 28 GHz (Japan, Korea, United States); and 39 GHz (China, United States). These wide (GHz +) bands provide the promised step function in capacity but come with technology standards through poor propagation characteristics that force deployment of much smaller cell sizes, overlay networks, and other nontraditional deployments. Later in this chapter, we will discuss what these nontraditional form factors can look like and how they force the industry to rethink implementations along the lines of O-RAN standards and concepts.

Licensed private spectrum

Deployment of private networks requires availability of spectrum to deploy the network on. In fact, spectrum availability typically defines the deployment format. There is no harmonized spectrum for private networks across the globe. By implication, equipment choice, RF regulation, and cost will end up varying by region. Government harmonization efforts are under way but are not expected to solve this challenge in the short-term future. We look at three examples: United States, Germany, and Japan.

United States

In the United States, the FCC is opening 150 MHz of "CBRS" spectrum in the 3.5 GHz (3550–3700 MHz) band as shown in Fig. 1–5. This spectrum is intended for use by three tiers of users:

- Incumbent users can retain the rights to use the band for military and Wireless Internet Service Provider purposes. WISPs will continue to have incumbent access to the 3650–3700 MHz band under the terms that are currently in place, with no modifications needed to deployed equipment.
- Priority Access Licensed (PAL) users (yellow) are allocated in the 3550–3650 MHz spectrum. Spectrum auction is done on a per-county basis (~ 3200 in the United States).

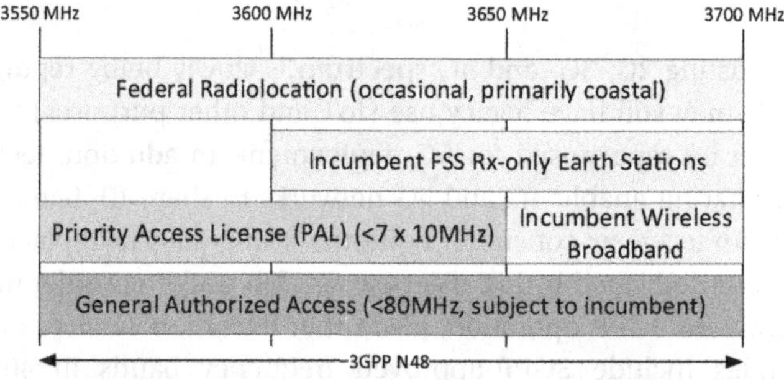

FIGURE 1–5 CBRS spectrum options.

In each market, there will be 7 × 10 MHz TDD PALs, with a maximum of four channels assigned per licensee. PAL users will mostly drive deployments of public networks.

- General Authorized Access (GAA) (green). All other users can register with a spectrum allocation system (SAS) when SAS determines that the spectrum is not in use by incumbent or PAL users with higher access priority. Channels throughout the overall band (3550–3700 MHz) are available for GAA: access to 70 MHz, which is shared with PAL users and subject to availability (PAL users have priority over GAA users), and access to the remaining 80 MHz is reserved to GAA access (but PAL users can also use this part of the band under the GAA provisions). GAA use requires spectrum sharing using coexistence techniques (see next). GAAs will mostly be used for private networks.

This new regulation creates a deployment framework for 4G and 5G in this band which was previously underutilized.

Spectrum allocation system and related components

The SAS (Fig. 1–6) is an automated frequency coordinator that coordinates all radio equipment that operate in the CBRS band. There are currently four SAS Administrators: CommScope, Federated Wireless, Google, and Sony. SAS Administrators coordinate spectrum, not only among all their managed transmitting devices but also among each other.

There are several parameters fed to the SAS, including CBSD-ID, location, height, antenna characteristics, and transmit power levels.

There are several types or categories of equipment allowed to operate in the CBRS band:

- A User Equipment (UE) device
  - operates <23 dBm Effective Isotropic Radiated Power (EIRP) (typically a handset)

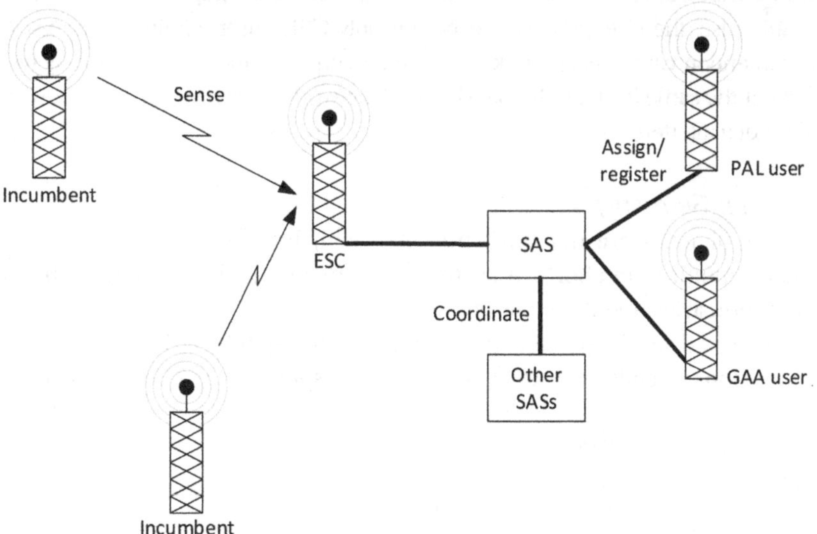

FIGURE 1–6 SAS components. *SAS*, spectrum allocation system.

  - not required to be registered to a SAS
- Category A CBSD (CBRS Device)
  - operates <30 dBm EIRP (per 10 MHz)
  - required to be registered to a SAS
- Category B CBSD (CBRS Device)
  - operates <47 dBm EIRP (per 10 MHz)
  - required to be registered to a SAS.

There is the CPI (Certified Professional Installer) requirement for all Category B CBSDs as well as Category A CBSDs that are (1) installed above 6-m Height Above Average Terrain and (2) unable to geolocate.

The ESC is a network of sensors that detects incumbent activity in the CBRS band along the coast and transmits this information to a SAS, which determines which part of the CBRS spectrum can be used by the two other tiers—PAL and GAA. PAL and GAA users are not allowed to use CBRS channels while they are in use by incumbent users.

Coexistence

The SAS is responsible for spectrum allocation among users. PAL users are prioritized with GAA users getting access to remaining spectrum. SAS authorization is required for both PAL and GAA users. GAA users only need to register but do not need a license. SAS operates in real-time, but in practice, allocation is assumed stable over time.

SAS assigns the spectrum for GAA, but it does not coordinate access among GAA users. Coexistence methods such as IEEE 802.19.1 need to be deployed to guarantee access among GAA users. Given technology neutrality in CBRS, these methods need to be working across standards.

An enterprise may not allow other CBRS users to install their equipment within the area it controls, and in this case, the enterprise would be the only CBRS user within its premises and could manage any contention within its network. However, enterprises and other users may want multiple CBRS networks at the same location. In that situation, network-coexistence methods are necessary to manage and reduce interference.

Europe (example: Germany)

European RF frequencies are not uniformly allocated, but defined on a per-country basis, with ongoing harmonization efforts. A current list is maintained, for example, by the Alliance for Connected Industries and Automation.

Germany allocated local licensed spectrum in 3700–3800 MHz band range to industries for their applications already in 2019 (Fig. 1–7). This spectrum is made available for private

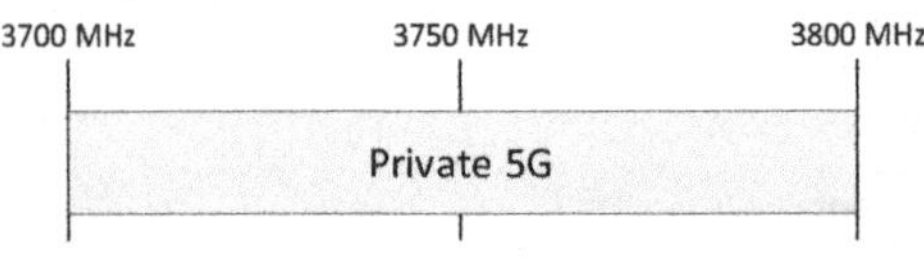

FIGURE 1–7 German private 5G bands.

network use through a transparent and low-cost process by the Bundesnetzagentur (Federal Network Agency, or BNetzA). The BNetzA oversees the telecommunications industry in Germany.

Applications for spectrum are done in "chunks" of 10 MHz spectrum, with fees defined according to the following formula since December 2019: fee $= 1000 + B \times t \times 5 \times (6a_1 + a_2)$.

- 1.000 is the base fee in €,
- B is the bandwidth in MHz (min. 10 to max. 100 MHz),
- t is the time of allocation in years (e.g., 10 years), and
- a is the area in km^2, differentiation between settlement and transport areas (a_1) and other areas (a_2) -> a_1 applies to industrial use and a_2 applies to farming and forestry.

There are no explicit definitions/limitations on transmit power, but the regulations stipulate that the local network operators are responsible for a reasonable deployment that does not cause interference.

Note that in Germany, Broadband Fixed Wireless Access (BFWA) has been allocated to the 5.8 GHz spectrum (5755–5875 MHz) band and as such does not share the private network spectrum as it does in the United States.

The 26 GHz band strategy for local networks is being decided during the second half of 2020.

Japan

Spectrum for local 5G has been released in Japan at the end of 2019 as shown in Fig. 1–8. This is intended for local government and enterprise use. Target is to realize "Society 5.0" through state-of-the-art technologies such as IoT, AI, robots, and self-driving vehicles, and their incorporation into all industries and sectors related to lifestyle.

Unlicensed spectrum

As mentioned earlier, mmWave spectrum is a key to unlocking 5G potential. Licensed mmWave spectrum for 5G is mainly in the 24–26, 28, and 39 GHz bands. Release 17 is intended to bring 5G support to the unlicensed 60 GHz band that was previously dedicated to 802.11ad/ay standards (with limited commercial success). Another option is to use unlicensed spectrum in <7 GHz bands. 3GPP defined the use of unlicensed spectrum already for 4G/LTE. In 5G, NR-Unlicensed

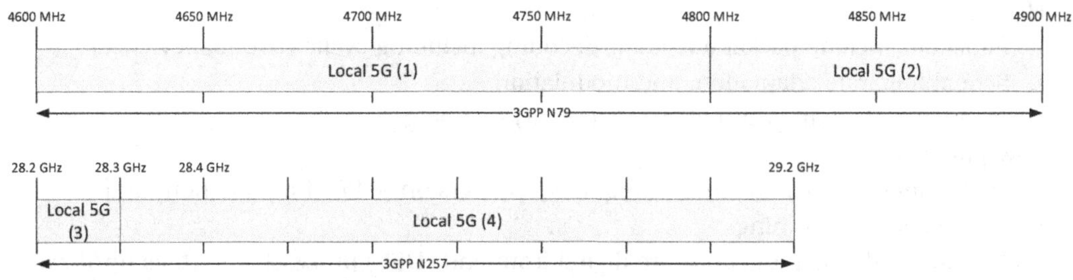

FIGURE 1–8 Japan Government/Enterprise 5G bands.

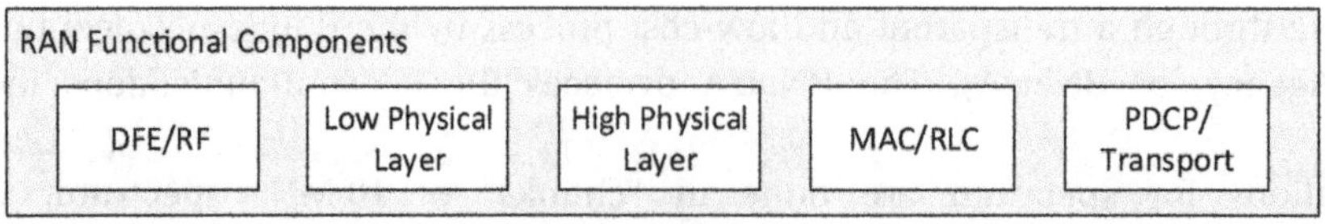

RAN Functional Components

DFE/RF | Low Physical Layer | High Physical Layer | MAC/RLC | PDCP/ Transport

FIGURE 1–9 RAN functional components. *RAN*, radio access network.

(NR-U) standards define the use of what traditionally would be Wi-Fi spectrum for use with the 5G/NR waveform in a coexistence scenario. This allows Wi-Fi bands to supplement data throughput. This is known as LAA (License-Assisted Access) or "anchored NR-U" and this technology was not widely picked up for LTE—it is unsure how popular a LAA will evolve into NR-U will prove to be. Technology challenges to address when supporting NR-U include:

- spectrum sharing with existing users including Dynamic Frequency Selection, Listen-Before-Talk, and Transmit Power Control techniques and
- low transmit power support which constrains coverage to (largely) indoor operation, just like Wi-Fi.

Spectrum in the 6–7 GHz band is an obvious candidate for NR-U deployment. This spectrum was first opened in the United States but is opening in other areas of the world as well.

Traditional base station architectures

At a high level, these systems include the following functional components that are described in more detail elsewhere in this document and are shown in Fig. 1–9.

- User plane PDCP and transport
  - Transport functions include IP/IPSec and GTP tunneling to carry the user IP packet between the RAN and the core network.
  - PDCP key functions include air interface ciphering, packet (re-)ordering and retransmission, header compression.
- User plane MAC/RLC
  - RLC functions include packet segmentation and retransmission.
  - MAC layer functions include wireless air interface scheduling, packet concatenation and multiplexing, and HARQ retransmission.
- High-physical layer
  - Functions include packet encoding/decoding including cyclic redundancy check.
  - Scrambling, rate adaptation, and modulation.
  - Frequency-domain grid (resource element) mapping.
- Low-physical layer
  - 3GPP functions include time↔frequency conversion (OFDM modulation) and optionally beamforming.
  - Non-3GPP-defined functions are digital front end (DFE) processing such as up/down conversion and filtering, crest factor reduction, and digital pre-distortion.

- RF
  - Converts the digital baseband IQ signal into a high-power radio frequency signal that is transmitted over the air (and vice versa).
- Control plane.

All-in-one base station

Built and deployed in the 1980s and 1990s, 1G and 2G systems were architected as a system-in-a-box where all components described above were integrated into a single unit, with coax cables toward the RU that is mounted on top of the RF tower—conceptualized in Fig. 1–10. The base station was located at the bottom of the cell tower, together with power supplies, air-conditioning, backhaul transmission equipment, and other components. These cell sites are bulky in nature and associated with macro cell towers, as typical in 1G and 2G networks.

The internal architecture of the all-in-one base station was highly proprietary and not visible to the network operator. Given its all-in-one nature, there was no possibility to mix and match components from different vendors.

Note that the very concept of the cell site (and the handover of a voice call between cell sites) was only established in 1G, deployed in the 1982, with the Advanced Mobile Phone System in the United States.

3G/4G/5G macro cell

Starting from 3G standard support in the 2000s and migrating to 4G/LTE in the 2010s, large vendors such as Ericsson, Huawei, and Nokia started to disaggregate the base station by splitting the radio head from the baseband through a semistandardized interface: Common Public Radio Interface (CPRI) in the case of Ericsson and Huawei or Open Base Station Architecture Initiative (OBSAI) in the case of Nokia (Figs 1.11 and 1.12). Even through CPRI and OBSAI are standardized interfaces, the implementation was made highly proprietary through use of optional features. As a result, the disaggregation achieved the goal of

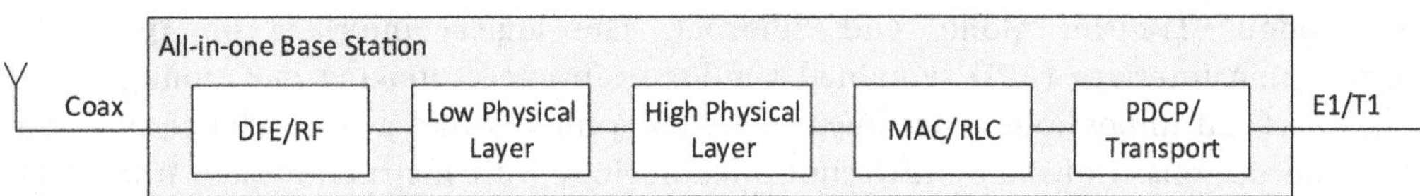

FIGURE 1–10 All-in-one base station.

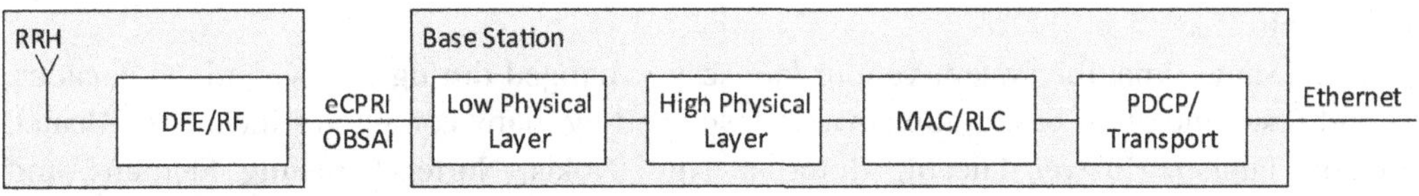

FIGURE 1–11 Macro base station architecture.

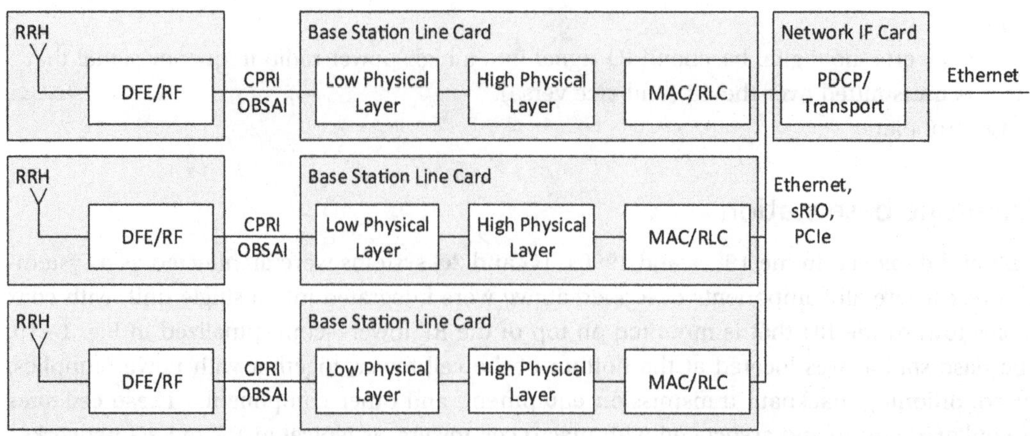

FIGURE 1–12 Multisector base station architecture.

removing high-cost (RF power coax) cabling but did not allow for vendor mix and match between baseband and radio head.

Note that both CPRI and OBSAI standards use proprietary serialization/deserialization standards and application-specific (silicon) devices for their implementations. This proprietary nature eliminated standard products from being used for large portions of the base station and forced the use of ASICs and field programmable gate arrays—both of which are highly proprietary and costly.

As system performance requirements increased over time, the base station architecture evolved to separate into two main components: The Network Interface Card (NIC) and the Line Card, as shown for a three-sector base station.

This evolution allows better scalability to increase performance from a single sector to three sector, six sector, and above. It also defines a new interface between the Line Card and the NIC. Physical interfaces used included PCIe, serial Rapid IO (sRIO), Asynchronous Transfer Mode, and Ethernet. The logical interface or Application Programming Interface (API) remained vendor proprietary, making decoupling of NIC from Line Card impossible. Even lower level components had a wide choice of implementation options. Consider MPU/controller devices with instruction sets from ARM, MIPS, PowerPC, or x86, DSP architectures from Analog Devices (Tigersharc), Freescale (StarCore), or Texas Instruments (C66 family of devices), as well as many proprietary architectures and implementations. This broad technology portfolio was clearly not sustainable.

At the same time, the system vendor landscape changed during the 3G and 4G decades. 3G and first-generation base stations were developed by many companies including, Alcatel, Ericsson, Fujitsu, Huawei, Lucent, Motorola, NEC, Nokia, Nortel, Samsung, Siemens, and ZTE.

Facing intense competition, the industry went through a sequence of mergers during the 3G/4G period, with the "famous five": Ericsson, Huawei, Nokia, Samsung, and ZTE capturing ~95% of the infrastructure market.

5G base station architectures

In the previous chapter, we discussed traditional base station architectures as deployed for macro sites supporting 2G, 3G, 4G, and 5G air interfaces. As we indicated earlier though, the 5G standard is designed to support flexibility, which means that the deployment form factor options increase. This chapter shows some relevant examples.

Integrated small cell

The integrated small cell (ISC) in many ways is a massively size-, power-, and cost-optimized version of the all-in-one base station that we described earlier and repeated in Fig. 1–13.

Defined as a "base station in a box" and implemented as a "base station on a chip," the ISC is targeting two markets:

- Coverage limitations. User retention is key to a wireless operator. By implication, user satisfaction needs to be achieved even if this implies additional financial expense. For those users who have coverage issues (dropped calls, too few bars, and low Internet throughput), an ISC can provide in-home solution at a reasonable cost. Backhaul from the ISC to the operator network is provided through (third party) wired Internet connectivity such as DOCSIS/cable, DSL, or other traditional wireline access means. A "femtocell gateway" forms the interface between the ISC network and the core/macro network, allowing firewalling and other gateway functionality to be implemented for network abstraction.
- Capacity enhancement. High-traffic areas such as the often-used example of the coffee shop impose a high user-count and traffic rate that can be beyond the capacity of the traditional macro base station, especially when the location of the high-traffic area is at/near the cell edge.

First-generation ISC (femtocells) targeted WCDMA and to a wider extent, LTE deployments with small user counts (up to 16) and ~100-Mbps throughput in an indoor form factor with passive cooling. RF output power per antenna is ~250 mW.

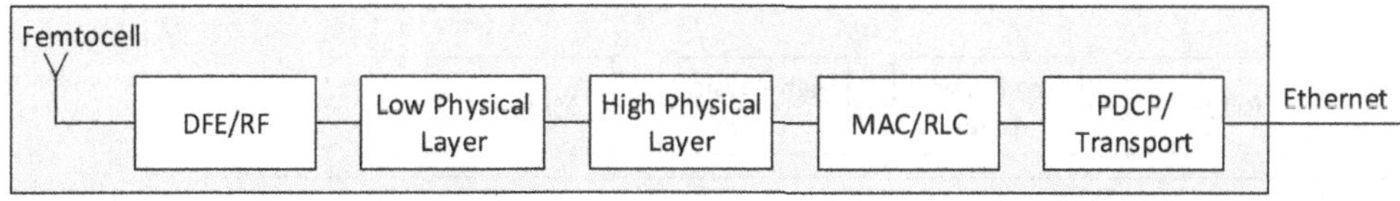

FIGURE 1–13 Integrated small cell architecture.

The femtocell concept has never been adopted as widely as initially anticipated and as a result, the number of chipset/products supporting this market has diminished over time. Reasons for lack of market traction include the following:

- Interference with macro base station. Not a topic for coverage limited networks but a major issue when targeting capacity enhancement solutions.
- Consumer value proposition. While the femtocell provides a net benefit to the operator, the operational cost is carried by the consumer (electricity, wireline backhaul, dealing with "yet another box" in the home).
- Alternative solutions such as Wi-Fi solve (pieces) of the problem in an often more cost-effective way, especially when macro coverage is "good enough" to provide voice support and data can be offloaded to Wi-Fi.

The ISC is expected to play a larger role in 5G deployments because of the physical characteristics of the deployments in higher frequency bands (C-Band, mmWave). These frequency bands have impaired propagation characteristics that dictate smaller cell sizes. In addition, certain target 5G markets such as private networks are ISC markets by definition: regulators stipulate low RF power and implied small cell size.

Pico/micro cell

The pico/micro follows much the same concept of physical implementation as the ISC, but at a larger scale, supporting typically multiple concurrent RF bands and potentially multiple operators as shown in Fig. 1–14. RF output power/antenna can be higher as the pico/micro cell is owned and managed by the wireless network operator and hence subject to different regulations.

Capacity enhancement is the first and foremost reason for pico/micro cell deployment for obvious reasons. The higher output power of the pico cell allows much less dense deployment compared to ISC but still provide the required capacity enhancement. Shopping malls, dense urban deployments, and similar are targeted by pico/micro cells. These cells are typically connected directly to the operator network.

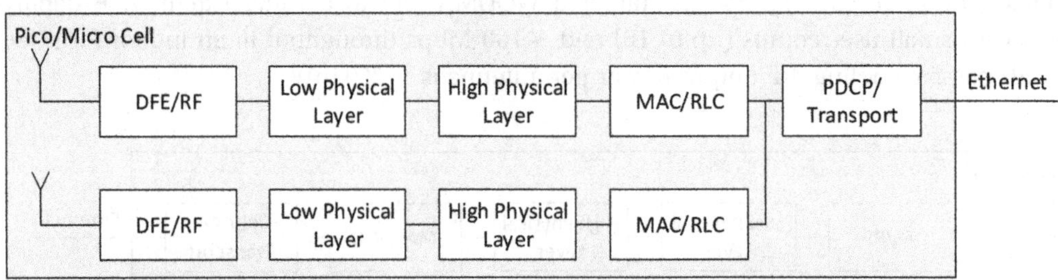

FIGURE 1–14 Pico/micro cell architecture.

Distributed antenna systems

Distributed antenna systems (DAS) as shown in Fig. 1–15 are a different physical implementation but targeting the same goal as ISC or pico/micro cells and that is coverage and capacity enhancements. Rather than implementing the complete RAN functionality in a single-cost/power-optimized box, these systems rely on a central base station (potentially reused from a macro base station system design) connecting to multiple/many (low-power, low-cost) RUs that provide RF coverage over a wider geographical range such as a mall or a sports arena:

The DAS design reuse from traditional macro base stations implies that the solutions are inherently compatible with macro systems and benefit from the same hardware/software enhancements over time. This makes DAS systems an interesting value proposition to the system vendors that can now leverage existing engineering work to a large extent. Ericsson Radio Dot and Huawei Lampsite are commercially successful examples of DAS systems.

Massive multiple-input, multiple-output

Given that below 6–7 GHz, there is only limited amounts of spectrum available, and that deploying physical sites is expensive with regards to install and maintenance, it makes sense to extract highest potential value from existing (macro) network sites. Massive MIMO is one of the techniques to enable this.

Use of MIMO for capacity extension to move from single antenna deployments to two, four, or eight antennas is nothing new: both 3GPP and Wi-Fi systems employ these techniques since many years. What massive MIMO does is to increase the antenna count to much higher numbers (e.g., 32 or 64), to enable a step function in potential throughput of the system, without changing the system architecture—the traditional "macro base station" split

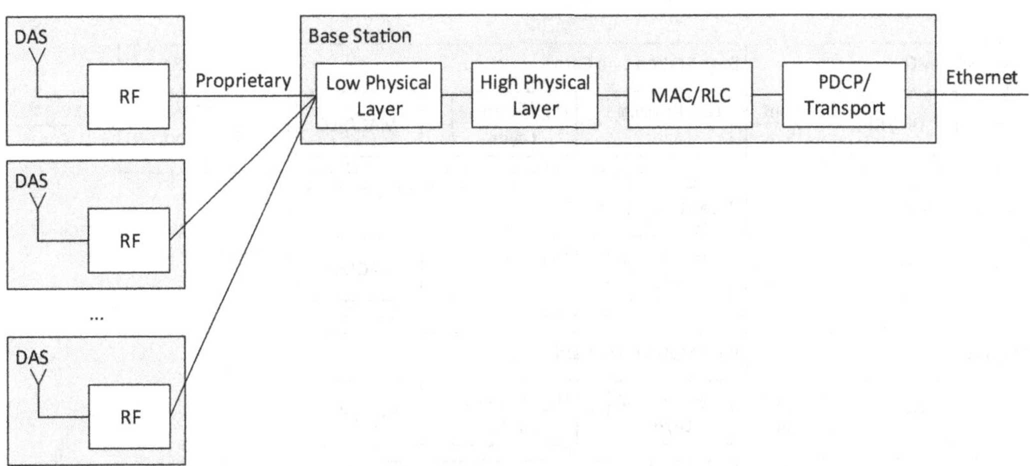

FIGURE 1–15 Distributed antenna system architecture.

between base station and (now, more complex) radio head is maintained as shown in Fig. 1–16.

Challenges to make massive MIMO work include the following:

- Raw compute requirements and compute precision, especially in the radio to enable signal processing from a large amount of antennas.
- System scalability—cost targets dictate that each antenna path needs to be built from relatively low-cost components so that the antenna scaling factor does not make the system prohibitively expensive.
- Power consumption associated with the above.
- Time and frequency synchronization between all the antenna elements.

C-(as in: centralized) radio access network

The concepts of RAN centralization are not new, as we have shown earlier. In its simplest form, a C-RAN system can be implemented with the same/similar components as a traditional macro base station, just by increasing the amount of processing power dramatically, targeting deployments in dense urban environments where there is not enough space to deploy individual macro cells as shown in Fig. 1–17.

Benefits of the C-RAN architecture have been discussed earlier so we will not repeat them again. The intention here is to highlight the scale of deployment form factors in 5G, from femtocell to large-scale C-RAN.

Functional splits

The previous discussions on form factors open the discussion on "functional splits"—what is the correct functional partitioning between different functional units (products), what is the

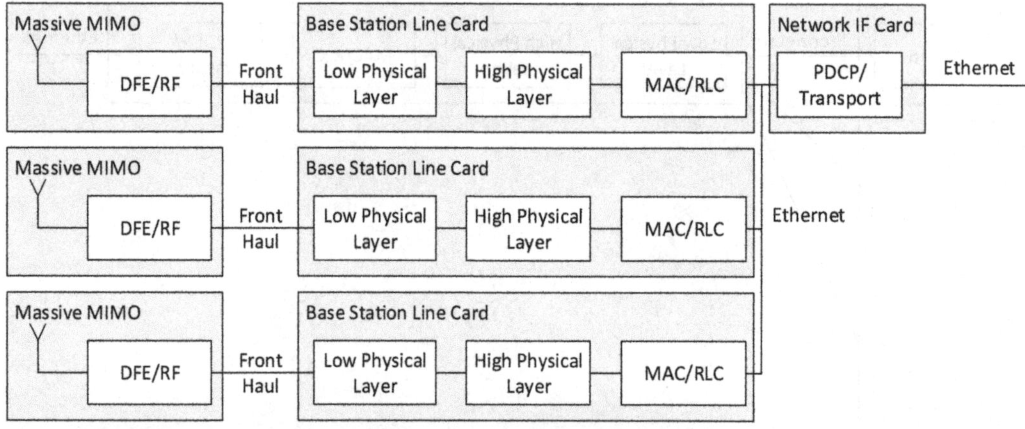

FIGURE 1–16 Massive MIMO architecture. *MIMO*, multiple-input, multiple-output.

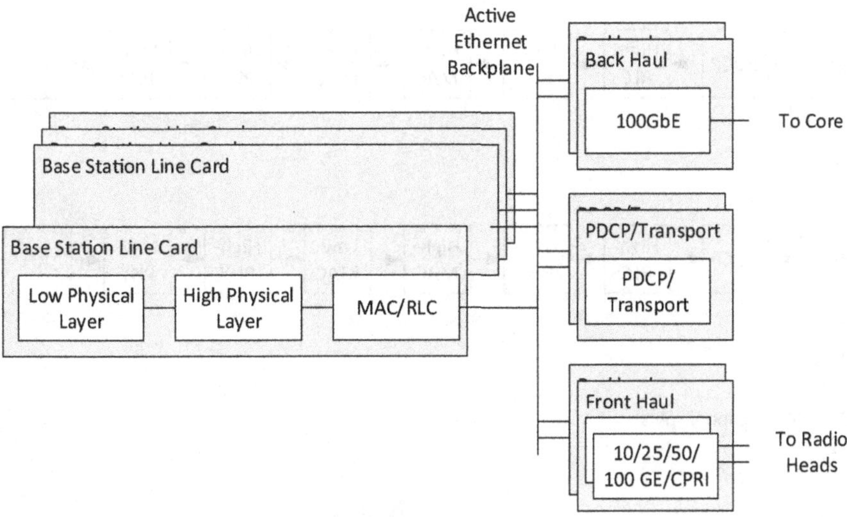

FIGURE 1–17 Centralized RAN architecture. *RAN*, radio access network.

correct interface between these units, and where are the protocols/interfaces defined in an open and interoperable manner?

Given high product development cost and timelines, the challenge is now to define appropriate "split" point in the 3GPP processing stack as well as interoperable physical and logical interfaces that allow a minimum set of physical products to target as broad as possible an O-RAN market.

This chapter discusses these topics, introducing the O-RAN concepts of central unit, distributed unit, and radio unit.

Distributed versus centralized processing

Splitting up processing chains into different physical components/locations is a well-studied topic with extended coverage in literature. In various previous C-RAN-related publications, functional splits are presented by different names and focusing on different directions.

3GPP standards started to define these functional splits in 38.801,[2] to *formalize* potential protocol split options, without *dictating* an implementation. Functional splits are intended to allow functions to run in a central location (e.g., the IT room in an enterprise deployment) versus distributed in the radio heads (the antenna units that are connected to the baseband unit). 3GPP functional split options are shown in Fig. 1–18.

The picture shows that 3GPP allows for any functional split. Practical implementations zoom in on specific use cases and involve trade-offs that center around the following four issues:

- Fronthaul interface throughput and latency requirements. As more functionality is pushed toward a central location (lower level split), fronthaul throughput increases and latency requirements get tighter.

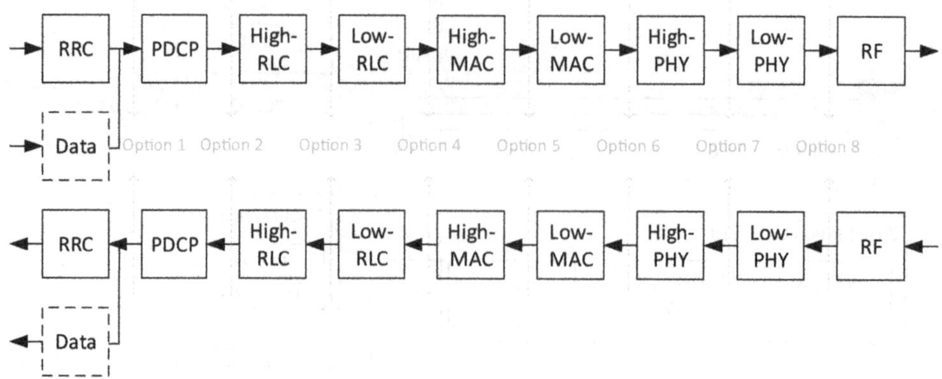

FIGURE 1–18 3GPP functional splits.

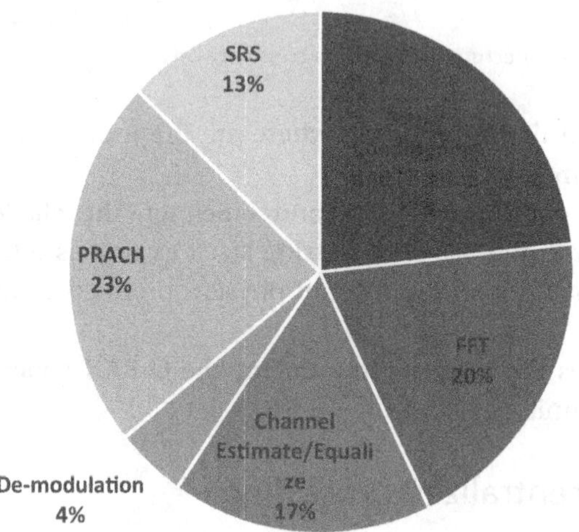

FIGURE 1–19 Compute complexity for physical layer processing in 4G/5G standardized use case.

- Implementation complexity. There is a benefit to a low-complexity radio head (size, weight, power, and design efforts are low) which tends to be supported with a lower level (Option 7/Option 8) split.
- Feature support. Centralization of processing can allow for more advanced features to be implemented such as Coordinated Multipoint Processing.
- Resource sharing. As we discussed, one of the prime objectives of early C-RAN initiatives was resource pooling/sharing given that peak versus average use of RF resources.

Let's discuss each one of this in a bit more detail for each functional split, observe Fig. 1–19:

Split 8. This is the traditional 3G/4G/5G macro base station split that we discussed earlier with the full scope of 3GPP physical layer specifications implemented in the base station. In this implementation, the only digital functionality left in the radio head is up/down sampling and DFE including signal conditioning and filtering which are simple in nature (but computationally potentially complex as we will discuss later). The radio head is little standard specific. Interface to the radio head historically has been proprietary: even though CPRI and OBSAI standards are publicly available, their "optional features" effectively make them proprietary, blocking interoperability between base station and radio head vendors. The fronthaul link carries baseband IQ samples in a digital form, in a time division multiplexing format, which provides time/frequency synchronization to the radio head as well. Given the baseband interface, the throughput of the fronthaul is constant and defined by the antenna/carrier configuration: bandwidth (BW) = antenna/carrier count (A × C) × bitwidth (IQ) × signal rate/carrier (SR). This bandwidth is equal for uplink and downlink.

An alternative physical implementation for this interface is Ethernet through standards such as (more about that later) CPRI over Ethernet or eCPRI. Moving to Ethernet removes the proprietary aspects of the standard but does not fix the high bandwidth requirements associated with baseband interfacing.

Partial mitigation of the bandwidth challenge can be achieved by sample compression, either in CPRI/OBSAI interfaces or on Ethernet interfaces.

Split 7. In the LTE air interface, only ~60% of available subcarriers are used for communication (1200 out of 2048 in a 20 MHz carrier), with the remainder carriers defined as guard carriers. Combined with cyclic prefix-associated overheads, we see that roughly half of the fronthaul bandwidth can be eliminated by moving the frequency↔time conversion (FFT functionality) to the radio head. In addition, note that the FFT function is computationally (as counted in multiply–accumulates) intensive though highly regular and well understood with optimized implementations widely available. When combined with DFE functions such as signal conditioning/filtering and RACH (Random Access Channel) processing, this constitutes the bulk of the compute requirements in the physical layer chain:

Within the Option 7 split, we define three subcategories of incremental offload to the radio head as shown in Fig. 1–20.

Option 7-1 strictly offloads the (i)FFT per the description given earlier. Incrementally, we can offload resource element mapping and beamforming components to the radio head to make *Option 7-2*. To understand the benefit of this option, consider massive MIMO systems where the number of logical layers is (much) smaller (say, 8 layers) than the number of physical antennas (say, 64 antennas). See Fig. 1–21 for the downlink (transmit) chain only—but assume equivalent for the uplink.

Compute and IQ sample IO bandwidths scale with number of layers before the beamforming operation and with number of antennas after the beamforming operation. In an IO-constrained scenario where the number of physical antennas is higher than the number of logical layers, Option 7-2 becomes an obvious partitioning candidate.

Stretching this even further, we note that in the downlink (only!), operations of layer mapping and precoding are relatively of low-complexity and need low computational effort. Moving to *Option 7-3* takes advantage of this by sending unmodulated bits from the base station to the radio head, thus lowering IO bandwidth even more. This option only applies to the downlink because in the uplink direction, channel estimation/equalization (which are the equivalent functions in the receiver) are still done in the

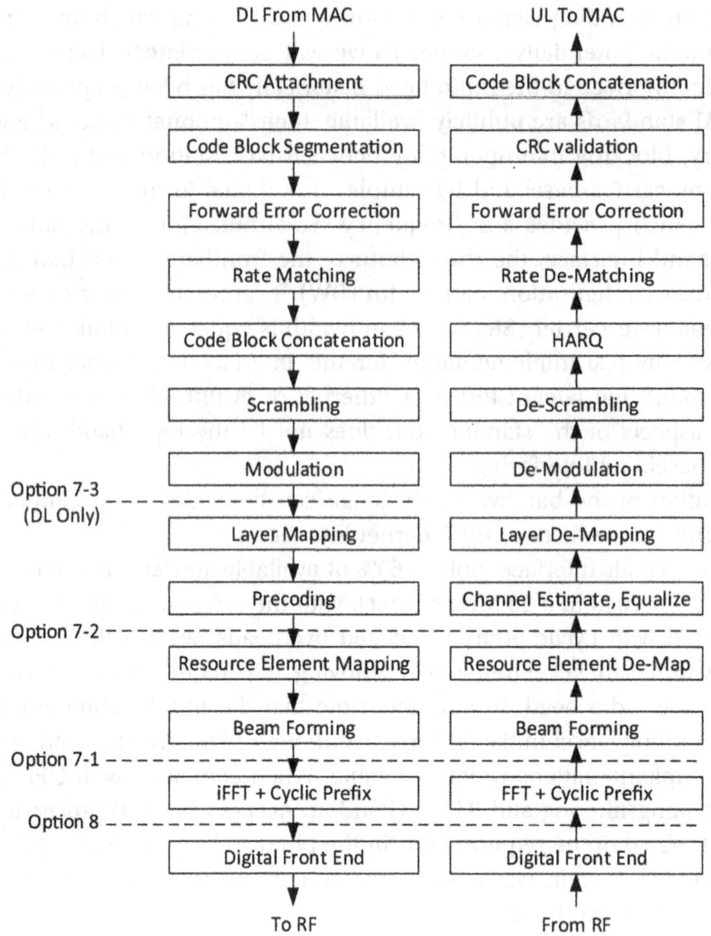

FIGURE 1–20 Option 7 and Option 8 splits.

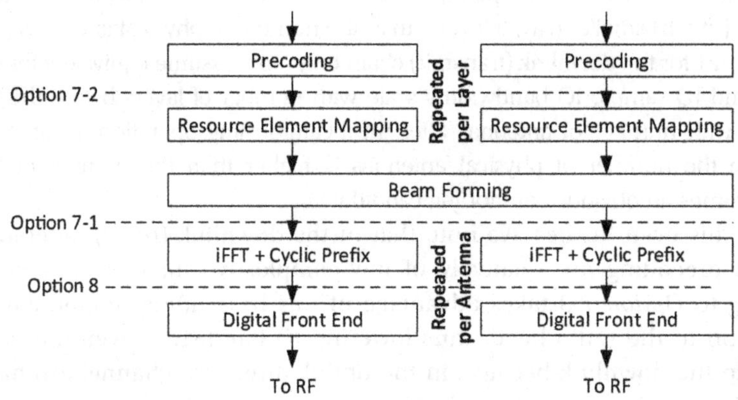

FIGURE 1–21 Option 7 splits.

(centralized) base station. As we will see later, in use cases with asymmetrical throughput requirements (more downlink than uplink), this can be a good candidate.

Remainder split options (6.1) are above the physical layer as shown in Fig. 1–22.

Option 6 is a MAC/PHY split where messages between both stacks are exchanged over a standardized, network-based interface (nFAPI being a key candidate, see next). This removes any physical layer processing from the centralized location, thus making it very software- and virtualization-friendly. This split is also very natural from a software ecosystem point of view: physical layer and MAC/RLC/PDCP stacks are often provided by different software vendors, and the FAPI interface concepts are well established. This interface builds on both factors, thus making implementation complexity relatively low—it is an easily understood interface.

Options 5, 4, and 3 similarly split the upper stacks to offload more and more processing to the radio head. However, given that the IO bandwidths within the Layer-2 stack are relatively similar (so little benefit to be had from a fronthaul capacity point of view) and that MAC/RLC stacks are typically developed in a tightly coupled manner, these options have not proven to be very popular in the industry.

A more well-adopted functional split is defined as *Open 2* which divides the L2 stack between PDCP and RLC processing domains. Compare this to the functional split between the NIC and the Base Station Line Card in the traditional 3G/4G/5G macro base station to understand why this is a very logical partitioning choice to allow different scaling of PDCP and MAC/RLC components.

Fronthaul throughput

To contrast the functional split options against each other, compare some popular split options in the chart given in Fig. 1–23 that takes two typical use cases and calculated theoretical fronthaul throughput for a 100 MHz wide 5G/NR RF channel.

Takeaways from this chart that are discussed previously include:

- Massive MIMO benefits significantly from Option 7-2 where layer-level information is transferred rather than antenna-level information.
- Option 7 reduces bandwidth sufficiently from Option 8 to warrant moving the FFT function away from the centralized unit.
- Options 7-3, 6, and 2 all cut fronthaul bandwidth by an order of magnitude compared to more low-layer split options.

As we will show later in the discussion on ISCs, the fronthaul throughput requirements (and the limitations on available throughput) impose hard limitations to centralization.

Centralized processing features: Coordinated Multipoint Processing

Cellular networks do not consist of just a single base station talking to multiple clients but have several base stations covering a geographical area—as large as a country or as small as an enterprise or industrial site. This creates the problem of interference: when a client is physically between two base stations, it will receive two signals with similar signal strength. Similarly, this client would transmit a signal to two base stations simultaneously, potentially

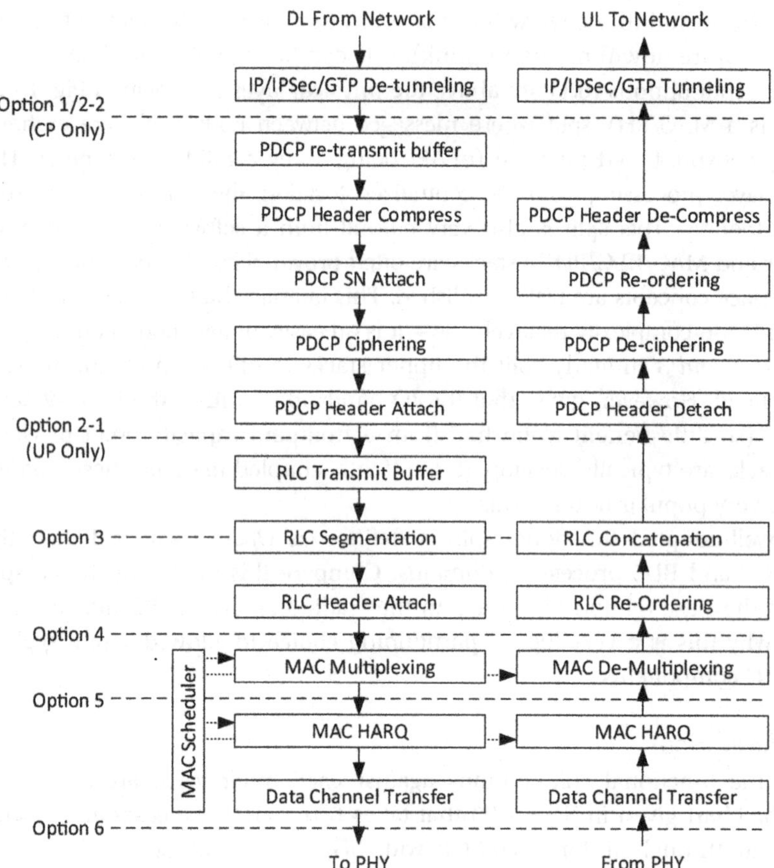

FIGURE 1–22 Options 2, 3, 4, 5, and 6 splits.

causing interference to another user trying to do the same. This is shown in Fig. 1–24 where the colored bands indicate the frequency used to communicate between the client and the base station for that specific base station/sector.

Frequency reuse techniques are commonly used in cellular networks to mitigate this type of interference, but they come at the expense of lower spatial bandwidth efficiency. In a generic frequency reuse scheme, the total available bandwidth is divided into subbands and used by different cells in a way that no two neighboring cells would use the same subband. See Fig. 1–25 for an example dividing up the frequency into three bands and (of course) three sectors. Note how the cell-edge user's communication path to each of the three base stations is done on a separate frequency band, as indicated by the different colored bands.

A variation of this scheme is called fractional frequency reuse which improves the spatial frequency bandwidth efficiency by reusing a large portion of the available bandwidth for their inner cell users while assigning the rest to the borderline users.

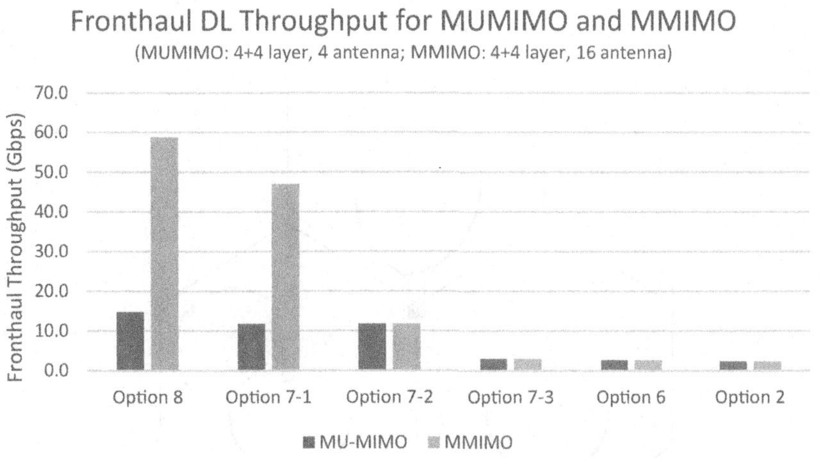

FIGURE 1–23 Fronthaul downlink throughputs for common functional split options.

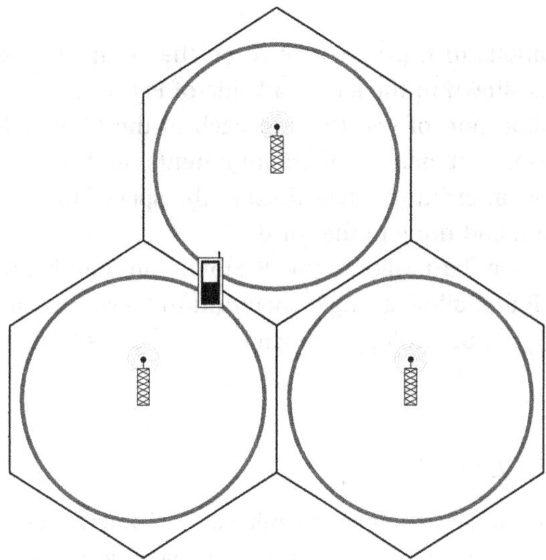

FIGURE 1–24 Single-frequency band operation.

Frequency reuse as described above is widely deployed in macro base stations today. Its main drawback is the available spectral bandwidth and the static nature of the spectral division into smaller bands. In addition, note that the base station needs to support three sets of antennas and associated electronic components, making it complex to scale this solution down to very low cost.

A more flexible approach is shown in the figure given next, which is still using frequency reuse to mitigate interference, but making the reuse scheme fully flexible by allowing each of

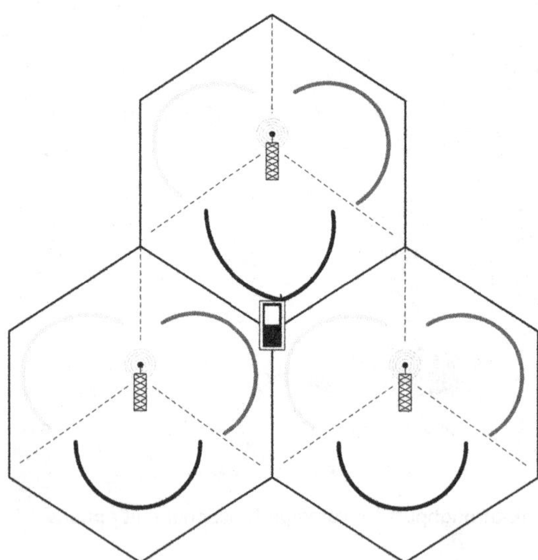

FIGURE 1–25 Sectorized frequency reuse.

the three shown base stations to transmit/receive on the complete available spectrum (single-frequency network) as shown in the left-hand side of Fig. 1–26.

In this scheme, the allocation of spectrum to each of the base stations is done dynamically, based on traffic load, user bandwidth requirements, and so on. The example on the right-hand side shows—as an example—two-third of the spectrum allocated to one base station, one-third to a second, and none to the third.

Note that the 3GPP standard allows for flexibility in bandwidth allocation through mechanisms such as BWP that allow a single spectrum to be split up into multiple segments, if required even with unique numerology, transmit power, and so on.

Coordinated Multipoint

Coordinated Multipoint (CoMP) is intended to mitigate cell interference and improve network capacity, specifically for cell-edge UEs—hence improving system-level coverage. Consider the "single-frequency band operation" shown earlier where a cell-edge user is considering an interference scenario when communicating to/from multiple base stations at once. Instead of mitigating interference, multiple base stations [or technically: Transmit Receive Points (TRP)] could be operating to transmit/receive *the same signal, at the same time* to/from this single user, where the signals to/from both base stations are combined to create a better signal-to-noise (SNR) ratio to the user and hence improved coverage. This is called CoMP.

Note that there is not a single CoMP technique, but separate techniques for uplink and downlink. In both directions though, CoMP can either imply communication to/from a single TRP but in coordination with other transmissions (coordinated scheduling or

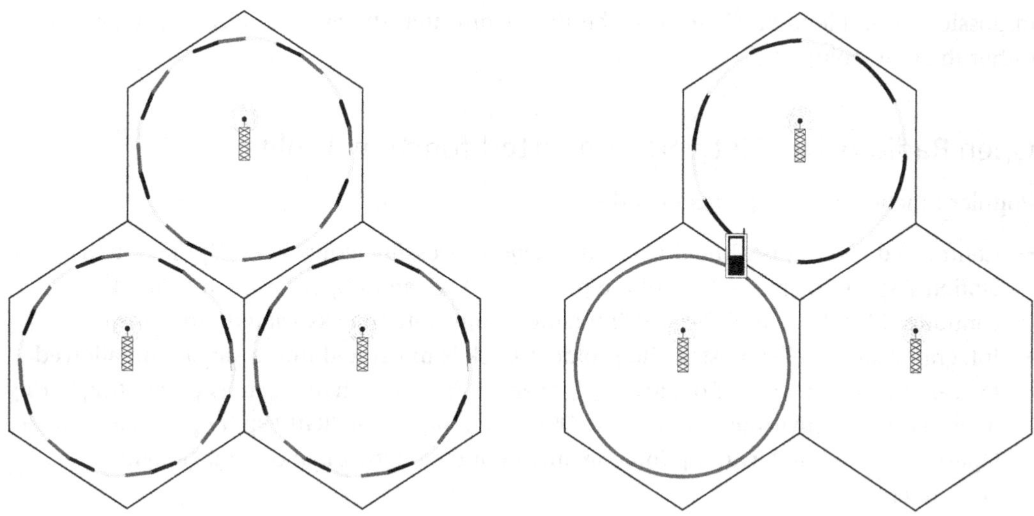
FIGURE 1–26 Dynamic spectrum allocation.

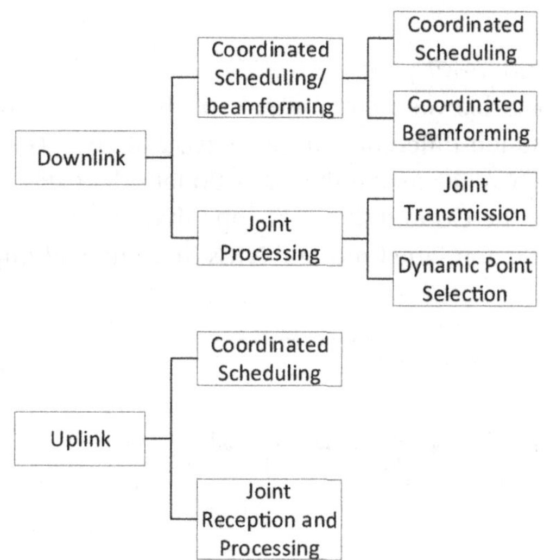

FIGURE 1–27 Coordinated Multipoint implementation techniques.

beamforming), or by doing joint transmission where multiple TRPs are communicating to the same UE to improve signal quality. See Fig. 1–27 for a summary.

It is obvious that techniques such as CoMP imply centralized processing or deep cooperation between TRPs, including tight restrictions on timing. Moreover, we can turn this argument upside down: without centralization, many CoMP techniques become tough or impossible to implement. CoMP may be the feature that drives centralization of processing rather than ISC solutions.

Open Radio Access Network-supported functional splits

Popular functional split options include:

- Centralized versus distributed and radio head processing along 3GPP Option 2 and Option 7 splits where the bulk of the physical, MAC, and upper layers are run in centralized locations and the RUs implement only antenna-associated functions.
- Integrated small cell—most of the protocol stack is integrated into a single unit allowed for very fine granularity of deployment, over nonideal backhaul such as preexisting wiring in an industrial environment or cable/DSL-based backhaul. 3GPP split options are either Option 1 or Option 2, with optional inclusion of core network and management components.

For example, an in-home femtocell as deployed to fix coverage issues, or an industrial deployment that has limited backhaul capacity from preinstalled wiring.

Both options are discussed in more detail in the following sections.

Central unit/distributed unit/radio unit

Supporting functional splits between components of the base station is nothing—we have seen being an implementation differentiator since 2G, 3G, and 4G networks. So, what is so important about the O-RAN Alliance and their functional splits? (Fig.−1.28).

First, the split between the CU and the DU is one that follows the functional split we saw between the Macro Base Station Line Card and NICs in traditional implementations.

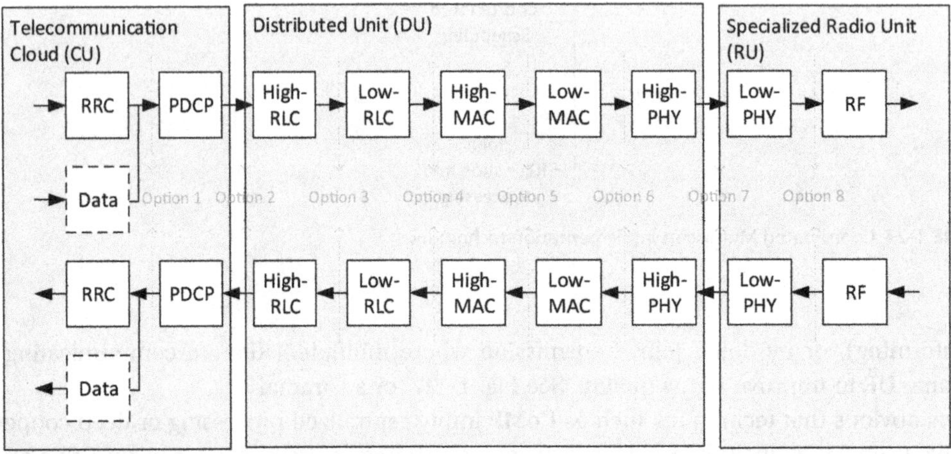

FIGURE 1–28 Central, distributed, and radio unit splits.

The CU main function is the PDCP function (user plane or UP), but with this function typically comes IP/IPSec backhaul tunneling and of course support for the GTP to transport packets to/from the core network and other CUs over the X2 interface as we show next. This PDCP function is envisioned as a software-centric function (with ciphering being a conveniently forgotten CPU intensive component as we will show) running on generic server platforms. CU functionality is not strictly correlated to physical time: the is no reliance on real-time software components and specific operating system/hardware support to achieve real-time deadlines (e.g., PREEMPT_RT features in Linux). Also, the CU scaling is not correlated to number of deployed radios, sectors, antennas, or other physical components. This makes the CU a component that naturally migrates to a server/software-centric solution that scales performance (throughput in Gbps) by adding CPU cores and IO interfaces, leveraging Moore's law.

The DU does scale with the number of deployed radios as each radio requires its own MAC/RLC and high-physical layer stack implementation. Each radio is transmitting and receiving aligned to physical time which imposes real-time deadlines on both MAC and physical layer. Both these aspects (1:1 relation between DU/RU and real-time requirements) force the DU implementation to be more embedded in its implementation—tighter coupling between functionality and resources the functionality is running on, even when it is a software-centric implementation. This explains why the DU is separated from the CU, to allow it to run in a different environment, less cloud/virtualization centric.

The second main split chosen by the O-RAN Alliance is the "Lower Level Split" (LLS) between DU and the RU along (more or less, see next) the 3GPP Option 7-2 dividing line. The balance between fronthaul throughput, implementation complexity, and other aspects is believed to be best struck with this functional split, allowing for an interface that can be supported by a wide ecosystem of vendors.

Physical implementations can (and do) of course combine multiple functions into a single hardware box: there is no obstruction to combine CU and DU functions onto a single hardware board or even a single (multicore) chip where the physical interface becomes a logical one.

Option 7−2x

As we said earlier, the split between DU and RU is key and involves a trade-off between two competing interests: on the one hand, the desire to keep RU complexity low to save power, cost, and implementation complexity, and, on the other hand, the desire to reduce the fronthaul bandwidth. The xRAN standard (later picked up by the O-RAN alliance) defined a new split point called "7−2x," or also called the Lower Layer Split/CU plane ("lls-CU"). This split is nearly the same as the 3GPP Option 7-2 but allows precoding/digital beamforming either to be implemented in the DU or in the RU, depending on implementation choice.

RUs that require precoding done in the DU are called "Category A" and RUs that support precoding in the RU are called "Category B." From the example of massive MIMO, we showed earlier, we can see that Category B is an obvious choice for systems where the antenna count is higher than the logical layer count.

There are some specific complexities involved in supporting Category B RUs, such as the transport of the necessary control information between the DU and the RU.

Relevant 3GPP standards

The O-RAN Alliance key interface description is the LLS one that defines the interface between DU and RU: the fronthaul. 3GPP standards are more important in the mid- and backhaul portion of the network that connects the CU and DU components to the 4G and 5G core network as shown in Fig. 1–29.

These standards include:

- the Higher Level Split between DU and CU, or the F1 interface as defined by 3GPP and
- the E1 interface between the CU User Plane (CU-UP) and CU Control Plane (CU-CP).

Relevant standards documents for both are defined in 3GPP 38.47x series. We will discuss these standards in more detail elsewhere in this document.

nFAPI

We have already mentioned that the functional split 6 connects the base station and the radio head over a Ethernet network interface at the MAC/PHY boundary. Although neither defined by 3GPP or by O-RAN, there is a standard worth mentioning in this regard: nFAPI or "Networked Femto Application Programming Interface." nFAPI builds on the femto API that has been the industry-standard interface for connecting L2 (MAC/RLC/PDCP) stacks and L1 (physical layer) stacks from different vendors, typically in an ISC/femtocell environment. This standard has been available from the Small Cell Forum[3] since the late 2000s when femtocells started entering the market. It has enabled relatively pain-free interoperability

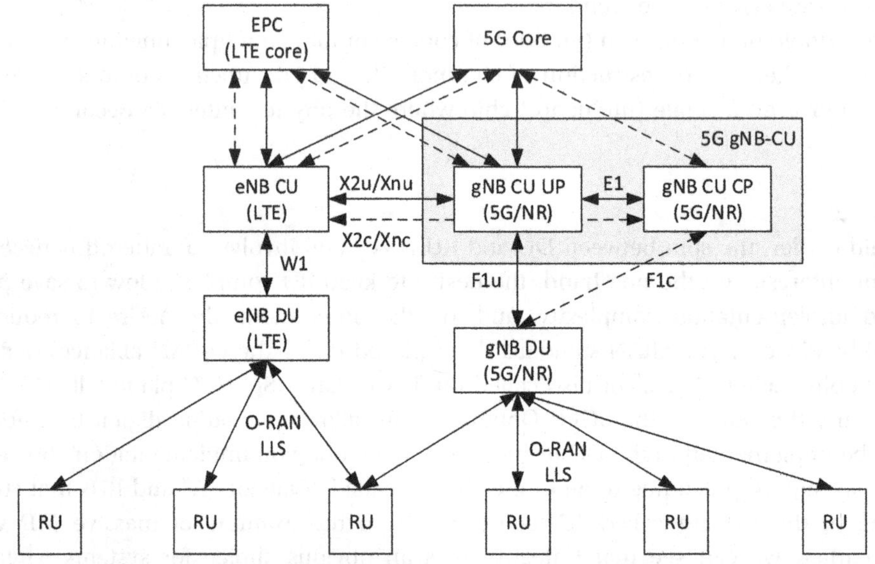

FIGURE 1–29 3GPP-defined mid- and backhaul standards.

between software vendors and proven an enabler of ecosystem approaches in base station implementations.

As the name says, the nFAPI standard extends this "vendor-independent API" concept to a networked version to enable O-RAN and similar implementations with a centralized/virtualized MAC layer and a network connected physical layer. The standard is defined for both 4G (4G-nFAPI) and 5G (5G-nFAPI).

Other relevant standards

Recent contributions to IEEE standards show increasing focus on Time-Sensitive Networking (TSN). TSN is part of the 802.1Q series of standards focusing initially on the use of Ethernet as a transport mechanism in industrial, time-sensitive environment. Being Ethernet-based, it is a Layer-2 technology. It uses a centralized scheduling scheme to coordinate scheduling and ensure deterministic, low latency. TSN is the new standard for real-time networking.

Building on TSN work, a recent (2018) contribution based on joint CPRI and IEEE standardization work is the 802.1CM standard that defines a "TSN for fronthaul" profile that maps to the use of Ethernet in the fronthaul (5G) network. This standard supports the various fronthaul splits and defines timing (latency) requirements that match to 3GPP/O-RAN standards. We will discuss this standard in more detail elsewhere in this document.

Like many other components network management in 3GPP has been traditionally implemented using proprietary operations support systems and protocols. These protocols need to be mapped to more open, cloud-based, and dynamically changing environments, calling out for standardization of DU and RU operational control (orchestration) standards. The NETCONF protocol, using the YANG modeling language, are traditionally used for this purpose in wireline (router, switch) remote management and control, allowing operators to move from command line interface to API-based control—and with that, automation. These APIs provide abstraction between the DU/RU implementation and its operation and are key to interoperability between RAN components in a multivendor environment. Given the use of YANG in wireline packet orchestration, this network control is a very natural way to allow a natural extension to the wireless world.

Integrated small cell

Referenced in O-RAN WG7 as the "Integrated gNB-DU," this is a platform that where all functionalities of an O-RAN DU (O-DU) and an O-RAN RU (O-RU) are performed in the same box. This box adheres roughly to the same functional split and functionality as the ISC discussed earlier, with either a split-1 or split-2 (PDCP is implemented externally) toward the operator network or a split-6 implementation to support some of the advanced centralized features such as joint scheduling.

The ISC is a much more "closed" solution where high- and low-physical layer are typically tightly coupled and running on colocated hardware. The ISC is also expected to be a small form factor (low power and cost), driving for a cost-optimized (ASIC-like) hardware solution with highly proprietary software.

Table 1–1 Categorization of nonideal backhaul (3GPP 36.932).

Backhaul technology	Latency (one-way) (ms)	Throughput	Priority
Fiber Access 1	10–30	10M–10 Gbps	1
Fiber Access 2	5–10	100–1000 Mbps	2
Fiber Access 3	2–5	50M–10 Gbps	1
DSL Access	15–60	10–100 Mbps	1
Cable (DOCSIS)	25–35	10–100 Mbps	2
Wireless Backhaul	5–35	10–100 Mbps typical, <1 Gbps max	1

This goes against a lot of the "software-driven" principles of O-RAN system but shows the strength of the O-RAN standards: O-RAN defines interfaces rather than dictating an implementation, allowing for tightly integrated solutions to exist in the market.

The main benefit of the ISC solutions is the support for a nonideal backhaul to the operator network. See Table 1–1 (copied from 3GPP 36.932).

The table shows the two big challenges with nonideal backhaul: latency is too high to maintain HARQ roundtrip time (more about that later), and throughput is not high enough to support physical layer functional splits (e.g., 3GPP splits 7 or 8).

An additional example is that of the industrial environment, which is a key target market for 3GPP systems. Industrial environments (the factory floor) often are prewired with an Ethernet/TSN network that is capacity-limited, say to 1 Gbps CAT-5 wiring. The near-10 Gbps throughput numbers required for a CU/DU/RU split are simply not supported by such cabled backhaul. Asking the customer to outfit the factory floor with cabled networking to support a new wireless network can be too much to ask.

Note that in either implementation (CU/DU/RU or ISC), core network software components can be integrated into the CU component of the system or hosted as a cloud service in a local (Metro Edge Compute) or remote server. In private/small-network deployments, the core network is supplied by the same vendor as the radio system and thus no interoperability challenges are expected.

A real-life example: enterprise 5G networking

Say, the task at hand is to look at dimensioning and design a system to support a 5G private network. First thing is to establish requirements which is a process that covers the systems architecture team, marketing, and other customer facing organizations as well as hardware and software teams. We need to establish the use case for the networking by answering a set of questions:

Consider the following topics discussed in the subsequent sections:

Network installation and maintenance. Who will run the network?

- Self-install versus professional install. A professional install comes with high installation cost which can be prohibitive to customer pickup. But a self-install capable system needs to be much more "foolproof," and this can have low-level impact on hardware and software requirements: Consider whether the front, mid-, and backhaul ports in the system are allowed to be associated (labeled!) with specific physical interfaces on the hardware or the customer should "just" be allowed to plug any wire into any port. This can impact hardware (need for a switch?) and software (a control layer that assigns internal and external IP addressing and routing tables appropriately). There is obvious boot flow and automated software install impact.

- Who is the system integrator? As in so many IT-related areas, 5G/private network equipment is made by combining hardware and software from multiple vendors. This combining and integration effort is the job of the system integrator which is the vendor that combines all the hardware/software and system integration aspects and delivers (and optionally installs) a complete "kit" of equipment to deploy a site—as well as to guarantee successful operation. The equipment itself can either be highly custom and optimized or made from a combination of commercial software stacks running on relatively general-purpose hardware. The latter "white box" systems are becoming more and more popular due to ease of development and support of a wide range of deployment options. When considering system integrator choice, consider what the target features of the system will be:
  - network services;
    - mobility/handover support;
    - interoperability with commercial operator networks;
    - QoS requirements such as voice, video, and industrial/URLLC; and
    - integration with other applications, such as (Metro) Edge Computing, virtualized services, and application integration.
  - Client equipment
    - modem cards/clients with proven interoperability with the deployed network;
    - general-purpose (EQ mobile phones) or application-specific (URLLC/industrial Ethernet bridges); and
    - SIM cards.
- Indoor versus outdoor deployment which impacts both RF aspects (output power/EIRP) as well as form factor (passive or active cooling, hardware noise, moisture, and dust resilience) and other related topics. This sets the boundaries around the system that can be deployed.
- Regulatory requirements—location, transmit power, sensing, and registration are unique per geography as we discussed. What are the RF testing requirements that need to be finished and how do they impact design and hardware timeline? And obviously, the RF specifications define the target system requirements.
- Security requirements that are often overlooked. Does the system have a secure-boot requirement, or can it be assumed in a safe-enough environment protected by other factors? What is the process for applying critical software patches without disrupting user experience?

- System optimization point. It is impossible to design a system that both scales to every possible performance point while being cost optimized over all those performance points at the same time. What do we pick as functional and performance minimums and maximums?

Who are the users of the network?

- Internal only or visitors and guests? Internal use only implies that the test environment can be greatly simplified: the breadth of connected device(s) (types) is very different as this changes the scope of feature development and test.
- What are the performance requirements? There is no need for wide bandwidth operation if the client does not support this. Sometimes, we really need to optimize for latency rather than bandwidth, and so on.

What are the applications running on the network and what are derived key performance indicators?

- Application definition. As we show elsewhere, industrial (5G URLLC) deployments are quite different from consumer (5G eMBB) ones with regards to throughput (industrial: less strict) as well as latency and reliability (industrial: stricter). Do we need to support local (edge) compute or only provide a wireless network/pipe without concern for other application presence? Do we need to colocate a core network or is this handled elsewhere? A well-defined application/use case allows the system architect to derive requirements from the elements discussed next rather than having to establish them individually.
- What are throughput and latency requirements? Throughput is defined as both single-user peak throughput and combined (network-level) aggregate performance over all users. The two numbers do not need to be the same and can have system-level impact (see our discussion on "Elephant Flows" later).
- The target user count includes definition of both connected users who are visible to the control plane of the system but are not transmitting or receiving data at all times (e.g., an idle phone) as well as active users who are defined as users that are actively transmitting and/or receiving data.
- What are the coverage requirements? Providing a 5G alternative to Wi-Fi coverage may require much be lenient RF coverage as compared to trying to build a world-class system with associated equipment and install cost. Consider RF planning to be a part of product definition.

Total cost of ownership

Maybe the most important of all, this is split between network cost, network maintenance, and end-user equipment cost:

- Network cost—this is the cost associated with installation of the physical network.
    - the centralized/IT room units (if any)
    - wiring
    - TRPs—how many RUs will be needed. To be estimated from physical area, link budget, etc.
    - labor
- Network maintenance cost
    - spectrum licensing—typically a yearly cost
    - software/equipment maintenance licensing to the system integrator
- User device cost
    - including SIM card or similar (eSIM) provisioning and management.

Use-case example

In our private network example, we establish some requirements:

- USA CBRS deployment: 3.x GHz operation.
- Self-install, which implies category A radios (<30 dBm EIRP/10 MHz).
- Optimization for small and medium enterprises: as per the 2012 Commercial Buildings Energy Consumption Survey showed 98% of commercial buildings in the United States are 100,000 square feet or less (or roughly 10,000 square meters or less).
- Small and medium businesses have (by definition) less than 1000 employees, which gives some indication of required connected user count.

RF aspects

We now look at some of the RF aspects that define the next level of dimensioning. Spectrum allocation needs to be matched to the throughput requirements—and by implication, network planning needs to include an assessment of required system throughput. The RF band selection is mostly defined by regulatory aspects as defined earlier in this document and local to each country.

Other aspects include definition of the required RF bandwidth (e.g., 10/20/40/.../100 MHz) which defines the peak throughput that the system can achieve to an individual user. Aggregate system capacity can be increased through RF bandwidth expansion (but can be financially costly) or by cell densification: use of more TRPs in the system. The latter has an obvious impact on the system installation cost. Note that, depending on regulatory aspects, (maximum) RF output power can be limited, in our case to 30 dBm EIRP.

RF propagation obviously limits the signal power that is conveyed from transmitter to receiver. Fig. 1–30 shows the RF propagation per standard 3GPP models.

The propagation model can be used to predict the received signal power at the client by calculating the SNR at the receiver and comparing this against the required SNR for successful demodulation, even if the calculation is high level/rough and used only for initial dimensioning with refinements made later. The receiver-required SNR depends on the quality of

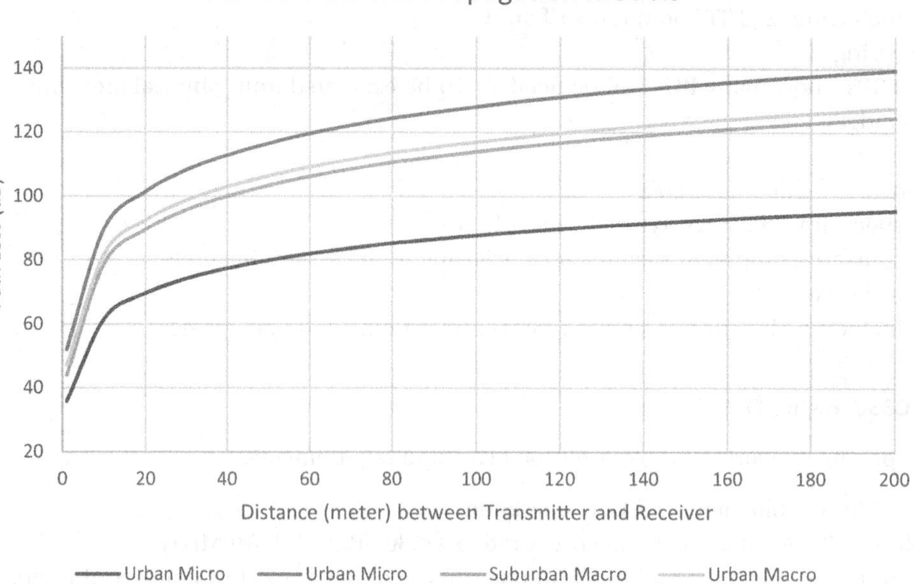

FIGURE 1–30 3GPP propagation model example.

the receiver implementation but is typically −5.5 dB for QPSK, 5–15 dB for 16QAM and 15–20 dB or better for 64QAM.

The required signal level S is now computed as: S = noise figure (say, 5 dB) + thermal noise (−174 dBm/Hz + 10*LOG(used bandwidth)) + SNR.

For example, for 10 MHz bandwidth and 20 dB required SNR for 64QAM decode: S = 5−174 + 70 + 20 = −79 dB.

Given an assumed 30 dBm EIRP, the acceptable path loss is 30 − (−79) = 109 dB or ~100 dB when taking safety margin. Assuming a pessimistic propagation model, this is ~20-m radius—or better when reducing safety margin, required signal strength, or advanced algorithmic support including CoMP.

We take the 20 m radius to define the TRP count: with some safety margin added, each TRP can support ~1000 m^2. Our 10,000 m^2 application use case implies an optimization point of 10–15 TRPs, assuming placement of TRPs will never be perfect.

Next question is key: do we want to develop a CU/DU/RU deployment architecture? This implies higher bandwidth and more latency constrained fronthaul Ethernet and typically results in dedicated wiring between the CU/DU and the RUs. An ISC deployment can leverage existing networking infrastructure with higher latency and lower bandwidth but comes at challenges for supporting advanced algorithms (centralized processing/CoMP) as shown in Fig. 1–31.

Given cost being a prime consideration, we assume RF spectrum utilization being the key and hence a drive for centralized processing. As a side benefit, we note that CU/DU/RU

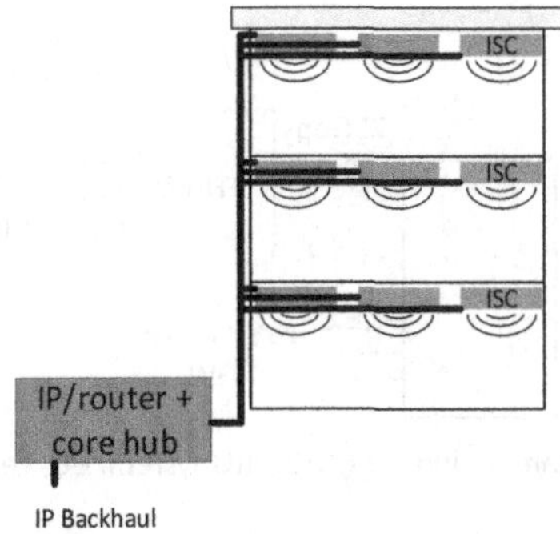

FIGURE 1–31 Integrated small cell deployment architecture.

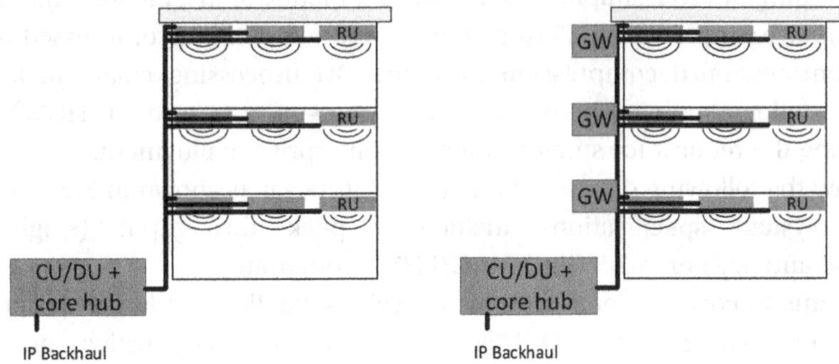

FIGURE 1–32 CU/DU/RU deployment architecture.

processing architecture allows for more choices for COTS hardware platforms and associated (somewhat) reduced time-to-market.

In this CU/DU/RU deployment, the centralized unit implements the MAC/upper layer stacks and physical layer functions in addition to the core network components. By implication, the bandwidth between centralized unit and DU is higher than for the ISC. A potential bandwidth/cabling overhead mitigation is the use of a "fronthaul gateway" which can reduce cabling overhead (star vs hub topologies) but can also support advanced functions such as CoMP. The deployment is shown in Fig. 1–32.

To dimension the system to the next level of detail, we consider RF aspects and fronthaul bandwidths/wiring associated with the CU/DU/RU architecture. Let's assume that maximum/peak over-the-air bandwidth is ≤100 MHz and we can have two to four antennas at each RU.

Calculating the bandwidth associated with two to four antennas at 100 MHz RF (per Chapter 3), we see that when using IQ compression, a 10 Gbps Ethernet link (which can be supported by both electrical and optical links) supports this bandwidth. Let's assume a fronthaul gateway (GW in Fig. 1–32 on the right-hand side) aggregating RU traffic from ≤8 RUs into a single 25 Gbps Ethernet link using downlink joint transmission and uplink joint

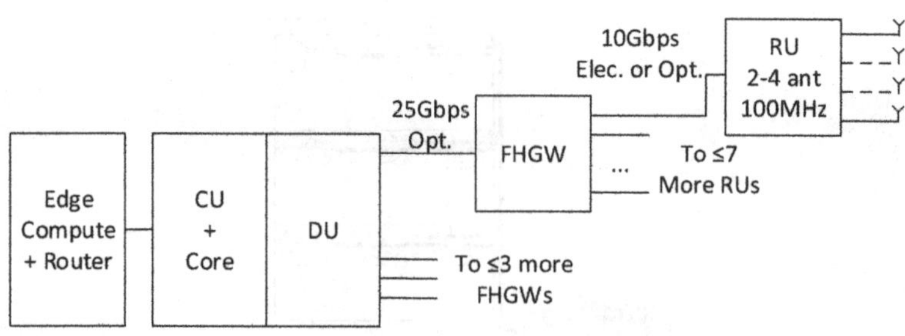

FIGURE 1–33 Connectivity picture for example indoor CU/DU/RU system. *CU*, central unit; *DU*, distributed unit; *RU*, radio unit.

reception. This makes each set of RUs connected to a single GW look to the DU as a single "logical cell." Scaling out to multiple logical cells is a matter of instantiating more links from DU to FHGWs, for example, using 25 Gbps optical fiber carrying uncompressed IQ so to offload eCPRI compression/decompression from the DU processing chain, making it more lightweight on implementation. The overall system can support up to 4 (FHGW) × 8 RUs = 32 RUs, satisfying the request for small-to-medium enterprise deployments.

We can draw the following diagram for a system proposal as shown in Fig. 1–33.

Additional system specifications include a peak throughput (single user) of 100 MHz × 2–4 antennas or ≤2.5 Gbps MAC/PHY throughput.

This book aims to cover systems like this as well as smaller and larger designs from ISC up to data center scale centralized O-RAN systems. We cover connectivity and bandwidth calculations; component choices in CU, DU, and RU; and dimensioning of each component for software, IO, and hardware processing load.

Summary and conclusions

We showed how the O-RAN standard evolved from various other initiatives with similar goals around extending SDN from the wireline space into the wireless domain. The first enabler of 5G is newly available spectrum being made available around the world, both in the traditional licensed operator space and private spectrum for industrial and enterprise use.

This nontraditional 5G spectrum is driving new deployment options, ranging from ISCs through mmWave and massive MIMO, which pushes for implementations that can scale from the very small to the very large. This drives for an ecosystem of vendors and standardized interfaces between various components in the RAN domain.

We show how O-RAN Alliance and standards including 3GPP, IEEE, and others are defining standardization of functional splits of various units that make up the RAN, as well as the interfaces between these units. This introduces the concepts of the central unit, distributed unit, and radio unit as well as the Integrated Small Cell. Between them, these units make up the bulk of the O-RAN-based RAN deployments.

It is best to understand the various aspects involved in system design by going through an example. We pick one of an indoor, enterprise deployment that provides sufficient coverage for small- and medium businesses and use this example to define the set of questions that the system architect needs answers to in order to define the "sandbox" she or he gets to play in. This starts defining which boxes need to be build, connectivity between them, performance targets, and so on. We will use this (simple!) example where required throughout the remainder of this book as a use case where relevant to show dimensioning aspects.

References

[1] O-RAN Alliance. Available at https://www.o-ran.org.

[2] Study on new radio access technology: Radio access architecture and interfaces. *TR, 38* (801), 2016.

[3] Small Cell Forum. Available at https://www.smallcellforum.org/.

Further reading

"*Study on new radio access technology: Radio access architecture and interfaces, V14.0.0 (2017–03),*" 3GPP, Sophia Antipolis, France, Rep. TR 38.801, 2017.

802.1CM Time-Sensitive Networking for Fronthaul (802.1CM). Available at https://1.ieee802.org/tsn/802-1cm/.

2

System components, requirements, and interfaces

This chapter outlines the key functionality that encompasses the O-RAN system: the central unit (CU), distributed unit (DU), radio unit (RU), and integrated small cell (ISC). We also discuss the key interfaces between these components and include relevant standard references. The objective of this chapter is not to provide a deep-dive training on 3GPP—this is done very well in many books and online tutorials. We give an overview of key functions implemented in each of the functional layers from 3GPP as shown in Fig. 2–1 and delve deeper into those aspects that impact connectivity (relevant interface standards) and dimensioning.

Next-Generation Radio Access Network overview and terminology

Let's first discuss some 5G terminology. The whole 5G network, including base stations, core network components, and interfaces between them, is called the 5G system and is shown in Fig. 2–2. Note how the name for the base station itself evolved from BS (Base Station) in 1G to BTS (Base Transceiver Station) in 2G, NB (NodeB) in 3G, eNB (evolved NB) in 4G to gNB (next-generation NodeB) in 5G.

The 5G air interface is called New Radio (NR or 5G/NR), and the radio access network is called the Next-Generation Radio Access Network or NG-RAN. Base Stations (also called NG-RAN Nodes) that build up the 5G NG-RAN can either be defined as gNB when they provide a 5G/NR interface to the User Equipment (UE/client) as we discussed before, or ng-eNB if they provide an E-UTRA [4G Long-Term Evolution (LTE)] interface to the UE. The core network components are called 5G Core Network or 5GC.

As if this is not confusing enough, we also have specific terminology for the interfaces between various components. The link between the NG-RAN and the 5GC is defined not only as the NG interface—but also as the N2 [control plane (CP)] or N3 [user plane (UP)] interface by the 3GPP 5G system architecture group documents. The combined N2 and N3 interface set is the equivalent to the S1 interface in 4G/LTE language. The link between gNB and NG-eNB is defined as the Xn interface, the equivalent of the X2 interface in 4G/LTE. These relationships are shown in Fig. 2–2, copied from 3GPP 38.300.

Before we delve deeper into the communication links between the various components, we first look at the underlying protocol stacks (Fig. 2–3). Communication is done through a/any Internet Protocol (IP)-based network: per 3GPP 38.410, UP traffic is carried through a

Open Radio Access Network (O-RAN) Systems Architecture and Design. DOI: https://doi.org/10.1016/B978-0-323-91923-4.00006-9

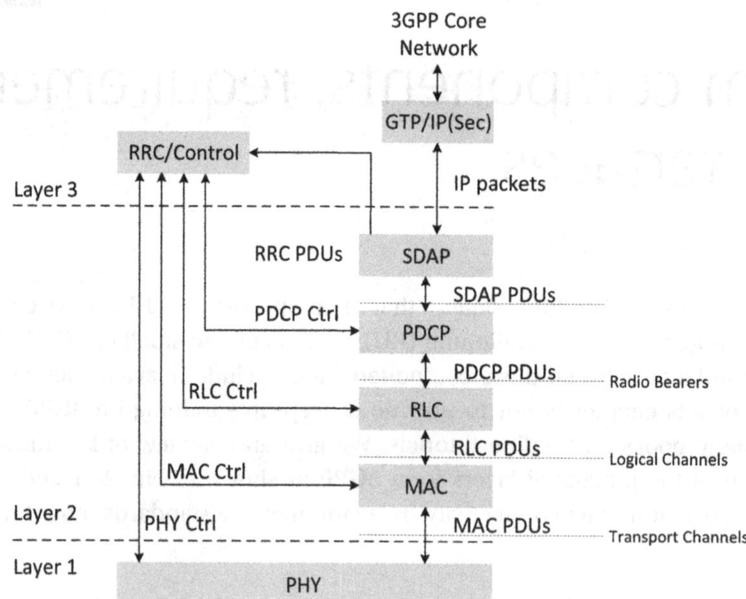

FIGURE 2–1 3GPP protocol split and terminology.

GTP-U tunnel (3GPP 29.281) defined as NG-U. CP traffic is carried over Stream Control Transmission Protocol (SCTP) as defined in NG-C standards:

In this protocol stack, the NG-C/U Physical Layer can pretty much be any implementation that supports packet (frame)-based transport (3GPP 38.411 standards defined—this standard is near-empty). The Data Link Layer is also open to any implementation, if it supports IPv4 or IPv6 at IP layer (per 3GPP 38.412 for CP, 38.414 for UP). DiffServ marking support is a required feature.

The explicit separation of CP and UP is a key aspect of 5G standardization. We delve into the various interfaces that enable this separation later. More on 5G RAN Quality of Service (QoS) is explained later, and we dedicate more space to DiffServ concepts in a separate section at the end of this chapter.

Over the air between Base Station and UE, the protocol stacks for both UP and CP are shown in Fig. 2–4. These protocol stacks are what the O-RAN CU, DU, and RU implement.

Within the NG-C, there are (few) UP and (many) CP, with delineation between UP and CP being a key design target for 5G standards as compared to the 4G Evolved Packet Core (EPC), to actively support implementing them in different physical locations and software environments, including cloudification (Fig. 2–5).

UP functions are implemented in the UP function (UPF):

- UPF. Packet routing and forwarding and inspection, QoS handling (enforcement), acts as external Protocol Data Unit (PDU) session point of interconnect to data network (DN),

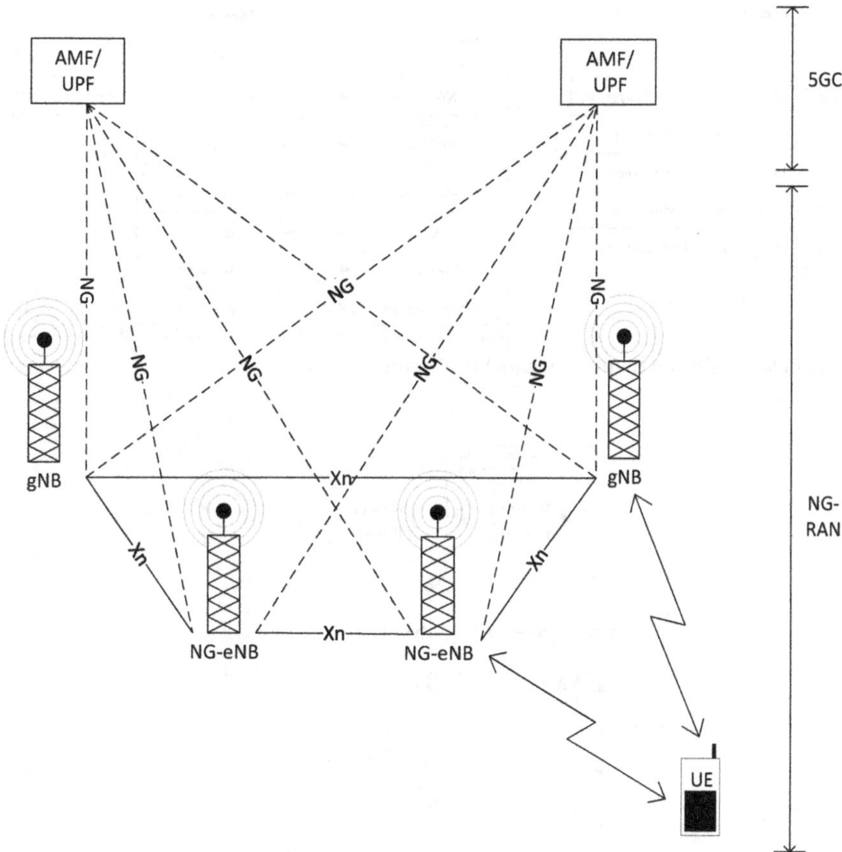

FIGURE 2–2 NG-RAN (copied from 3GPP 38.300). *NG-RAN*, Next-Generation Radio Access Network

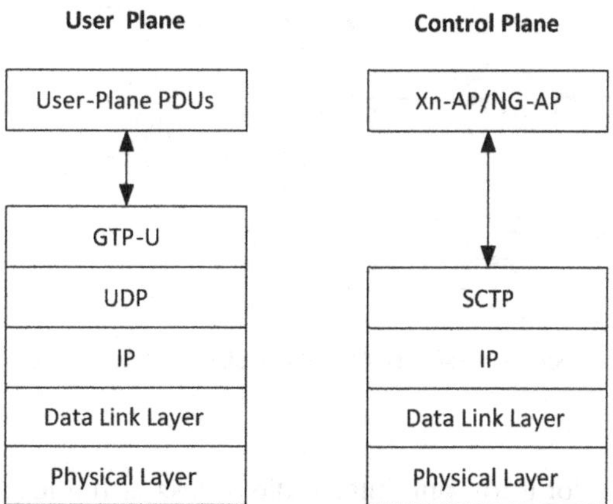

FIGURE 2–3 User plane (left hand side) and control plane/signaling plane traffic.

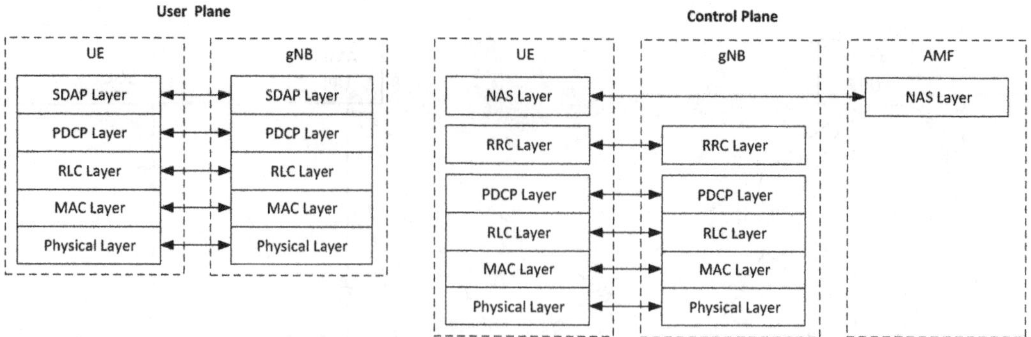

FIGURE 2–4 User plane (left) and control plane (right) 5G protocol stacks.

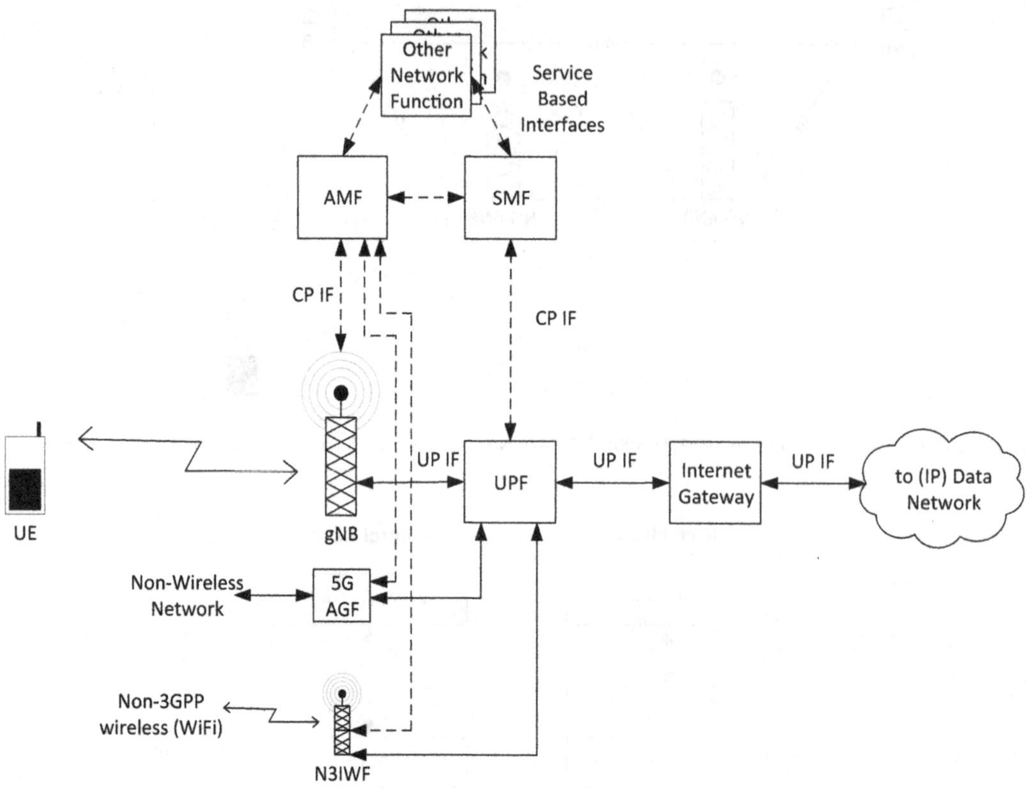

FIGURE 2–5 CP/UP Split in 5G network components. *CP*, control plane; *UP*, user plane.

and is an anchor point for intra- and inter-radio access technology mobility. The UPF is explicitly design to support multiple access networks including wireline and other wireless interfaces, targeting fixed-mobile convergence.

CP functions are implemented in the following components:

- Access and Mobility Management Function (AMF). Termination of Non Access Stratum (NAS) signaling, NAS ciphering and integrity, registration/connection/mobility management, including roaming, access authentication and authorization, and security context management. This function terminates the CP protocol to the UE as we showed before.
- Session Management Function. Session management (establishment, modification, and release), UE IP address allocation and management, Dynamic Host Configuration Protocol (DHCP), termination of NAS signaling related to session management, DL data notification, and traffic steering configuration for UPF for proper traffic routing.
- Policy Control Function (PCF). Unified policy framework, providing policy rules to CP functions, access subscription information for policy decisions in user data repository.
- Authentication Server Function (AUSF). Authentication server to the UDM
- Unified Data Management (UDM). Generation of Authentication and Key Agreement (AKA) credentials, user identification handling, access authorization, and subscription management. The UE encryption key (UESUCI) is stored here.
- Application Function (AF). Application influence on traffic routing, accessing NEF, interaction with policy framework for policy control.
- Network Exposure function (NEF). Exposure of capabilities and events, secure provision of information from external application to 3GPP network, translation of internal/external information.
- NF Repository function (NRF). Service discovery function which maintains NF profiles and available NF instances.
- Network Slice Selection Function (NSSF). Selecting of the Network Slice instances to serve the UE, determining the allowed NSSAI, determining the AMF set to be used to serve the UE.

The way the CP works is that required CP functions are invoked when necessary to provide specific network-related functions. For example, during UE registration, messages are exchanged between UE and AMF for session establishment, which involves communication to the UDM to ensure this UE is allowed on the network, where UDM invokes AUSF services. Next, a data connection (QoS flow in 5G which is the equivalent of a bearer in 4G language) needs to be built between the UE and the DN on the UP. For this purpose, the UE initiates a PDU session establishment request to the AMF (via the gNB). Once this exists by means of an established GTP-U tunnel between UPF and gNB, the gNB configures a data radio bearer (DRB) that translates this GTP-U-tunneled traffic into over-the-air traffic to the UE and so on.

Again, the point here is not to delve into 3GPP protocol stacks as has been done in plenty of other online and offline documentation. We want to show instead how 3GPP targets of openness and disaggregation align with those of O-RAN. Disaggregation and software centricity allow for scalability where the core network (both CP and UP) components can either be software elements that reside as a function in an enterprise network or a dedicated hardware function that resides in the operator network.

Wired networking analogy

The concept of CP/data plane (DP) separation has been around since a long time in wired networking—notably the early 2000s with the advent of network processors. Software-defined networking (SDN), for example, defined by the Open Networking Foundation (ONF), is about separation of CP and DP operations. Separation of these components from a software/API perspective allows them to (potentially) run in different hardware, physically colocated or explicitly not, and in fact nonvirtualized or virtualized. As such, also the concept of standardization of the CP/DP API, through initiatives such as the Network Processor Forum has been standardized in the wired networking world for long time (http://www.oiforum.com/public/NPF_IA.html). A fair question at this point is: "why should people adopt SDN explicitly?" One piece of the answer is: SDN allows the replacement of control that is standard, application-dependent, often distributed with control that is generalized across standards, applications, as well as more centralized. This means that SDN matters to entities who can benefit from centralized control and who are able to develop application-specific control SW. Motivations often involve simplicity of network management, but even the management becomes network-specific: SDN provides for a standardized interface between CP and data plane, as such "exposing the network" to networking applications developers.

OpenFlow, an SDN standardization vehicle driven by the Open Networking Foundation (ONF), is a concept that is closely associated with SDN. OpenFlow defines an API level interface between (today) Layer-2 (L2) switching functions executing in the DP and higher level CP functions, allowing for separation of DP control and monitoring. By defining a dedicated DP component, the OpenFlow concept accepts the notion that certain functions are indeed best executed on an optimized hardware platform such as a network processor. The OpenFlow concept introduces centralization of the CP using the SDN principles, allowing (but not forcing) for the CP to move to a remote (centralized/cloud) environment.

Central unit

As we discussed in the chapter 1, 3GPP (38.801) defines multiple functional splits between functionality in the CU and DU. Referring to Fig. 2–6, we assume what is defined as Option 2 which defines the Radio Link Control (RLC)/Medium Access Control (MAC) to be allocated to the DU and the PDCP and Radio Resource Control (RRC) to the CU. In addition, following CP/UP split principles, the RRC CP piece is separated from the Packet Data Convergence Protocol (PDCP, UP). The Service Data Adaptation Protocol (SDAP) function is associated with the PDCP and defined in 3GPP 37.324. This component is responsible for mapping between QoS flows and radio bearers. More information about this is present later.

The PDCP layer (defined in 3GPP 38.323) is responsible for transfer of user data between SDAP layer and RLC. In the transmit direction, the PDCP defines a unique sequence number to each payload packet and adds its own header to the packet that includes this sequence number to be used by the remainder processing steps in the PDCP layer. For one, a "SDU discard timer" is associated with the packet, allowing the PDCP layer to track successful delivery of the packet to the receiving entity after which the associated buffer can be returned to its pool.

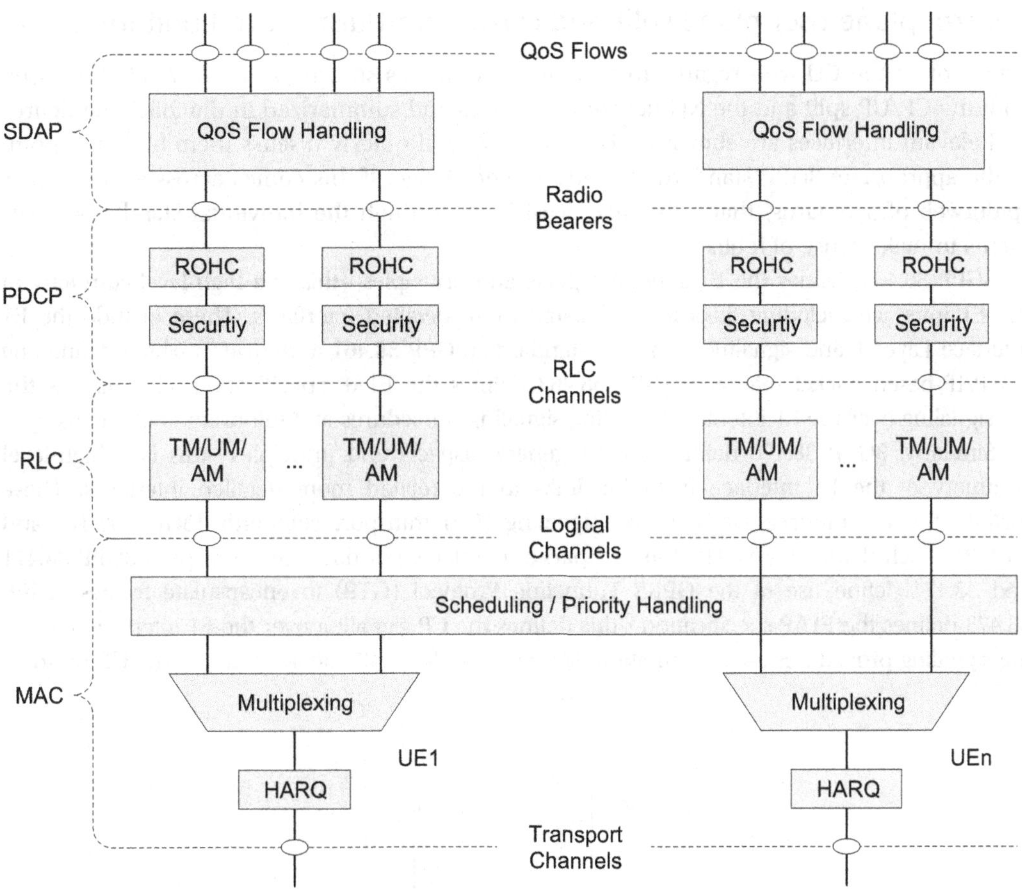

FIGURE 2–6 3GPP Layer-2 protocol stack and interfaces (replicated from 3GPP 38.300).

The PDCP layer includes (optional) Robust Header Compression (RoHC) to compress IP/UDP and/or other header types, reducing the effective payload size of the packet that is transferred over the air. After header compression, packet ciphering and (optional, for control bearers) integrity protection are applied to the packet, using AES, SNOW, or ZUC algorithms. The packet is then routed to the appropriate RLC/MAC layer (see the discussion on dual/multiconnectivity later).

On the receive side, the sequence number is used to detect reception of duplicate packets as well as to ensure in-order delivery of the packets to the next layers.

There are two types of radio bearers: DRBs for data channels and signaling radio bearers (SRBs) for control channels. The RRC protocol is responsible for signaling and data radio bearer establishment, configuration, maintenance, and release. This includes the same steps for additional carriers associated with dual connectivity (DC, more about that later) operation. The RRC is also responsible for broadcasting system information (SI) and handling mobility.

Control plane user plane split and central unit/distributed unit interface

Let's look at the CU with regards to its interfaces first, as shown in Fig. 2–7. We discussed both the CP/UP split and the NG network interfaces and summarized in the following figure.

Relevant interfaces are shown in the figure. We will quickly discuss them here and point to the appropriate 3GPP standards for further knowledge. If this comes across as a complex spiderweb of standards, that is because it is! Read through the individual standards documents to make sense of it all.

3GPP 38.460 defines the E1 general aspects and principles—this is a high-level summary of the E1 interface including links to the related more detailed interfaces. These include the E1 interface Layer 1 and signaling transport standards (3GPP 38.461 and 38.462) which define any SCTP/IP-based packet interface. 3GPP 38.463 defines the E1AP specification—this defines the CP signaling over the E1 interface including signaling procedures and information elements.

Similarly, 3GPP 38.470 defines the F1 general aspects and principles—this is a high-level summary of the F1 interface including links to the related more detailed interfaces. These include the F1 interface Layer 1 and signaling (CP) transport standards (3GPP 38.471 and 38.472) which define any SCTP/IP-based packet interface. For data (UP) transport, 3GPP 38.471 and 38.474 define use of the GPRS Tunneling Protocol (GTP) to encapsulate frames. 3GPP 38.473 defines the F1AP specification—this defines the CP signaling over the F1 interface, including signaling procedures and information elements. 3GPP 38.425 defines the generic UP protocol

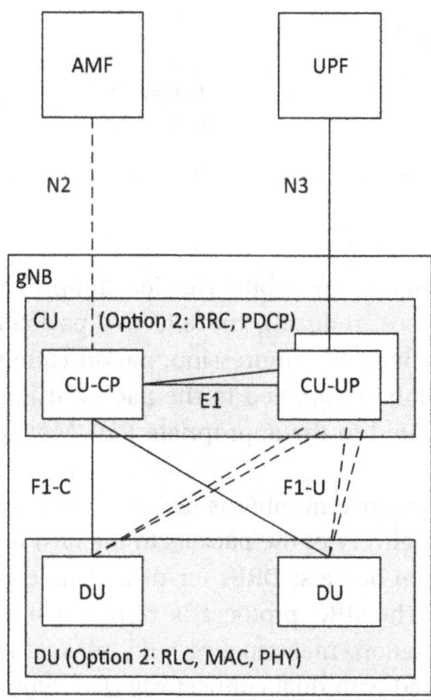

FIGURE 2–7 CU-associated interfaces. *CU*, central unit.

that also covers the F1 interface as well as the Xn (and X2 for LTE) interfaces. This standard references 3GPP 29.281 for the definition of GTP. 3GPP 38.425 carries control information related to user data flow management of radio bearers: the UP procedures. Each instance is unique per radio bearer/GTP tunnel and instantiated together with the associated GTP tunnel.

3GPP 38.420 defines Xn general aspects and principles, for both CP and UP aspects including mobility management, DC/multiconnectivity aspects, and other procedures. These include the Xn interface Layer 1 and signaling (CP) transport standards (3GPP 38.421 and 38.422) which define any SCTP/IP-based packet interface. 3GPP 38.424 defines the use of the GTP to encapsulate UP frames. 3GPP 38.425 carries control information related to user data flow management of radio bearers: the UP procedures. Each instance is unique per radio bearer/GTP tunnel and instantiated together with the associated GTP tunnel. 3GPP 38.415 manages the PDU session UP which is required when PDU sessions need to be transferred between Xn- and NG-related interfaces for handovers.

3GPP 38.410 defines NG (N2, N3 above) general aspects and principles for both CP and UP aspects. These include the Xn interface Layer 1 and signaling (CP) transport standards (3GPP 38.411 and 38.412) which define any SCTP/IP-based packet interface. 3GPP 38.413 defines the NG application protocol which is used to signal messages over the N2 interface. 3GPP 38.415 manages the PDU session UP which is required when PDU sessions need to be transferred between Xn- and NG-related interfaces for handovers.

O-RAN specifications as written by O-RAN WG5 around these interfaces do not redefine any 3GPP interface specifications but define profiles (such as information elements to be supported) and expected message sequences that are expected to be supported by an O-RAN compliant vendor.

Internet Protocol (IP) and Internet Protocol Security (IPSec)

The Data Link Layer specifications for all transport networks such as NG and F1 interfaces define the use of any frame-based protocol, or in the bulk of the cases: IPv4 or IPv6. What is implicit in this definition is that when the backhaul is not physically secure, support for Internet Protocol Security (IPSec) becomes an implied requirement that can consume a large amount of compute/processing resources when requirements go to 10s or 100s of Gbps throughput associated with UP processing. Given the point-to-point nature of these links, the vast number of operators looks to deploy a single-tunnel mode implementation: it gives simplicity. However, other scenarios can ask for multiple tunnel support, including:

- traffic type separation including QoS-class-based differentiation or user plane/control plane separation;
- multiple 3GPP service deployments, such as mass-market and public safety concurrent deployment;
- Mobile Virtual Network Operator deployments that require separation on per-operator level, but also base station sharing scenarios (single cell tower shared across multiple nonvirtual operators);
- Internet offload models; and

- core network deployment models can force traffic differentiation (again, user plane and control plane separation being an example).

As a small but important last point, note that IP fragmentation is required to be supported on all the F1, NG, and other links.

Quality of Service and related concepts

The purpose of QoS is to guarantee sufficiently good customer experience on a network that has limited capacity. QoS is measured though metrics such as packet loss, throughput, latency, jitter availability, etc. Some applications (say, voice traffic) are intuitively understood to be sensitive to packet loss and latency, while others (large downloads) can sustain latency but need to have high (average) throughput, and so on.

QoS in 3GPP (both 4G and 5G but we will focus on 5G first and cover changes from 4G after) is interesting because it carries aspects from wireline QoS and wireless QoS (the latter being the scheduler algorithm that we discuss extensively later). The wireline piece of QoS is defined by 3GPP to be based on DiffServ concepts. A short introduction to DiffServ is added to the end of this chapter to explain its basic concepts.

QoS in 5G centers around the concept of QoS flows. What is a QoS flow? This is defined by 3GPP as the finest granularity of QoS, defined within a PDU session, and identified by a QoS flow Id (QFI) (Fig. 2–8). In the downlink direction (from UPF to UE), classification of traffic [mapping from DiffServ code point (DSCP) and/or application detection] is done by

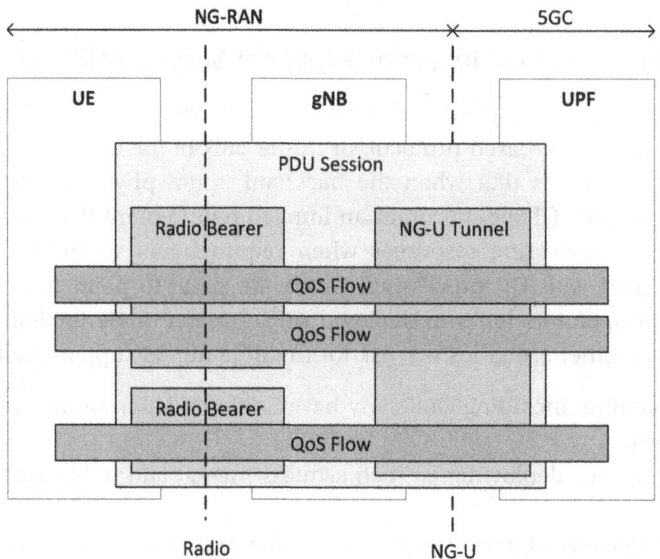

FIGURE 2–8 QoS flows in 5GC and NG-RAN (reproduced from 3GPP 38.300). *NG-RAN*, Next-Generation Radio Access Network; *QoS*, Quality of Service.

the UPF, and it is the UPF that assigns the QFI. In the uplink direction, this job is done by the UE. Over the NG-U connection, the 6-bit QFI field is carried inside the GTP-U extension header where the GTP tunnel carries the overall PDU session.

Fig. 2–9 explains the functional split between gNB and UPF elements where (in the downlink direction) the UPF components classify ingress IP flows to PDU session and QoS flow (QFI). The gNB SDAP layer maps these QoS flows to DRBs that are subject to wireless air interface scheduling and transmission over the air.

Each PDU session is carried in a GRPS tunnel and identified by the Tunnel Endpoint Identifier (TEID). The TEID unambiguously identifies a tunnel endpoint in the receiving GTP-U (GPRS Tunneling Protocol—User) or GTP-C (GPRS Tunneling Protocol—Control) protocol entity. The receiving side of a GTP tunnel locally assigns the TEID value for the transmitting side to use. The TEID values are exchanged between tunnel endpoints using GTP-C messages [or RANAP (Radio Access Network Application Part) in the UTRAN (UMTS Terrestrial Radio Access Network)].

See in Fig. 2–10 for how these pieces fit together in the overall protocol stack.

The QoS flow context that is pointed to by the QFI contains a set of QoS-related parameters that includes:

- 5G QoS Identifier (5QI). Defines a set of preconfigured QoS characteristics.
- Allocation and Retention Priority (ARP). Defines traffic priority level (PL—lower number indicates higher priority) and indicates whether this flow can be preempted or preempt other flows.
- For GBR QoS flows:
  - Guaranteed Flow Bit Rate for both uplink and downlink
  - Maximum Flow Bit Rate for both uplink and downlink

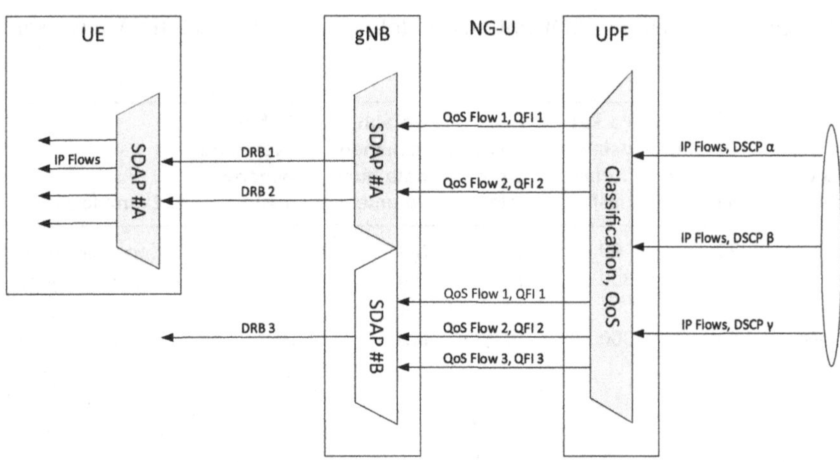

FIGURE 2–9 Mapping between IP flows, QoS flows, and DRBs (downlink). *DRBs*, data radio bearers; *IP*, Ingress Protection; *QoS*, Quality of Service.

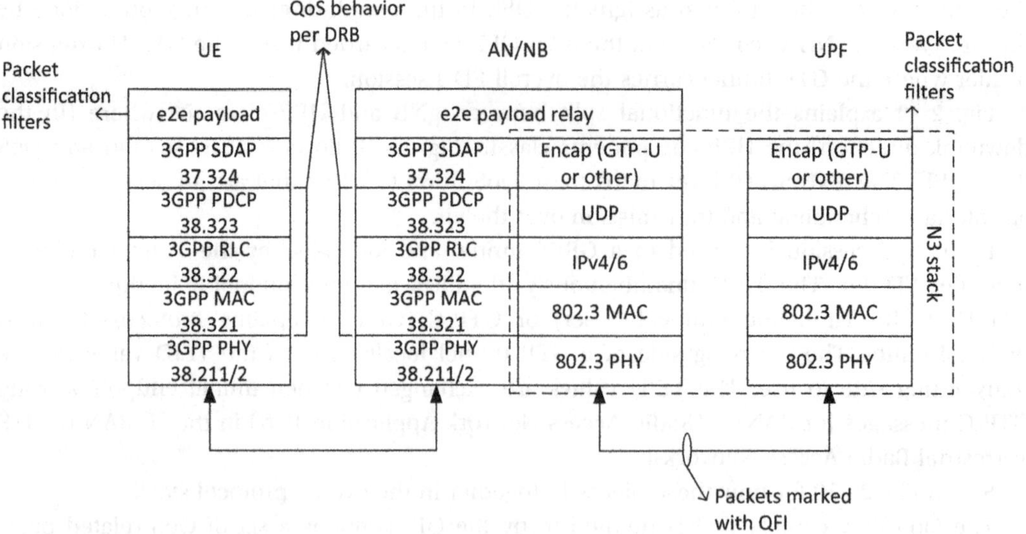

FIGURE 2–10 Relationship between radio bearer (DRB) over the air and QFI backhaul QoS definition 3GPP). *DRB*, digital radio bearers; *QFI*, QoS flow Id; *QoS*, Quality of Service.

- Maximum Packet Loss Rate for both uplink and downlink
- Delay-Critical Resource Type
- Notification Control

- For Non-GBR QoS flows:
  - Reflective QoS Attribute
  - Additional QoS Flow Information

The following table defining 5QI options is defined in 3GPP 23.501. A few examples are copied next:

5QI value	Resource type	Default priority level	Packet delay budget (ms)	Packet error rate	Default maximum data burst volume	Default averaging window (ms)	Example services
1	GBR	20	100	10^{-2}	N/A	2000	Conversational voice
4	GBR	50	300	10^{-6}	N/A	2000	Nonconversational video (buffered streaming)
6	Non-GBR	60	300	10^{-6}	N/A	N/A	Video (buffered streaming), TCP-based (e.g., www, e-mail, chat, ftp, p2p file sharing, progressive video, etc.)
79	Non-GBR	65	50	10^{-2}	N/A	N/A	V2X messages

(*Continued*)

(Continued)

5QI value	Resource type	Default priority level	Packet delay budget (ms)	Packet error rate	Default maximum data burst volume	Default averaging window (ms)	Example services
86	Delay-critical GBR	18	5	10^{-4}	1354 Bytes	2000	V2X messages (advanced driving: collision avoidance, platooning with high Levels of Automation (LoA))

The 5QI equivalent in 4G/LTE is defined as the QoS Class Identifier (QCI) which is also a scalar value associated with a bearer that defines the traffic type and expected service. A comparison between 4G/LTE and 5G/NR parameters is shown in Table 2–1.

The mapping between traffic categories and DiffServ code points is implementation-specific but done based on 5QI and PL (per 3GPP 38.414). Recommendations have been published by IETF.[1]

Reflective Quality of Service

Optionally supported by UEs, the reflective QoS concept tells the UE to monitor the downlink for the QoS that has been applied by the RAN to the traffic. It then applies the observed QoS parameters in the uplink for the traffic it generates toward the RAN.

Packet Data Convergence Protocol (PDCP)

In earlier versions of Global System for Mobile communication (GSM) and 3GPP standards, PDCP was only used for the packet data bearers, and the circuit switched bearers connected directly from the host to the RLC layer. Because LTE is all-packet, this is now a place for higher layer functions that sit above the packing and unpacking that goes on in the RLC. PDCP functions in the UP include encryption/decryption, RoHC header decompression, sequence numbering, and duplicate removal. PDCP functions in the CP include encryption/decryption, integrity protection, sequence numbering, and duplicate removal. There is one PDCP instance per radio bearer.

Table 2–1 4G and 5G QoS parameters.

	5G/NR	4G/LTE
QoS identifier	5G QoS Identifier (5QI)	QoS Class Identifier
QoS granularity	QoS flow	EPS bearer
Flow/bearer identifier	QoS Flow Identifier (QFI)	EPS bearer ID (EBI)
Reflective QoS	Supported through Reflective QoS Indicator (RQI)	Not supported

Dual and multiconnectivity

The concept of multiconnectivity means that a single UE/client device is connected to multiple radios (RUs, TRPs) at the same time. DC was first introduced in Release 12 specification for LTE, assuming a UE to be connected to two eNBs at the same time, where the eNBs are connected over the X2 link. The concept of a Master and Secondary eNB (MeNB and SeNB in 4G or MgNB and SgNB in 5G) was established to define a single base station (the master) that maintains the CP connectivity, where the UP connectivity can be provided by both eNBs. Consider the scenario with the MeNB as a macro cell and a SeNB as a small cell as a typical deployment scenario in Fig. 2–11.

3GPP defines different multiconnectivity options that allow deployment in various operator core network and base station configurations, with the target of allowing a mix-and-match approach between 4G or 5G Core Networks and 4G and 5G RAN (Base Stations). The purpose is to allow the operators to incrementally build a network deployment, where the network and RAN components do not need to be swapped out all at once. Keep in mind that the Option numbering terminology for deployment options as we are discussing here is distinctly different from the Option 1, 2, 3, . . ., 8 functional splits between CU, DU, and RU.

To explain the options, we start with the straightforward ones, options 1 and 2 as in Fig. 2–12.

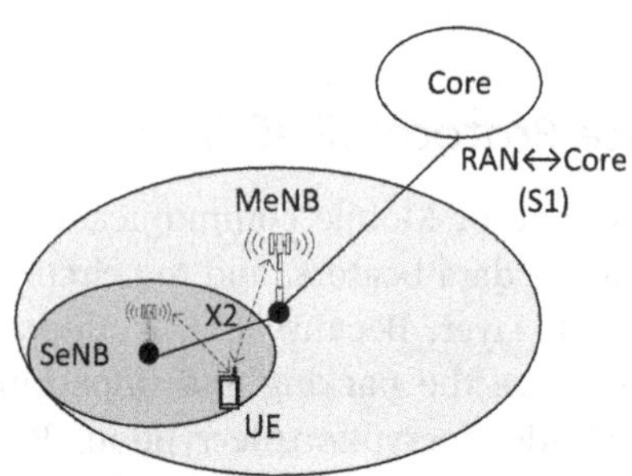

FIGURE 2–11 LTE dual connectivity with MeNB and SeNB. *LTE*, Long-Term Evolution; *MeNB*, Master eNB.

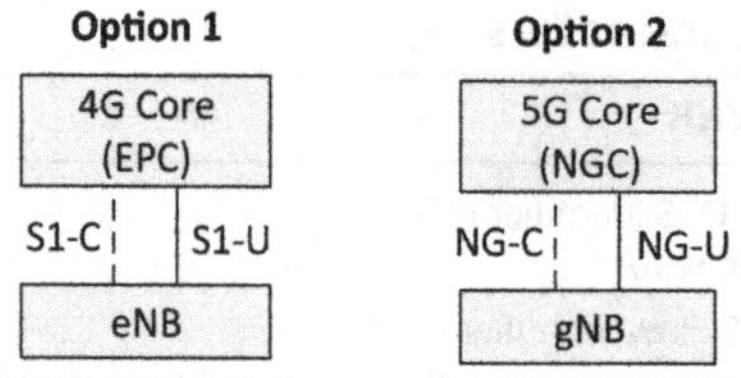

FIGURE 2–12 Deployment options 1 and 2.

These options show an LTE-only (LTE eNB and LTE EPC) and a 5G-only (5G gNB and 5G NGC) deployment with 4G and 5G standards defining C-plane and U-plane connectivity between the two.

Options 3, 3a, and 3x represent Non-Stand Alone (NSA) deployments where the core network is LTE (EPC) and both 4G/LTE and 5G/NR RANs are used for wireless access. The difference between the suboptions defines the interconnectivity between the core and the RAN components as well as between the 4G and 5G RAN components. In Option 3, the LTE eNB is connected to the EPC with non-standalone NR. The NR UP connection to the EPC goes via the LTE eNB (Option 3) or directly (Option 3a/3x), refer to Fig. 2–13. Option 3x is a combination of 3 and 3a in which user data traffic flows from EPC to gNB from where it is delivered over the 5G/NR air interface or forwarded over the X2 interface to the eNB for transmission over the 4G/LTE air interface. Typically, the forwarded (LTE air interface targeted) traffic is voice or another low-bit rate application:

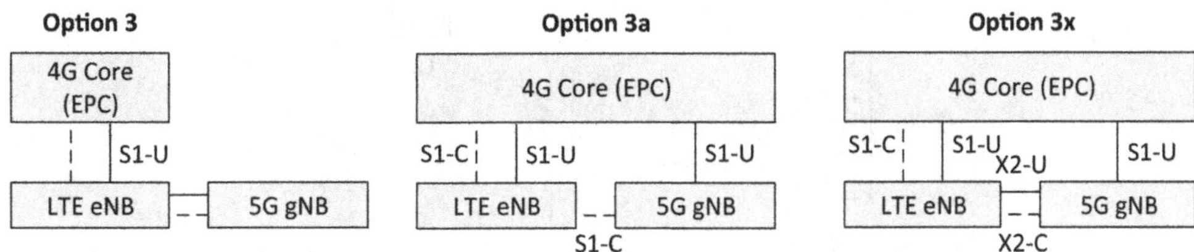

FIGURE 2–13 Deployment options 3, 3a, and 3x.

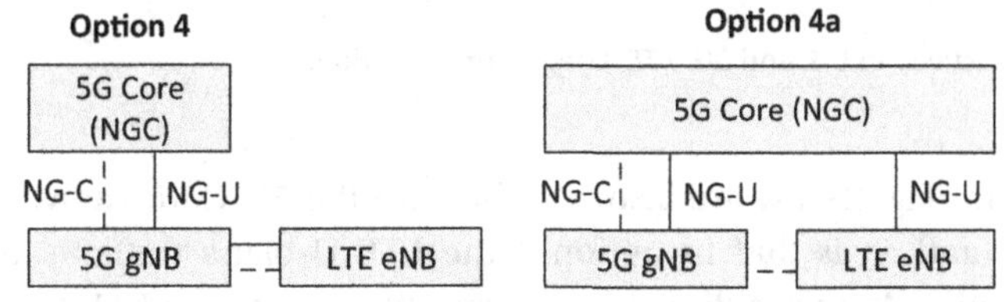

FIGURE 2–14 Deployment options 4 and 4a.

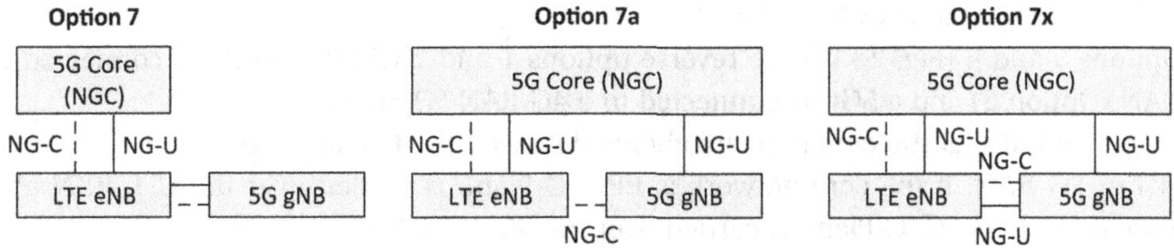

FIGURE 2–15 Deployment options 7, 7a, and 7x.

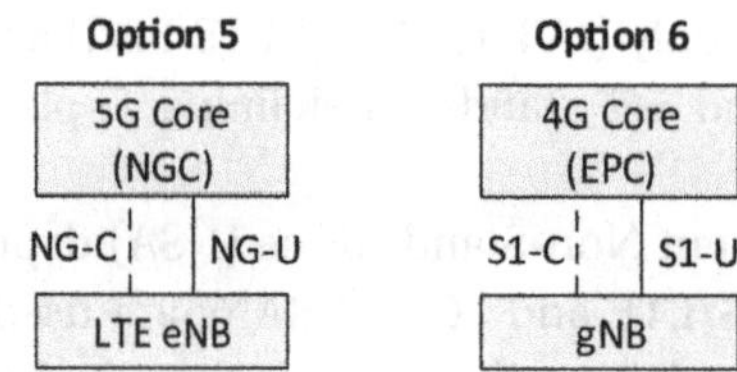

FIGURE 2–16 Deployment options 5 and 6.

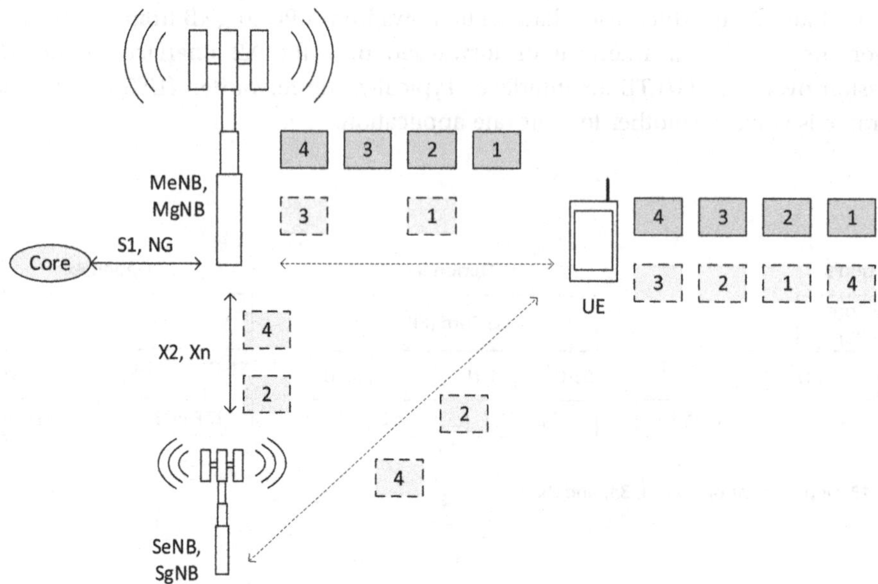

FIGURE 2–17 Dual connectivity in LTE and 5G. *LTE*, Long-Term Evolution.

Options 4 and 4a (Fig. 2–14) are also NSA but use the 5G Core Network (5GC). The difference between 4 and 4a is that in option 4 the LTE U-plane traffic is routed via the 5G RAN, whereas in option 4a this is directly sent from the 5GC to the LTE RAN.

Similarly, Options 7, 7a, and 7x (Fig. 2–15) are also NSA with a 5G NGC, but the difference between Option 4 and Options 7 is that the primary connection (and C-Plane) connection between core and RAN is to the LTE eNB.

Options 5 and 6 (Fig. 2–16) are reverse options 1 and 2 where an EPC is connected to a 5G RAN (Option 5) and a 5GC is connected to a 4G RAN (Option 6):

Keep in mind that the core network choice defines if the C-Plane signaling is carried over the 4G or 5G RAN. If the core network is EPC, C-Plane is carried over the LTE RAN and if the core network is 5G, C-Plane is carried over the 5G/NR RAN.

The concept of dual connectivity (Fig. 2–17) adds a level of complexity to the definition of the PDCP implementation to take care of buffering and packet distribution from MeNB to

Secondary eNB (SeNB). The concept is that an MeNB "feeds" enough packets to the SeNB to keep the SeNB↔UE link occupied while not overloading the SeNB with packets because (assuming the SeNB is a smaller cell size), in the case of mobility, the SeNB would no longer be able to maintain a connection toward the UE. Flow control between the MeNB and SeNB is a topic of differentiation between vendors. When implemented in different physical units (i.e., CU and DU), this communication link is defined in the 3GPP 38.425 standard.

Initial definition of DC in Release 12 for LTE assumed a single MeNB and a single SeNB and was designed for downlink traffic only. As standardization evolved, features were included to support uplink (Release 13), and in Release 14, the concept of multiconnectivity where the UE/client can connect to multiple nodes as the same time. This concept is carried forward in 5G. ENDC stands for E-UTRAN New Radio Dual Connectivity where the client device is connected to LTE and 5G networks at the same time, reducing handovers between the two networks and increasing (cumulative) available spectrum and therefore throughput.

Distributed unit

Given that DC uses a single PDCP entity to serve both MeNB and SeNB links, it is "natural" to make the split between CU and DU along the PDCP/RLC interface as is done in the "Option 2" logical split. This also eliminates the CP components to reside in the DU, which reduces the DU implementation complexity and allows for mix-and-match of CU and DU components in the network.

F1 termination

The DU needs to terminate the standardized F1-C and F1-U interfaces to and from the CU unit. This includes flow control as we described for the CU. Flow control limits payload buffering (and thus, DDR dimensioning) in the DU, keeping system cost and complexity down.

Radio Link Control and Medium Access Control

The 5G Layer-2 protocol stack is responsible for converting IP packets that are carried in QoS flows to transport blocks (TBs) that can be transmitted over the air, and vice versa. This protocol stack consists of four layers: SDAP and PDCP (associated with the CU) and RLC and MAC (associated with the DU). The latter is being discussed here. In these discussions, we use 3GPP terminology for the frames that go in and come out of the RLC and MAC layers (Fig. 2–18). Frames ingress to the RLC from the PDCP layer are called RLC SDUs and frames egress from the RLC layer are called RLC PDUs. These RLC PDUs that are ingress to the MAC layer are also called MAC SDUs and MAC output frames are called MAC PDUs. These MAC PDUs are the same as what is ingress to the Physical Layer: TBs.

Per 3GPP 38.322, RLC functions are defined as:

- transfer of upper layer PDUs according to transmission modes (transparent mode/TM, unacknowledged mode/UM and acknowledged mode/AM)

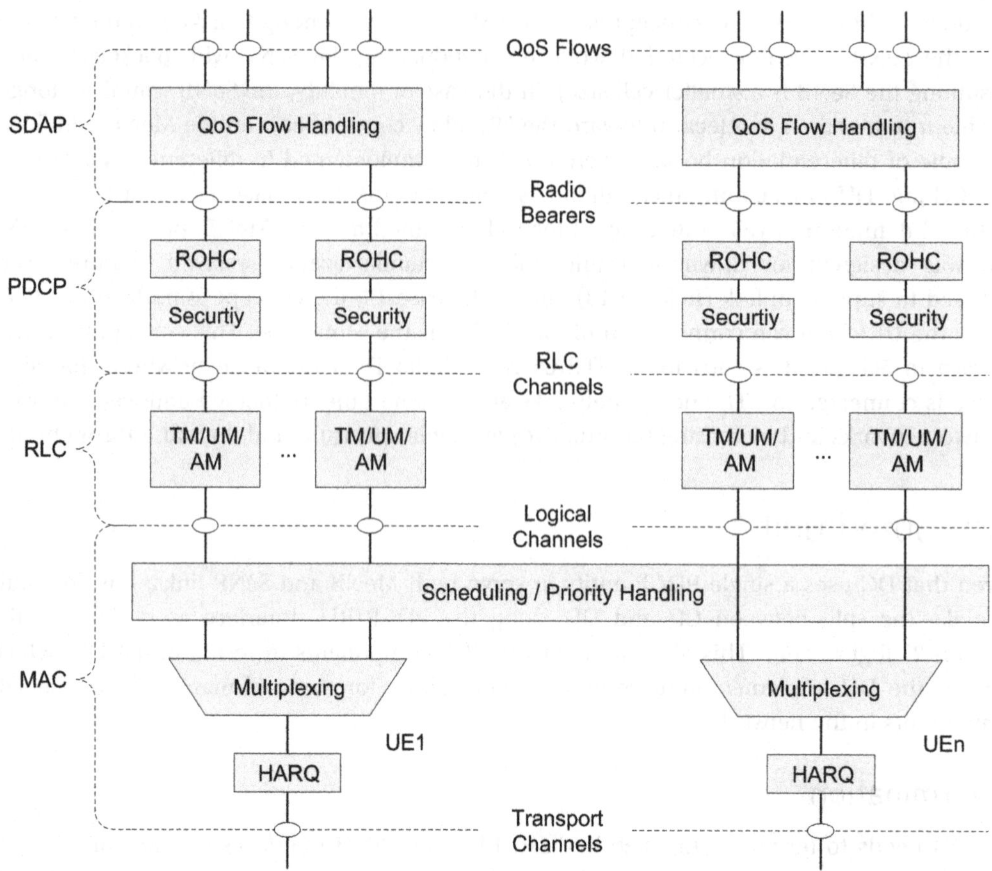

FIGURE 2–18 3GPP Layer-2 protocol stack and interfaces (replicated from 3GPP 38.300).

- error correction through retransmission in AM
- sequence numbering
- segmentation/resegmentation

RLC layer flows are set up, reestablished, and released by RRC layer commands and flows are unique per radio bearer/channel (and obviously, per UE).

The RLC transfer modes are defined by the type of channels that they are used for and the traffic characteristics associated with them. TM is associated with BCCH, CCCH, and PCCH. TM processing is effectively a bypass mode—there are no headers added and the RLC functionality is nil. UM processing supports fragmentation and reassembly. This function is needed to map the variable (IP packets after all) size RLC SDUs to the MAC PDU size that is determined by time/frequency resources that are dictated by Physical Layer limitations and allocated by the MAC scheduler algorithm. To enable out-of-order processing on the receiving side, the UM RLC protocol adds a sequence number and offset field and a couple of segmentation indicator bits in an RLC header that is attached to the PDU. AM operation is the same as UM operation but includes an acknowledgment mechanism through status reports that allows retransmission of dropped packets and thus reliable packet delivery to upper layers at the cost of increased latency associated with retransmission.

Selection between AM and UM operation is done during channel establishment and done based on the target QoS characteristics. UM is used when reliability is not required—for example, for voice and video traffic. AM is used when reliability is required, for example, for TCP-based flows such as web traffic.

Per 3GPP 38.321, MAC functions include:

- map between logical channels and transport channels and as such present logical channels to the RLC and upper layers
- multiplexing/demultiplexing of RLC PDUs onto MAC PDUs or TBs that are transmitted and received by the Physical Layer
- scheduling and priority handling
- error correction through Hybrid Automatic Repeat Request (ARQ) (HARQ)
- transport format selection

By far, the most complex and compute (and development!) intensive function is the scheduling/priority handling one (Fig. 2–19). Next level details on scheduler algorithm design are covered separately and extensively in chapter 7. The figure shows the high-level design with regards to inputs and outputs.

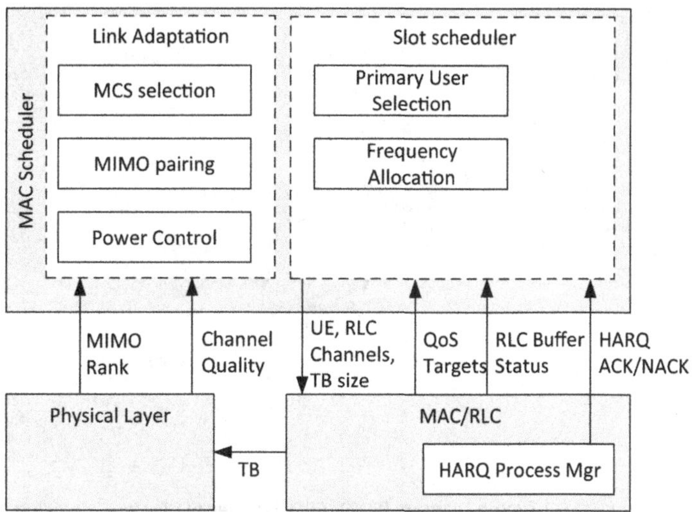

FIGURE 2–19 MAC scheduler interfaces. *MAC*, Medium Access Control.

Physical Layer TBs can be of any arbitrary size but are defined from a discrete set of options that is defined by:

- the number of layers
- modulation order
- code rate
- number of resource elements (allocation in frequency domain)
- physical transmission time (allocation in time domain)

3GPP 38.214 includes the procedures used for determining the (discrete) TB size from above mentioned parameters. The scheduler algorithm itself thus essentially defines layer/modulation/code rate/time + frequency resource allocations on a per-user basis. The scheduler operates every slot and outputs a list of allocated UEs and for each UE. These allocated resources together with an indication of which RLC logical channels the resources should be allocated to. This is input to the MAC PDU processing component that is responsible for composing the TB payload (= MAC PDU) by building MAC headers/subheaders and RLC PDUs. This MAC PDU processing component is responsible for calling the RLC layer sequentially to generate payload that composes the MAC PDU.

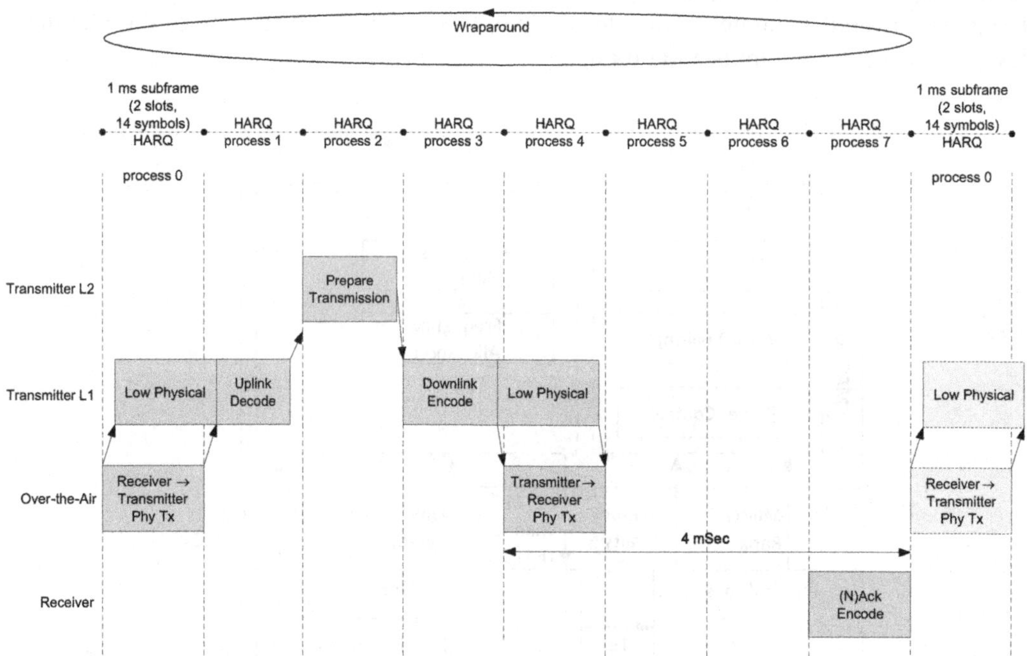

FIGURE 2–20 LTE synchronous HARQ Timing Diagram (FDD). *FDD*, Frequency Division Duplexing; *HARQ*, Hybrid Automatic Repeat Request; *LTE*, Long-Term Evolution.

Hybrid Automatic Repeat Request

HARQ is a combined Physical Layer (forward error correction) and MAC layer retransmission protocol that saves the information from a previously unsuccessful decode at the Physical Layer to combine this information with incremental retransmitted (redundancy) information in subsequent decode attempts. When the receiver fails to decode a TB [Physical Layer cyclic redundancy check (CRC) fails], it transmits a single HARQ Negative Acknowledgement (Nack) to the transmitter to indicate a request for additional information bits to try and do the decode operation again. HARQ is maintained by the MAC/MAC scheduler given that this entity is responsible for the time/frequency resource scheduling, including scheduling of HARQ resources. HARQ protocol in MAC is complemented by ARQ when operating in RLC AM to increase reliability of the end-to-end connection. HARQ can operate in synchronous or asynchronous manner. We first look at synchronous HARQ as used, for example, in the LTE uplink as shown in Fig. 2–20:

In the LTE synchronous HARQ implementation, the HARQ process ID is reused every eight subframes (shown as the wraparound in the figure above), assuming FDD operation. Time-Division Duplex (TDD) operation is predefined using a set of standardized timing tables but operates in similar concepts. In the example shown here:

- The transmitter prepares (at MAC level) a new transmission at time (counted as HARQ process ID) = 2.
- This TB is encoded by the Physical Layer at time = 3 and transmitted (low-PHY) at time = 4.
- Dictated by 3GPP standards, the receiver has four slots of time to prepare and transmit the associated Ack/Nack indication. This brings us to time = 8, which wraps around to time = 0 (there are 8 HARQ processes in the synchronous uplink for LTE).
- This Ack/Nack indication is decoded at time = 9 (or time = 1 in wraparound) for potential retransmission by the base station at time = 10 (or time = 2 in wraparound).

In the synchronous method of operation, the base station knows exactly when to expect the HARQ Ack/Nack indication and no explicit signaling is required: there is a fixed timing relationship between the over-the-air initial transmission (in above example, time = 4) and the transmission of HARQ Ack/Nack by the receiver, also over the air (in above example, time = 8 or wraparound to time = 0). The HARQ context (in 3GPP language: Process ID) is not signaled but inferred from the frame and subframe number.

Asynchronous HARQ is very similar from the functional point of view, but the timing aspects are different. In Asynchronous HARQ, where the HARQ process ID is assigned by the MAC scheduler and hence can be different for each (re)transmission. At the cost of explicit signaling of the HARQ process identifier, the retransmissions of a packet happen at any time compared to the initial transmission. Asynchronous HARQ as such allows more flexibility in scheduling. The mechanism for both the sender and receiver to know which HARQ process is used is the Downlink Control Information (DCI) indication. DCI carries the field called HARQ processor number. In LTE, only the DCI for downlink scheduling carries this field (since LTE DL use asynchronous HARQ) and the DCI for uplink scheduling does not carry this field (since LTE UL use synchronous HARQ).

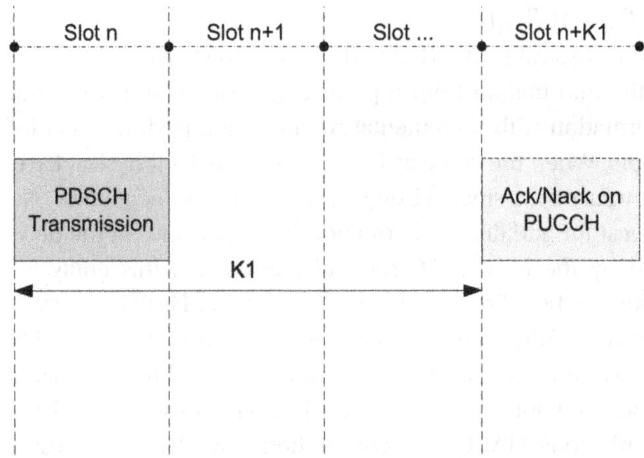

FIGURE 2–21 5G/NR asynchronous HARQ signaling. *HARQ*, Hybrid Automatic Repeat Request.

However in NR, both downlink scheduling DCI (i.e., DCI 1_0, 1_1) and uplink scheduling DCI (i.e., DCI 0_0, 0_1) carry the field HARQ processor number since they both use asynchronous HARQ. Given that the MAC scheduler controls what DCI are generated, the MAC scheduler can control allocation of both DL and UL HARQ process on a per-link basis. This gives the scheduler the tools to manage HARQ associated context store/restore bandwidth.

In LTE:

- Uplink HARQ is synchronous with 8 HARQ processes.
- Downlink HARQ is implemented asynchronous with 8 HARQ processes.

In 5G/NR:

- Uplink HARQ is asynchronous with ≤ 16 (RRC configured bound/number) HARQ processes. Note there is no explicit Ack/Nack signaling for UL HARQ in 5G/NR. Indication to the UE that it can start a new transmission is done by the base station through toggling the New Data Indicator signaling bit.
- Downlink HARQ is implemented asynchronous with ≤ 16 (RRC configured bound/number) HARQ processes.

Downlink HARQ timing in 5G/NR is signaled through parameter K1 by the base station to the client as shown in Fig. 2–21.

K1 defines the number of slots between the base station transmission and the expected Ack/Nack from the UE. In a so-called self-contained slot scenario, K1 can be as small as 0. K1 minimum capabilities are signaled by the client to the base station during connection establishment. K1 and HARQ timing capabilities on both client and base station side are defined by multiple parameters:

- Transmitter and receivers' Physical Layer capabilities:

- decoding, encoding latencies
- buffering capabilities
- modem sensitivity
- 3GPP specifications:
  - slot structures (UL/DL split, TDD structure)
  - Transmission Time Interval (TTI) duration: number of symbols per slot (e.g., mini slots), symbol duration/subcarrier spacing (SCS)
- Higher layers:
  - MAC scheduler
  - radio bearer management
  - max number of retransmissions
- Deployment scenario:
  - cell size/propagation delays
  - number of users per cell
  - propagation channel conditions

Note that supporting uplink synchronous HARQ has been limiting the deployment of O-RAN systems for 4G/LTE in conditions where the fronthaul is nonideal with regards to its latency. As the fronthaul interface latency goes up, supporting the strict 8-process synchronous HARQ requirement becomes impossible. According to Ref. [2], two strategies can be deployed to mitigate this challenge. The first one is called HARQ process interleaving where an HARQ process is not reused until the corresponding Ack/Nack has been received, rather than assuming the Ack/Nack reception to have taken place at a predefined time. This comes at the cost of throughput (we are not able to use all the allocated time resources in this way). The second strategy is referred to as "predictive HARQ," for example, backing off the target modulation and coding scheme to ensure higher rate of successful decode at the Physical Layer, thus not needing HARQ retransmission. In this type of scheme, the RLC (AM) layer ensures reliability through retransmission.

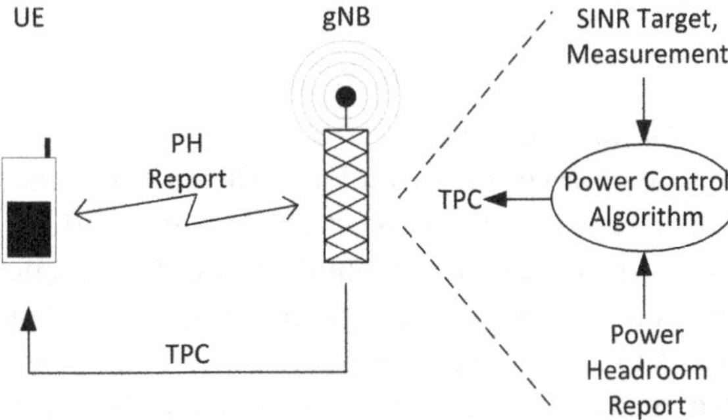

FIGURE 2–22 Closed-loop power control in 3GPP.

Power control

Receive side

3GPP systems, including 3G, 4G/LTE, and 5G/NR rely on closed-loop power control to aim for the received power from the multiple clients to the same at the base station. Keeping the received power (close to) the same for all clients allows the base station to decode the signals from multiple simultaneous transmissions such as are happening in an OFDMA waveform. For this purpose, the base station uses Transmit Power Control (TPC) commands (Fig. 2–22) to indicate to the client to increase or decrease transmit power, while the client reports back available headroom to the base station through power headroom reports.

The power control algorithm is proprietary to the base station implementation, but target performance equations are included in 3GPP 38.213.

Given that power control requires a round-trip (closed-loop) connection between base station and client, initial access cannot use this mechanism. For this purpose, initial access uses an open-loop power control mechanism that relies on the base station signaling target RACH preamble received power as well as base station transmitted reference signal power to the client for it to be able to estimate initial transmit power. A power ramp-up algorithm is employed beyond this to ensure that in less favorable radio conditions the base station can still end up detecting the RACH preambles.

Transmit side

3GPP systems have a limited range TPC. Two 3GPP requirements define the TPC:

- RE (Radio Equipment) power control dynamic range which is the difference between the power of a single RE and the average RE power for a BS—in effect the dynamic range of the RE level power control. Requirements are specified in 38.104 as maximum of +4 dB (up) and −6dB (down).
- The BS total power dynamic range is the difference between the maximum and the minimum transmit power of an OFDM (orthogonal frequency division multiplexing) symbol for a specified set of reference condition, which include transmission over a single RE versus transmission over a full OFDM symbol in the frequency domain. Requirements are specified in 38.141 and are up to 24.3 dB for a 100 MHz <6 GHz channel.

Discontinuous transmission and reception

A wireless data communication receiver uses a lot of DC energy, associated with digital signal processing (DSP), data conversion, analog demodulation, and so on. While data rates continue to increase, batteries only get incrementally more performance, putting more pressure on energy efficiency. Process node advancements help on the (mainly) digital side but there are additional savings to be had at protocol level, where the objective is to define a pattern of enabled/disabled regions during which the (mobile phone) receiver knows it can/cannot expect transmission from the base station. The objective is to turn off the radio for the most time possible while staying connected to the network. The radio modem can be turned off "most" of the time, while the mobile device stays connected to the network with reduced throughput. The receiver is turned on at specific times for updates.

For this purpose, both LTE and 5G/NR power save protocols include Discontinuous Reception (DRX) and Discontinuous Transmission (DTX). Both involve reducing transceiver duty cycle while in active operation. DRX also applies to the RRC_Idle state with a longer cycle time than active mode. However, DRX and DTX do not operate without a cost: the UE's data throughput capacity is reduced in proportion to power savings.

The RRC sets a cycle where the UE is operational for a certain time when all the scheduling and paging information are transmitted. The base station knows that the UE is completely turned off and is not able to receive anything. Except when in DRX, the UE radio must be active to monitor PDCCH (to identify DL data). During DRX, the UE radio can be turned off.

Semi-persistent scheduling

Semi-persistent scheduling (SPS) involves the base station dedicating resources for transmission and/or reception (independently) over multiple TTIs or slots. SPS is unique per cell and configured by the RRC layer by means of an indication of the UE identifier [Radio Network Temporary Identifier (RNTI)], periodicity, number of associated HARQ processes, and time/frequency resources (for the uplink case). The purpose of SPS is to reduce scheduling overhead associated with regular transmissions such as voice over IP.

Bandwidth adaptation/bandwidth part operation

5G/NR supports wide bandwidth operation, for example, in mmWave applications with carrier bandwidth of up to 400 MHz. This wide bandwidth may not be required by the

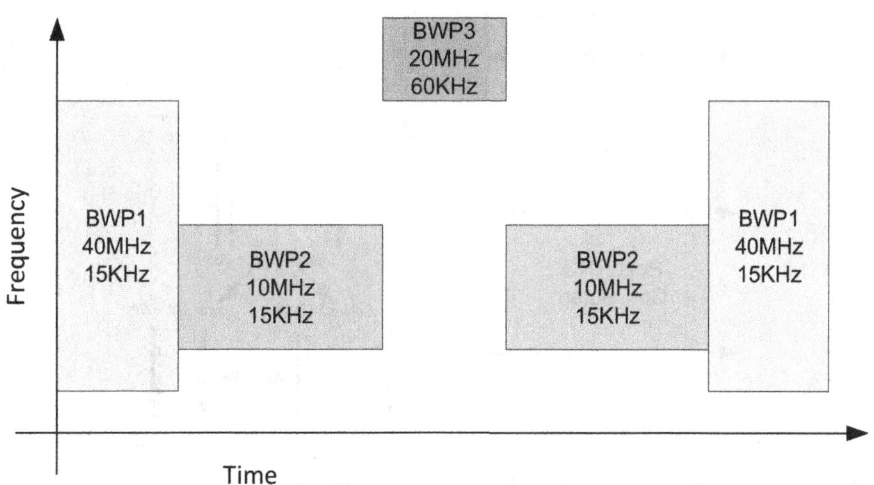

FIGURE 2–23 Bandwidth adaptation.

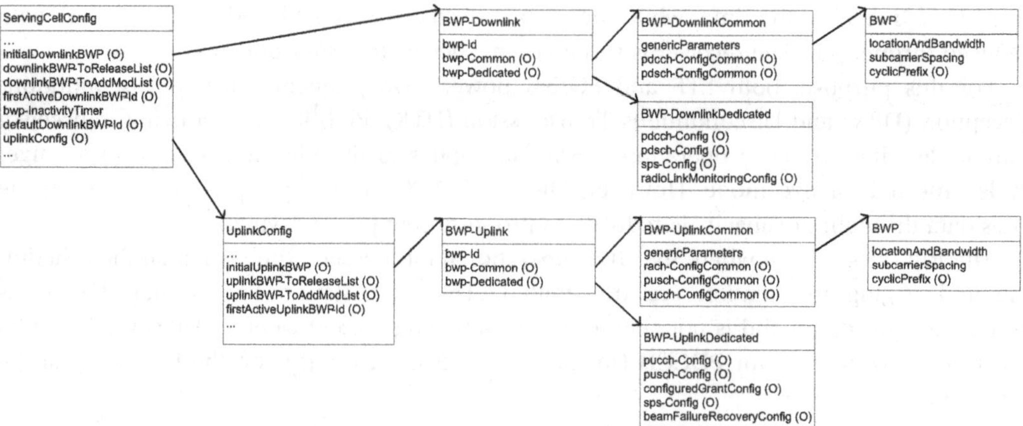

FIGURE 2–24 BWP-associated configuration information elements. *BWP*, bandwidth part.

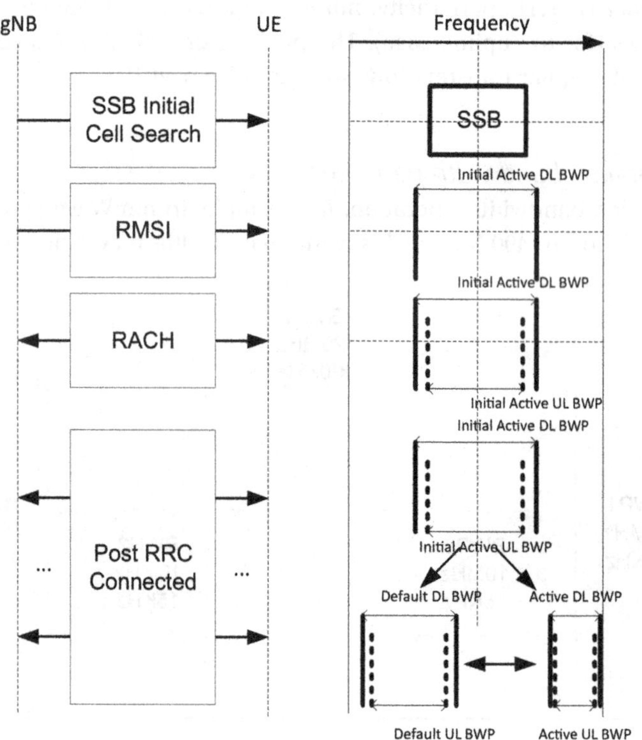

FIGURE 2–25 BWP configuration and operation. *BWP*, bandwidth part.

application, leading, for example, to unnecessary energy consumption caused by high sample rates and DSP processing requirements as well as inefficient RF utilization.

With bandwidth adaptation (BA), the receive and transmit bandwidth of a UE need not be as large as the bandwidth of the cell and can be adjusted: the width can be ordered to change (e.g., to shrink during period of low activity to save power); the location can move in the frequency domain (e.g., to increase scheduling flexibility); and the SCS can be ordered to change (e.g., to allow different services). A subset of the total cell bandwidth of a cell is referred to as a Bandwidth part (BWP) and BA is achieved by configuring the UE with BWP (s) and telling the UE which of the configured BWPs is currently the active one.

Each BWP can have unique numerology (e.g., SCS) and width (frequency). For example, as shown (copied from 38.300) in Fig. 2–23.

BWPs (defined per component carrier in the case of carrier aggregation) are communicated to the UE through RRC layer configuration of the BWP definition and ID through a "bandwidthPartId" (≤4 per UE) and configuration of location [frequency/physical resource block (PRB) offset], bandwidth, SCS, and cyclic prefix (CP) configuration. See outline of the associated information elements (IEs) per RRC specification (3GPP 38.331 and copied in Fig. 2–24).

When BA is configured, the UE monitors the control channel (PDCCH) on the active BWP(s). It does not have to monitor PDCCH on the entire DL frequency of the cell. A BWP inactivity timer (independent from the DRX inactivity timer described earlier) is used to switch the active BWP to the default one: the timer is restarted by the MAC entity upon successful PDCCH decoding and the switch to the default BWP takes place when it expires. On deactivated BWPs, the UE does not monitor the control channel and does not transmit on any channel. Fig. 2–25 shows the concept.

Carrier bandwidth part is a contiguous subset of the PRBs defined for a given numerology on a given component carrier.

Supplementary Uplink operation

Supplementary Uplink (SUL) operation is proposed in 3GPP 38.300 as a method to improve cell edge coverage, in a situation where the UE transmit power is (likely) lower than the base station power, giving reduced performance in the uplink as opposed to the downlink.

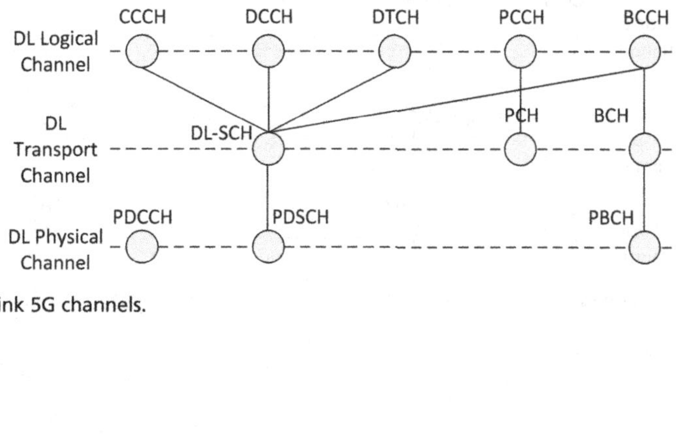

FIGURE 2–26 Downlink 5G channels.

The idea involves using a different RF band (at a lower RF frequency and thus with better propagation characteristics) for uplink transmission.

Several bands are defined for potential SUL use (n80,., n84 and N86), all of them below 2 GHz. SUL operation is designed to require only a single uplink MAC/Physical Layer instance to be operational at each point in time, even when the client is configured to support multiple uplink bands. This reduces (client) implementation complexity and cost.

Physical Layer

The Physical Layer is responsible for channel encoding including physical-layer HARQ; modulation; multiple-input, multiple-output (MIMO) (multiantenna) processing; and mapping of the signal to the appropriate physical time–frequency resources.

Channel mapping

The communication path all the way from the wireless interface (the air) and RLC layer is done through a set of so-called channels or communication pipes and interfaces that function at different levels of the stack. These channels are well-defined and become the anchor point for implementation and test. At the lowest level toward the air interface, a *physical channel* is defined by time/frequency resources in the OFDM grid that is used for transmission of a Transport channel that holds logical information coming from the MAC layer. The Physical Layer mapping function translates (maps) transport channels to physical channels. Transport channels are presented by the Physical Layer to the MAC layer where each transport channel has its own characteristics for transmission over the air interface (say, broadcasted or not or with specific targets for reliability and latency). Logical channels are presented by the MAC layer to the RLC. They are defined by the type of information they carry. By implication, multiplexing control channels to transport channels is done by the MAC layer.

Logical, transport, and physical channels in the downlink are shown in Figs. 2–26 and 2–27.

Logical channels (presented by MAC layer to RLC) are:

- Downlink:

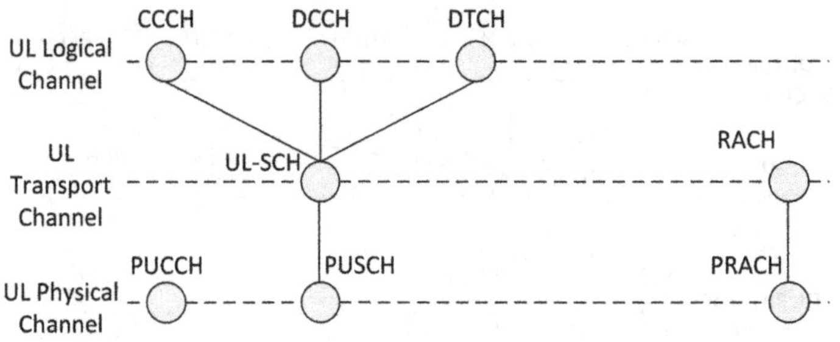

FIGURE 2–27 Uplink 5G channels.

- Broadcast Control Channel (BCCH)—used in DL only to transmit Master Information Block (MIB) that includes basic information required by the client to detect the cell site and the SI blocks that contain more RAN-related information including frame and radio access information needed to connect to the RAN.
- Paging Control Channel (PCCH)—used to page UEs for whom the location in the network is not known.
- Common Control Channel (CCCH)—used for signaling control information to and from pools of UEs, for example, including initial access-related control information,
- Dedicated Control Channel (DCCH)—used for signaling control information between a single UE.
- Dedicated Transport Channel (DTCH)—used for carrying user payload between the RAN and the UE.

- Uplink:
  - Broadcast Control Channel (BCCH)—used in DL only to transmit MIB that includes basic information required by the client to detect the cell site and the SI blocks that contain more RAN-related information, including frame and radio access information needed to connect to the RAN.
  - Paging Control Channel (PCCH)—used to page UEs for whom the location in the network is not known.
  - Common Control Channel (CCCH)—used for signaling control information to and from pools of UEs, for example, including initial access-related control information,
  - Dedicated Control Channel (DCCH)—used for signaling control information between a single UE.
  - Dedicated Transport Channel (DTCH)—used for carrying user payload between the RAN and the UE.

Transport channels (presented by Physical Layer to MAC) are as follows:

- Downlink:
  - Shared Channel (DLSCH)—used for bulk data transfer from the gNB to the UE.
  - Broadcast Channel (BCH)—used for transmission of parts of the BCCH SI, more specifically the so-called MIB.
  - Paging Channel (PCH)—used for transmission of paging information.
- Uplink:
  - Shared Channel (ULSCH)—used for bulk data transfer from the UE to the gNB.
  - Random Access Channel (RACH)—used for initiating access by the UE to the gNB.

Physical channels (presented by lower Physical Layer to the Upper Physical Layer) are:

- Downlink:
  - Physical Downlink Control Channel (PDCCH)—main responsibilities include transmission of control information that communicates to the UEs the scheduling information for both the downlink and the uplink shared channels.

Table 2–2 5G Numerology.

μ	$\Delta f = 15\ \text{KHz} \times 2\mu$	Cyclic prefix
0	15 KHz	Normal
1	30 KHz	Normal
2	60 KHz	Normal, extended
3	120 KHz	Normal
4	240 KHz	Normal
5	480 KHz	Normal

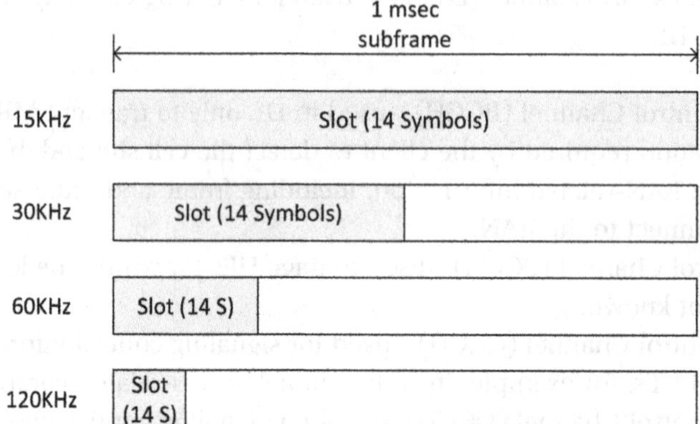

FIGURE 2–28 Frame structure.

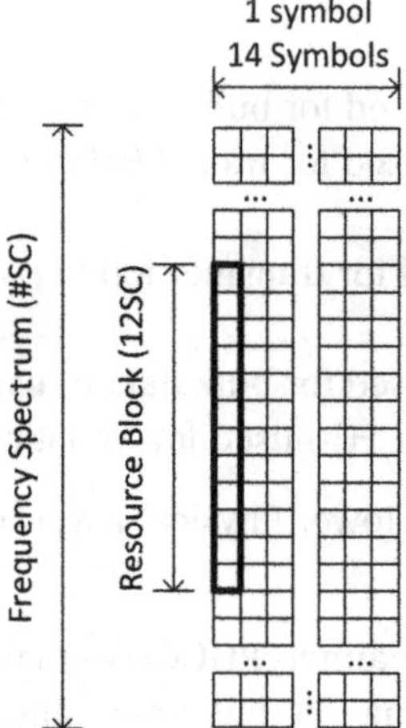

FIGURE 2–29 Resource block definition.

  - Physical Downlink Shared Channel (PDSCH)—this carries bulk data in time and frequency dimensions, sharing the channel capacity to different users.
  - Physical Broadcast Channel (PBCH)—this channel is used to communicate the MIB that is used as the first anchor for channel discovery from UE to gNB.
- Uplink:
  - Physical Uplink Control Channel (PUCCH)—used to carry uplink control data, or at least the portion of the uplink control data that is not transmitted through the PUSCH.
  - Physical Uplink Shared Channel (PUSCH)—counter to the PDSCH, this channel is used for bulk data transfer from UE to gNB.
  - Physical Random Access Channel (PRACH)—used for transmission of a channel access preamble (sequence) to initiate connectivity from UE to gNB.

Radio Network Temporary Identifier

RNTI stands for Radio Network Temporary Identifier and is used to identify a connected UE or group of UEs (paging) a radio channel or similar. It is the identifier that is used to identify either a UE within a cell (typically), a specific radio channel or a group of UEs (for paging, power control or system information transmission) by the gNB. The RNTI is used by the Physical Layer to identify which channels it needs to decode.

Numerology

A frame is defined as a 10-ms interval and is divided into subframes of 1 ms each. Air interface processing (including scheduling) is done on a group of 14 OFDMA symbols, which is defined as a slot. The physical time duration of a slot is defined by the parameter μ which defines the SCS, per Table 2–2.

For shared channel, 15, 30, and 60-KHz SCS is used in <6 GHz deployments where 120, 240, and 480-KHz SCS is used in >6 GHz deployments. Synchronization channels support 15/30 KHz SCS (<6 GHz) or 120/240 KHz (> 6 GHz) (Fig. 2–28).

5G also defines a so-called mini slot which is a sequence of 2, 4, or 7 symbols that are a smaller unit of scheduling used in URLLC systems, but that is not covered in-depth here.

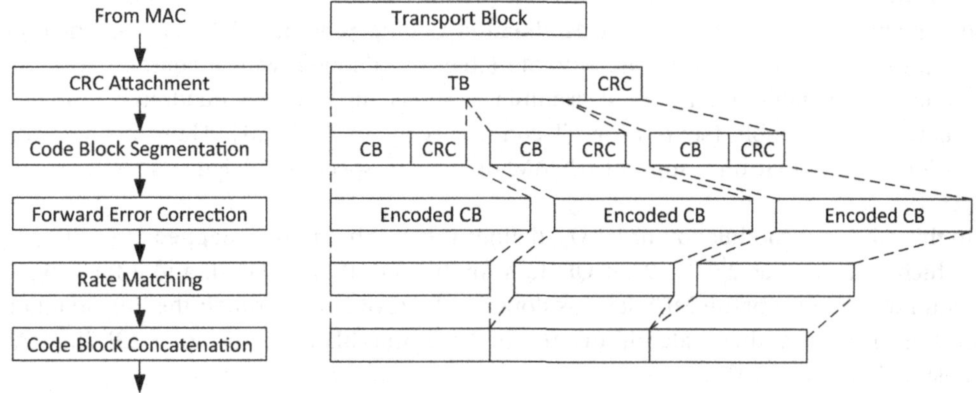

FIGURE 2–30 Channel coding and associated bit processing (3GPP 38.212).

In the frequency axis, subcarriers are grouped into PRB which is a group of 12 subcarriers as shown in Fig. 2–29.

A slot can be allocated to be all-downlink, all-uplink or a combined uplink + downlink slot (mixed slot), where the mixed slot configuration can be either static or dynamic (unique for each slot). The slot format is indicated through the Slot Format Indicator (SFI) which can be static or semistatic (configured through the RRC) or dynamic (configured through the DCI carried in the control channel).

• The Physical Layer processing chain is split into two pieces: channel coding (covered in 3GPP 38.212) and physical channel mapping and modulation (covered in 3GPP 38.211).

Channel coding (3GPP 38.212)

Channel coding transforms the MAC TB into a set of bits that can be mapped into the OFDM frame (channel) that is transmitted over the air. As shown in Fig. 2–30, this includes error detection (CRC) to be able to detect any errors in the wireless transmission, channel coding to add redundant information to reduce the impact of channel imperfections, and rate matching to reduce channel coding overhead and match the encoded TB size to the available space on the OFDM frame, as shown for downlink/uplink shared channels:

In a little bit more detail:

- Code block segmentation—this breaks the TB (or MAC PDU) as delivered by the MAC layer into smaller code blocks (CBs) with discrete size options. This stage also adds two CRCs: one at TB level and one for each CB.
- Channel (en)Coding—for the shared channel, this is done using LDPC coding. Control signaling is encoded with polar coding.
- Rate matching—channel coding increases the number of bits transmitted over the wireless channel by 3 ×. Rate matching is employed to reduce the coding overhead. Interleaving is also implemented in this step.

Overall code rate for the channel coding stage is defined as the *ratio* between useful bits (MAC PDUs or TBs) and total transmitted bit (useful + redundant, post rate matching). This ratio is defined in 3GPP 38.214 using Modulation Coding Scheme (MCS) tables, which predefine possible coding rate options. Multiple tables are defined, with a complex scheme to determine which table applies. One example is shown for the 256 quadrature amplitude modulation (QAM) table (supporting all code rates up to 256QAM). There are two other tables for systems that only support 64QAM or for low spectral efficiency systems such as URLLC (Table 2–3).

In this table, the modulation order Q_m defines the number of bits mapped to each IQ carrier which is defined as 2^{Q_m} or 2 for QPSK, 4 for 16QAM, 6 for 64QAM, and 8 for 256QAM. Selection of the appropriate MCS level is done by the gNB (DU) through the link adaptation algorithm. Typically, these algorithms try to limit the BLock Error Rate (BLER)/HARQ retransmission rate to 10%.

Table 2–3 MCS index table 2 for PDSCH (3GPP 38.214).

MCS index I_{MCS}	Modulation order Q_m	Target code rate R × [1024]	Spectral efficiency
0	2	120	0.2344
1	2	193	0.3770
2	2	308	0.6016
3	2	449	0.8770
4	2	602	1.1758
5	4	378	1.4766
6	4	434	1.6953
7	4	490	1.9141
8	4	553	2.1602
9	4	616	2.4063
10	4	658	2.5703
11	6	466	2.7305
12	6	517	3.0293
13	6	567	3.3223
14	6	616	3.6094
15	6	666	3.9023
16	6	719	4.2129
17	6	772	4.5234
18	6	822	4.8164
19	6	873	5.1152
20	8	682.5	5.3320
21	8	711	5.5547
22	8	754	5.8906
23	8	797	6.2266
24	8	841	6.5703
25	8	885	6.9141
26	8	916.5	7.1602
27	8	948	7.4063

Physical channel mapping and modulation (3GPP 38.211)

The processing steps in 3GPP 38.211 map the coded TBs to physical resources in the frequency domain before they are converted to time domain and readied for transmission over the air.

- Scrambling. Scrambling makes the transmitted bits less prone to interference by making them into a pseudo-random sequence through an XOR operation with pseudo-noise gold sequence generator.
- Modulation. Converts the scrambled signals to IQ values that are transmitted over OFDM subcarriers. Modulation options are QPSK, 16QAM, 64QAM, and 256QAM.
- Layer mapping maps the IQ samples onto layers, antenna ports, and virtual and PRBs. These steps convert the bit sequence to logical layers as needed for spatial multiplexing

that are mapped to antenna ports which are unique entities for every base station↔client communication and have unique reference signals.

MIMO is a key technology in 5G (and 4G). It introduces the concept of the antenna which (as opposed to representing a physical antenna) is a logical entity. It is associated with a reference signal that can be used by the UE to estimate the channel from that antenna port to it. Mapping to the physical channel can be done by 1: 1 or 1: many physical antennas. Each antenna port is associated with its own resource grid in the time and frequency domain and associated reference signals that are used for determining the channel state information associated with the port. If an antenna port is mapped to more than a single physical antenna, beamforming is (typically) used to focus energy to the user where the signal from multiple physical antennas—all carrying the same reference signal—to the UE. This beamforming is transparent to UE. Given that the reference signals are unique to the antenna port rather than the physical antenna count, the reference signals define the maximum number of concurrent transmissions. Separation of antenna port from physical antenna also explains the concept of the Type A and Type B RUs where the location of the port↔physical antenna mapping is likely to be done in the RU if the physical antenna port count is higher than the (logical) antenna port count.

Radio unit

An O-RAN radio unit is the logical entity that converts the frequency domain information that the DU produced into a time-domain signal that is transmitted over the air and vice versa. Major components include the RF component that generates (or receives) the analog signal, low-PHY functionality that includes time—frequency domain conversion and signal conditioning, and Enhanced Common Public Radio Interface (eCPRI) termination which converts IQ samples carried in Ethernet frames to IQ sample buffers that are processed by the low-PHY. The bulk of the RU complexity (and cost) is carried in the low-PHY and analog/RF components.

Generalized block diagram

Fig. 2—31 shows a generalized version of an RU block diagram, components of which we will discuss individually. The top half shows the transmit path and the bottom half the receive path, both for a single antenna instance. The example given next shows a zero-IF implementation, but this is only one of multiple design choices. We discussed IF and direct RF conversion previously.

Note that, more so than for many other components in the system, the physical implementation and the representation in the figure mentioned above do not need to match. Some elements of partitioning options are driven by the system requirements. The eCPRI and beamforming components, for example, are shared between all the antenna elements where the remainder [fast Fourier transform (FFT) up to antenna] components are (or better

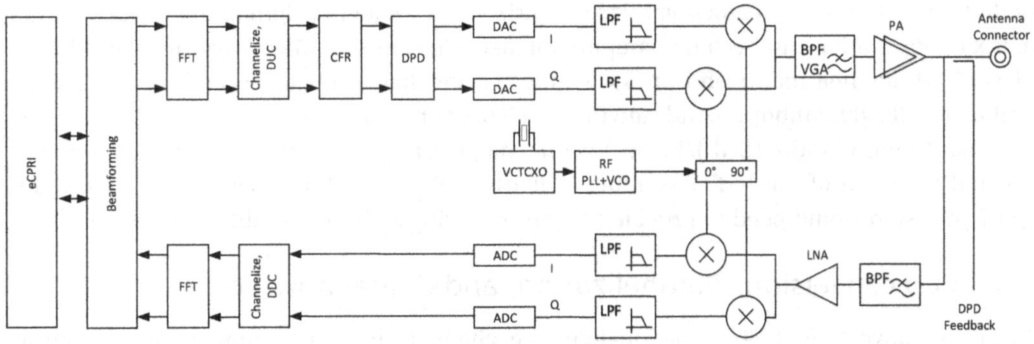

FIGURE 2–31 Generalized RU block diagram. *RU*, radio unit.

said: can be) individually implemented on a per-antenna basis. A logical functional split would colocate the eCPRI and beamforming components for all antennas into a single device, which can lead to scalability challenges in massive MIMO systems with up to 64 antennas. The remainder elements are independent but come with huge IO bandwidths—the IQ bandwidth associated with a single 100 MHz carrier is ~3 Gbps at baseband and higher after up-conversion. Note that the beamforming component is optional. The O-RAN fronthaul specifications define Type A and Type B RU options where the Type B RU includes beamforming in the RU that otherwise (Type A) would be in the DU. Type B RUs are intended for systems where the number of transmission layers is smaller than the number of physical antennas. For these systems, doing beamforming in the RU allows the fronthaul interface bandwidth to remain limited.

Typically, remainder processing functions are grouped or clustered into few as possible components to reduce IO bandwidth. Typically, this includes channelization and up/down-conversion up to and including crest factor (CF) reduction (CFR), digital predistortion (DPD), and additional filtering and sample rate adaptation components.

The analog-to-digital converter (ADC)/digital-to-analog converter (DAC) can be either on-chip (e.g., Xilinx RFSoC[3] or NXP Layerscape Access[4]) or connected over a JEDEC Standards (JESD) interface. JESD is a standardized IQ + synchronization interface that is designed for the purpose of ADC/DAC communication for telecommunication applications.

Remainder components are in the analog domain and can be either integrated with the ADC/DAC or implemented using discrete components, up to the power amplifier and Low Noise Amplifier (LNA) stages which are implemented as discrete components of their own or come pre-integrated in a front-end module. More information can be found in the chapter 4.

Digital beamforming and fast Fourier transform

The transmit and receive processing is straightforwardly defined as a (matrix) multiplication operation to apply the per user/antenna weights to the received IQ stream. The theoretical baseline processing required can be calculated from the required operation which is a matrix multiplication between an antenna × layer matrix and a frequency domain input vector. This matrix multiplication can become compute intensive for massive MIMO system. Consider an 8-layer × 64-antenna matrix and an 8 × 1 input vector that needs to be multiplied for each symbol at a 28,000 symbol/second rate for <6 GHz operation.

Signal-to-noise ratio (SNR)/Error Vector Magnitude (EVM) requirements define the required precision of the FFT. Assuming ~30 dB SNR required for 256QAM operation, the (i)FFT precision would need to provide margin by having SNR of ~45 dB or better.

Digital up conversion, channelization, and digital down conversion

Digital up-conversion (DUC) functionalities are channel filter, up-sampling, and mixing in the case where multiple separate digital carriers are transmitted over a single (analog chain and) antenna. The channel filter aims to filter out-of-band signal components to meet Adjacent Channel Leakage Ratio (ACLR) requirements. Up-sampling is done in the case of multiple carriers to bring all carriers to the composite (sum of all carriers) sampling rate. Mixing is to multiplex all carriers and is only present in the case of multicarrier.

This brings Tx signal to DAC sampling rate. Some filtering is done to remove the alias. The analog filter will also remove part of the aliasing so up-sampling filters are designed in conjunction of the analog filter so that both provide a good aliasing rejection.

Digital down conversion is the reverse step of DUC and used in the receiver path to bring the sampled signal down to baseband and provide antialiasing filtering beyond what was provided by analog filters. If needed, channelization is included to separate out individual RF carriers.

Peak-to-average power ratio reduction and performance improvement techniques

The OFDM waveform has gained a lot of popularity in the past two decades thanks to its robustness against multipath channel which is caused by the multiple reflections that occur on the radio channel of an air interface or even on the cable. The OFDM waveform (Fig. 2–32) breaks the channel in multiple quasi independent subchannels centered on subcarriers. These subcarriers are orthogonal and the QAM symbol that modulates each subcarrier can be safely recovered at the receiver end avoiding interference from neighbor subcarriers. One of the drawbacks of this OFDM modulation technique is the high peak-to-average power ratio (PAPR). Because of the independent modulation of each subcarrier, chance dictates that for some specific moments in time, a high number of subcarriers are co-summed in phase, thus causing a large peak in the time-domain signal that needs to be transmitted over the power amplifier (PA). These peaks, depending on the number of subcarriers, modulation pattern, etc. may fall in the range of 8...12 dB with respect to the average power level of the transmitted signal.

The PA in the transmitter needs to be designed to accommodate this PAPR. If we do not have any advanced techniques available, this can only be done by overdimensioning the PA target output by the amount of PAPR (8...12 dB as we said before). This overdimensioning

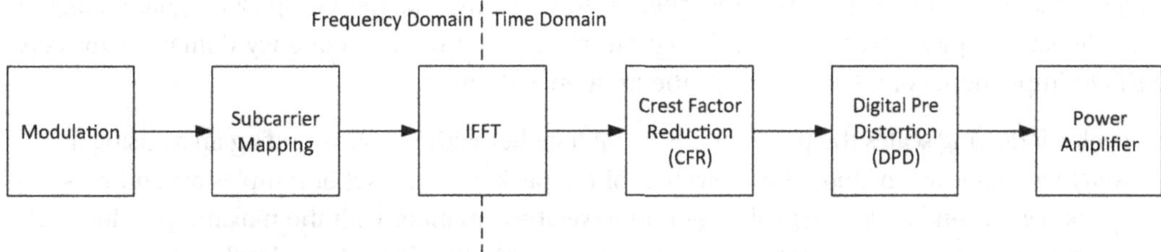

FIGURE 2–32 Generic picture of an OFDM transmitter. *OFDM*, orthogonal frequency division multiplexing.

can costly both financially (through deployment of more expensive components) as well as (DC) power consuming. Hence, it is ideally avoided. We discuss two relevant system components: CFR to reduce the PAPR and DPD to be able to operate the PA at higher power levels and thus be able to accommodate a higher PAPR.

Crest factor reduction

The point of CFR is to reduce the peaks in the time-domain OFDM signal and thus the PAPR, where we note that the CF (based on amplitude) is related directly to the PAPR (based on power) as:

$$CF = \sqrt{PAPR}.$$

Typically, CFR is applied to the baseband signal before other filtering, up-conversion, and DPD stages are executed. CFR can be implemented by multiple techniques, some of which (from Refs. 5,6) are discussed here:

- Clipping is the most intuitive way to implement CFR. If the signal is above a given reference (clipping) level, it is clipped to that value. In mathematical form:

$$X_{clipped}[n] = Clipfactor[n]X[n]$$

where

$$Clipfactor[n] = \begin{cases} 1, X[n] \leq A \\ \dfrac{A}{X[n]}, X[n] > A \end{cases}$$

with A being the desired clipping level. This clipping is typically followed by a filtering stage that smooths out the regions around the clipped area to reduce unwanted out-of-band components in the signal. Clipping and filtering works well with any signal type but is most efficient with contiguous signals composed of similar equal power carriers (e.g., 5 × LTE 20 MHz in the same band). Complexity depends on max allowed number of local peaks to cancel per pass. It is appropriate for medium–large instantaneous bandwidth (IBW) where performance can be traded off against complexity. Expect to need to tune parameters per composite signal configuration. Higher complexity versions of this algorithm can use time-to-frequency domain conversion (FFT) to implement complex filtering in the frequency domain.

- Peak windowing scales the peaks in the signal together with the surrounding areas using a weighted window function. After detection of the peak, define a set of samples around it as the "peak region" and scale this peak region in a weighted manner with the maximum reduction being enough to ensure that the reduced peak is within the desired amplitude range.
- The sigmoid method uses a modified sigmoid transfer to smooth out detected peaks, in mathematical form:

$$X_{sigmoid}[n] = Sigmoid[n]X[n]$$

where

$$Sigmoid[n] = \begin{cases} 1, |X[n]| \leq \frac{A}{\Delta} \\ \frac{A}{|X[n]|\left(1 + K\left(\frac{\Delta - 1}{K}\right)^{\frac{|x[n]|\Delta}{A}}\right)}, |X[n]| > \frac{A}{\Delta} \end{cases}$$

K is a shaping factor to control the smoothness of the transfer function and Δ is the shaping threshold above which the sigmoid function starts to be applied.

- Peak cancellation reduces the peaks in the signal by building a secondary "cancellation pulse" signal for each detected peak that is subtracted from the original signal. The cancellation pulse is shaped by a low-pass filter to ensure that it has few out-of-band frequency components.

In all implementations, CFR is a nonlinear function that introduces unwanted out-of-band signals. These typically are filtered out in remainder (digital) processing stages that are executed after CFR operation, including DUC that includes a filtering operation. CFR approaches are iterative where the signal can be processed in multiple "passes." Expect CFR to reduce the PAPR of the signal by ~3 dB for a typical 4G/LTE or 5G/NR signal.

Digital predistortion

There are two reasons to include DPD in the system: reduce unwanted (or out-of-spec) out-of-band emissions and improve the DC power efficiency of the RF subsystem.

Unwanted emissions are defined as the ACLR. They are caused by spectral regrowth due to the nonlinearity of the PA. One mechanism to lower ACLR is by reducing the PA output power as compared to its peak capabilities, or "back-off," but this comes at a financial cost associated with more expensive components and DC power consumption.

To look at the requirement for DC power reduction, we include the following chart which is strictly theoretical in that it calculates the associated DC power consumption for a PA as a function of the target output power of that PA, given a set of assumed PA efficiency levels. For example, 24 dBm output power (or 250 mW) corresponds to 2.5 W DC is the PA is operating at a 10% efficiency level. The chart is simply conveying this data rather than representing actual PA measurements. We include it to explain the reason for inclusion of DPD. As a rule of thumb, we assume that a PA without DPD operates at ~10%, whereas inclusion of CFR can improve efficiency to 15%–20% (3 dB reduced back-off as we discussed before) and inclusion of both CFR and DPD can achieve efficiency levels of 40%–50% or even better for state-of-the-art.

However, inclusion of DPD comes at a cost in terms of complexity and DC power consumption for the digital algorithm to implement DPD. The DC power consumed by the DPD algorithm can be measured in additional cost for the DSP device that implements the DPD algorithm. Say, we do a low-complexity DPD algorithm that improves the PA efficiency from 10% to 25% but its implementation comes at a DC power of ~1 W. Without DPD, the DC power associated with each PA subsystem and antenna port is ~2.5 W (250 mW/10%). With DPD, it is ~1.7 W (250 mW/15%) or reducing the DC power by ~800 mW. We see that the "cost" (1 W) and "benefit" are near. As a rule of thumb, for all systems mentioned above the very low-end (as we used in this example), the benefits of DPD outweigh the cost. In the very high-end with 10 seconds of Watt output power, aggressive use of DPD is the only way to keep system cost reasonable and DPD algorithm complexity can be the center-point of system design. The tradeoff is visualized in Fig. 2–33.

The function implemented by the predistortion system is shown in Fig. 2–34.

The target of the DPD predistorter is to implement the inverse function of the RF PA (considering (non-)linearity, memory effects, etc.), targeting improved PA efficiency as well as spectral compliance (such as ACLR).

The DPD subsystem is implemented through two software components:

1. The actuator—this component applies the DPD adaptation weights in the runtime system path of the system and hence performance critical.
2. The DPD adaptation algorithm—this component calculates the DPD adaptation weights and typically runs in a nontime-critical subsystem (e.g., on host ARM cores).

DPD is the block which implements a complex nonlinear function that approximates the inverse transfer function $G = H^{-1}$ of the PA baseband transfer function H. DPD implementations can be either memoryless or memory-based. Memoryless implementations assume the output to depend only on the current input as opposed to memory-based systems, where previous signal levels (last sample, the sample before that, etc.) are also taken into consideration. DPD algorithms look at both AM/AM distortion which defines the amplitude of the output signal as a function of the amplitude of the input signal as well as AM/PM distortion where the phase of the output signal is a function of the amplitude of the input signal. Implementation of the predistortion algorithm can be done by Lookup Tables (LUTs) and/or by a mathematical representation such as Volterra series.

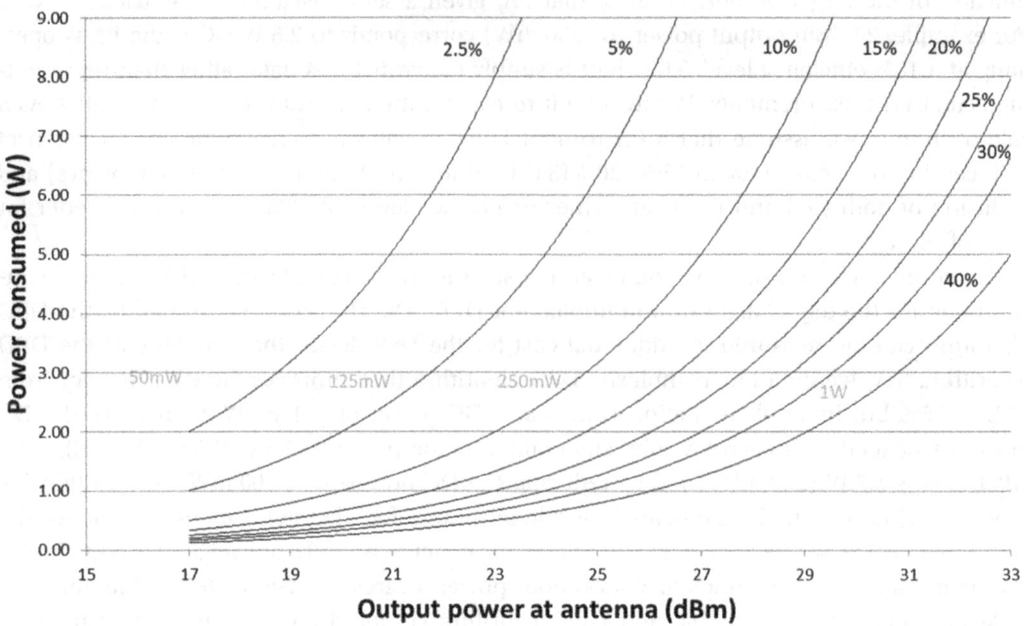

FIGURE 2–33 DPD justification for power efficiency. *DPD*, digital predistortion.

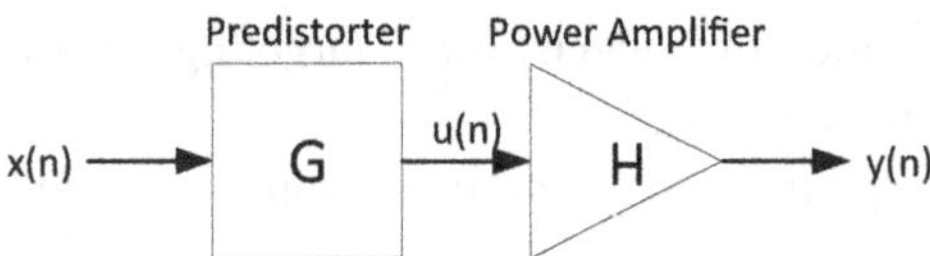

FIGURE 2–34 Digital predistortion concept.

Both implementations rely on characterization of the PA to establish the values that are used in the LUTs or Volterra series. This characterization can be done offline through measurement that are done on a test bench during development time, potentially augmented by an on-the-fly correction for temperature or other. For higher end systems, the characterization is done live by a feedback loop that measures the output of the PA and compares it against the DPD input signal to update DPD algorithms during system deployment (Fig. 2–35).

Assume an implementation using polynomial modeling (for more information, see later on in this document), complexity of DPD algorithms is typically accounted by the number of (multiply-accumulate) operations that is done on the sample stream that is input to the DPD algorithm—and multiplied by the rate of that sample stream to define the aggregate required multiply-accumulate count. Low-end DPD algorithms targeting indoor, <1 W output power

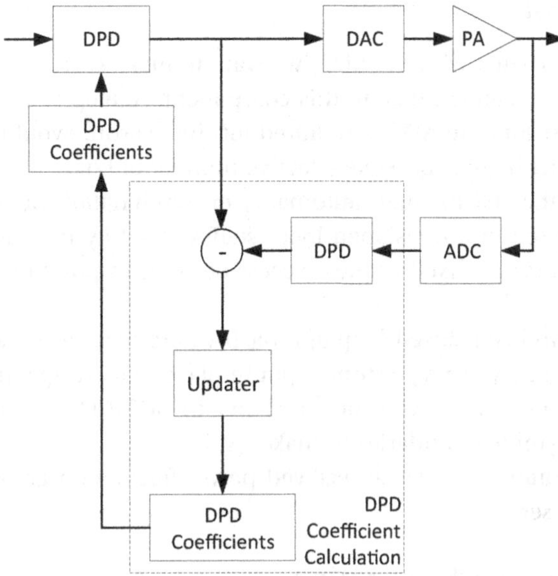

FIGURE 2–35 DPD feedback processing. *DPD*, digital predistortion.

levels can suffice with around a dozen operations per sample, whereas higher end system can use multiple dozens of operations on each sample. Keep in mind that the sample rate is not equal to the IBW or baseband rate (say, 122.88MSPS for a 100 MHz signal) but to a multiplier of the baseband rate to cover the out-of-band spectral components that need to be compensated. 3–5 × oversampling is a typical ratio seen in the field.

Note that the use of DPD has an impact on the performance requirements of the DAC used in the transmit path: unless it has sufficient performance [precision as counted in signal-to-interference and noise ratio and/or effective bits (effective number of bits)], the DPD algorithm will not be able to deliver a signal to the DPD with sufficiently low-noise floor over high-enough bandwidth (own channel and adjacent channels). In addition, feedback-based DPD algorithms need an associated ADC supporting enough performance to measure the generated distortion in the output signal.

Digital ↔ analog conversion (digital-to-analog converter/analog-to-digital converter) and analog components

After DAC a reconstruction filter (low-pass filter in the case of zero-IF interface systems) is needed to remove DAC aliasing images. The remainder signal processing steps include a modulator to bring (in our example) the baseband frequency to RF followed by filtering and amplification stages including the final stage: the PA. The reverse flow includes low-noise amplification, filtering, demodulation, and baseband antialiasing filtering to remove ADC aliasing images.

Analog gain control

Given the high cost associated with the ADC, we want to make sure that we make best possible use of the available dynamic range in this component. Analog gain control to one of the amplifier components before the ADC is required for this reason: avoid saturation (overflow) and underflow, providing highest possible effective number of bits.

Gain setting can either be manual, automatic, or combination of both. Manual means that the gain is configured in a quasiopen loop: statically set by the CP of the system. In a very static environment (think: fixed wireless access or point-to-point links), this can be good enough.

Automatic gain control is a closed-loop approach where receiver power is measured, and the gain is tuned accordingly. An algorithm is put in place to converge quickly enough to the appropriate value but does not overshoot/oscillate. In 3GPP/OFDM system, gain change should be applied on symbol boundaries to make.

Dynamic user scheduling can cause received power fluctuation across slots. By implication, a safety margin is set.

Time and frequency synchronization

Wireless 3GPP systems need to be both frequency- and time-synchronous across physical units (DU and RU), which implies a requirement for a board-level clock that is kept synchronous across the network. Time and frequency alignment are achieved through (a combination of):

- Synchronous Ethernet.
- GNSS/GPS.
- IEEE1588 (PTP).
- Radio interface-based synchronization (timing and frequency information is decoded over the air from other base stations).

Frequency synchronization requirements are defined in 3GPP as per the Table 2–4.

Synchronization accuracy of <1.5 μs to target a time alignment error across base stations of <3 μs [as stipulated by 3GPP 38.133 (cell phase synchronization accuracy)] as shown in Fig. 2–36.

Note that actually systems may require more tight timing requirements. Consider:

- massive MIMO and distributed Cooperative Multi-Processing (CoMP) where timing alignment will be (tens of) nanoseconds and
- O-RAN requirements which define timing jitter between DU and RU as discussed.

Most 3GPP systems incorporate a board-level oscillator (e.g., running at 30.72, 38.4, or 122.88 MHz) as well as a pulse per second source to perform both time and frequency synchronization functions. The board-level oscillator can be fed from a other frequency synchronous source or tuned through a software servo algorithm that runs on the Layerscape host processor.

Table 2–4 Time and frequency accuracy requirements per 3GPP.

Application	Frequency accuracy (ppb)	Phase/time accuracy
GSM	50	N/A
UMTS	50	N/A
LTE-FDD	50	N/A
LTE-TDD	50	≤ 1.5 μs
LTE-MBSFN	50	≤ 1.5 μs
LTE-A	50	≤ 1.5 μs
5G/NR	50	≤ 1.5 μs

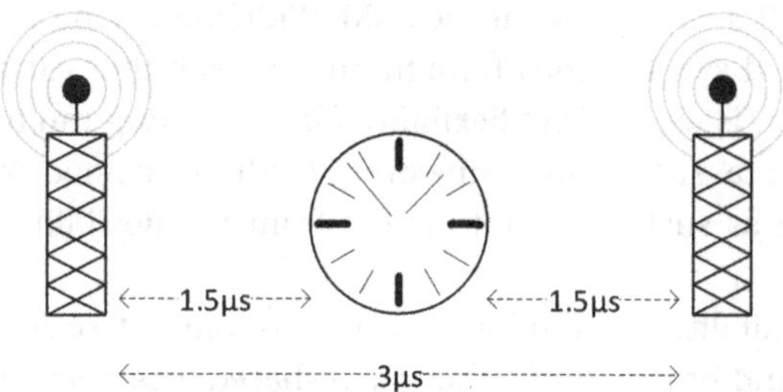

FIGURE 2–36 Timing alignment between radios.

PRACH

Preamble sequences are used as a method to obtaining synchronization between a client and a base station, both in 4G/LTE and 5G/NR implementations. The physical channel that carries the preamble (in the uplink only) is the PRACH. The preamble is a pattern (one out of 64 options) that is either randomly selected by the client (contention-based random access)—and assumed to be likely unique given its selection from a pool of 64—or allocated by the base station to the client (contention-less random access). The preamble is sent by client to base station over a PRACH which is a time/frequency allocation (PRACH occasion), obviously aligned with the defined bandwidth partition to the targeted client(s).

The preamble consists of two pieces: a CP and the preamble sequence itself. The sequence itself is a so-called Zadoff–Chu sequence which is a mathematical sequence designed to have a constant amplitude and to be orthogonal between sequences. Preamble formats can be split between long preambles and short preambles. Long preambles (length = 839 bits) are used in FR1/ < 6 GHz applications with a SCS of 1.25 or 5 KHz. They occupy 6 (1.25 KHz numerology) or 12 (5 KHz) numerology resource elements in the frequency domain. Definition of which long preamble format is to be used is done by the base station as part of the initial access (RACH) configuration. Short preambles (length = 139 bits) are used in both FR1/ < 6 GHz and FR2/mmWave applications with a SCS that is aligned to "conventional" subcarriers, so 15, 30, 60, or 120 KHz, allowing the same FFT to be used as for normal operation. They occupy 12 resource elements in the frequency domain.

Like for shared channel operation, PRACH processing is also offloaded to the RU as opposed to sending time-domain IQ samples from RU to DU. This is done both to reduce compute resources in the DU and to limit bandwidth between the RU and the DU.

Distributed unit/radio unit interface, Enhanced Common Public Radio Interface protocol overview

In LTE macro cell deployments, CPRI is the dominant form of sample transport between BBU and RRH. The CPRI standard defines a GSM/Wideband Code Division Multiple Access (WCDMA)/LTE optimized synchronous framing mechanism that carries user data, management, and control. There is extensively flexibility for supporting multiple carriers and different sample widths over a single link. However, CPRI is defined for (mainly) 3GPP I/Q transport use cases and as such does not benefit from the flexibility and economy of scale provided by Ethernet.

Given 5G trends, flexibility in partitioning and economy of scale, the interface between the DFE and the baseband processor is Ethernet/IP-based. Connectivity options include custom frame format, eCPRI or RoE. Ethernet-based I/Q transport depends on time-sensitive networking capabilities in the Ethernet standard (not discussed in this document) to guarantee frame timing.

As shown in the Fig. 2–37, different options in the RoE/eCPRI standards allow for mapping of split option 7 to standardized formats. Standardization efforts (e.g., xRAN Fronthaul Working Group) are ongoing to define Ethernet/packet format across eCPRI and RoE.

For illustrative purposes, Fig. 2–38 shows the protocol mapping for eCPRI or IEEE1914.3, including optional UDP/IP layers. TCP or SCTP is not considered a realistic protocol option given the strict requirements for latency that do not leave physical time for Ack/Nack and retransmission protocol support.

The eCPRI[7] standard protocol layering relies on UDP/IP/Ethernet addressing for source/destination routing. Time synchronization is provided through standard IP mechanisms such as IEEE1588. OAM and control and management also rely on standard IEEE/Ethernet mechanisms.

Both RoE and eCPRI use a custom header including a flow identifier (RoE) or physical channel ID (eCPRI) to define the antenna/port/flow/link that bits are carried for. Assuming large payload/frame sizes to be supported (few KB/OFDM frame), the dozens of bytes of Ethernet and protocol framing overhead constitute a very small percentage efficiency loss. Ethernet overheads for packetization and synchronization are estimated to be $\sim$10%. The eCPRI protocol is explicitly designed with strict separation of User and Control Planes in mind, allowing for a efficient hardware or software implementation like shown in Fig. 2–39.

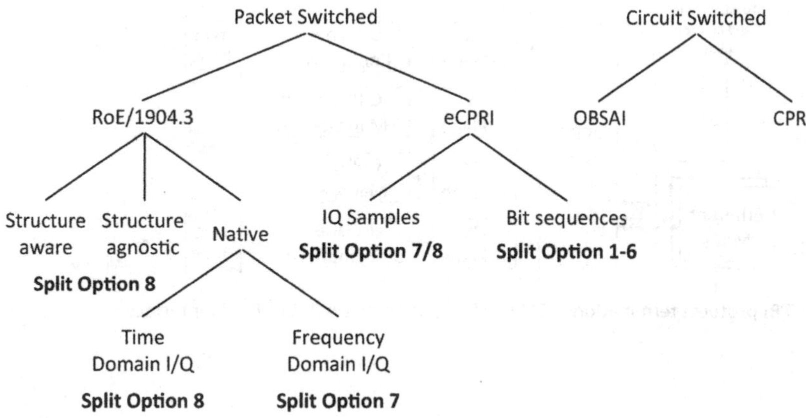

FIGURE 2–37 Ethernet-based sample transport mechanisms.

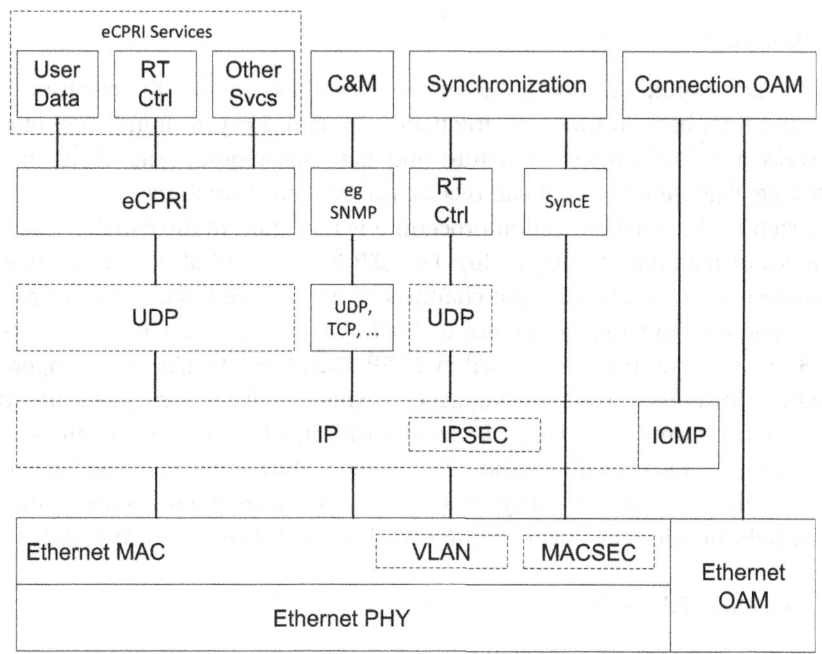

FIGURE 2–38 IQ sample protocol stack.

Initial access

Initial access is a complex topic in both 4G/LTE and 5G/NR systems, which touches on CU (CP), DU (MAC layer, PRACH), and RU (PRACH) components. We provide a high-level

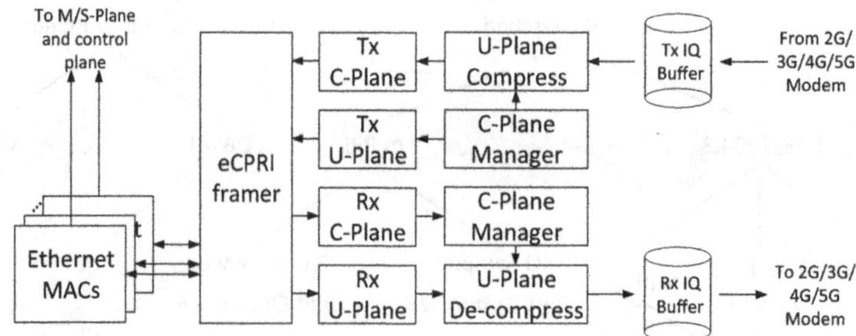

FIGURE 2–39 eCPRI protocol termination. *eCPRI*, Enhanced Common Public Radio Interface.

overview of the process to explain the relevant components, focusing on 5G/NR deployment —but note that 4G/LTE uses a comparable procedure.

Synchronization

Before starting any communication between client and base station, the client needs to synchronize to the RAN both in time and frequency domain. Keep in mind that, unlike Wi-Fi, 3GPP networks are synchronous in nature and that this requires the client to acquire a notion of timing alignment before it can receive (or transmit!) anything.

The first step in the synchronization procedure is reception of the Synchronization Signal Block (SS Block or SSB, refer to Fig. 2–40). The SSB is a handful of subcarriers that transmit primary and secondary synchronization channels as well as the Physical Broadcast Channel. The SSB is transmitted in a regular pattern defined by a burst set (one or more SSBs), a burst period and burst duration that are defined in 3GPP 38.213. For FR2/mmWave operation, SSB transmit pattern includes a beamforming aspects where each SS burst has a unique associated (base station transmit side) beam associated with it, where the client can assume beam reciprocity to identify the direction it should transmit to during initial access later on.

The SSB location in time (OFDM symbols 0, 1, 2, 3) and frequency (starting from subcarrier offset 0) help in initial time and frequency alignment. Content of the SSB is shown in Fig. 2–41.

Key components of the SSB are:

- Primary Synchronization Signal (PSS)—a Zadoff–Chu sequence that is derived from a portion of the cell identity (Cell ID). The PSS signal also identifies the slot boundary.
- Secondary Synchronization Signal (SSS)—an m-sequence that completes the Cell ID identification. 5G NR cells are identified by one out of 1008 IDs in 336 groups of 3 sectors each. The PSS contains the sector and the SSS contains the group, allowing the cell ID to be calculated as 3 × Cell ID (SSS component) + Cell ID (PSS component).
- Physical Broadcast Channel (PBCH)—this carries the MIB with additional timing and identification information in 31 information bits per Fig. 2–42.

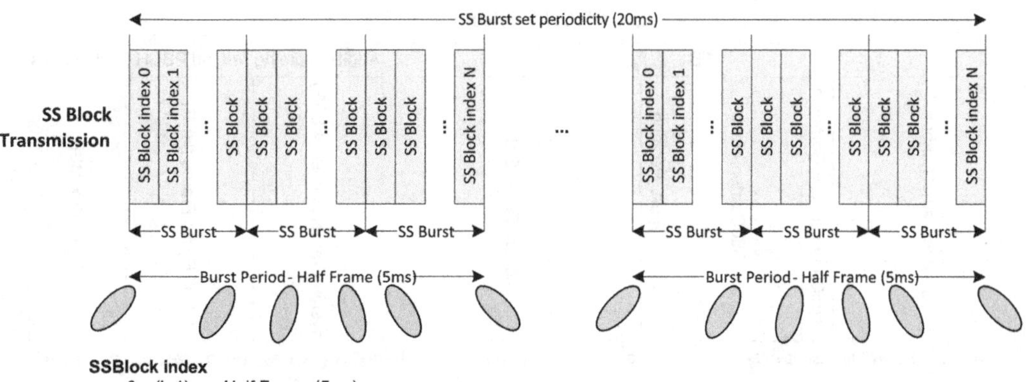

FIGURE 2–40 SS Block transmission.

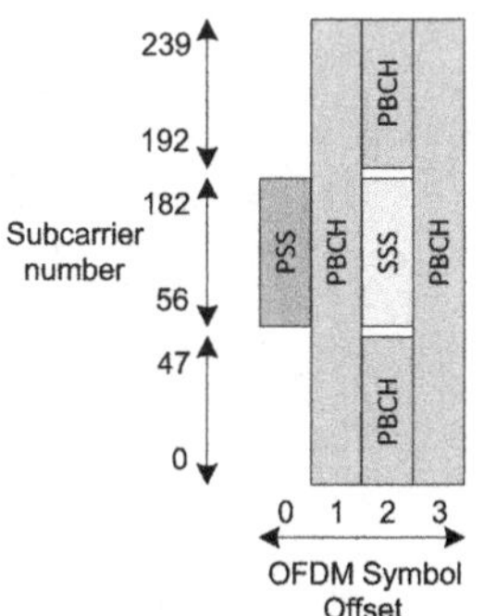

Channel	OFDM Symbol #	Subcarrier k relative to SS/PBCH block
PSS	0	56, 57, ..., 182
SSS	2	56, 57, ..., 182
Set to 0	0	0, 1, ..., 55; 183; 184, ...,236
Set to 0	2	48, 49, ..., 55; 183; 184, ..., 191
PBCH	1, 3	0, 1, ..., 239
PBCH	2	0, 1, ..., 47; 192, 193, ..., 1239
DM-RS for PBCH	1, 3	0 + v, 4 + v, 8 + v,..., 236 +v
DM-RS for PBCH	2	0 + v, 4 + v, 8 + v,..., 44 + v, 192 + v, 196 + v,..., 236 + v

$v = N^{cell}_{ID} \bmod 4$

FIGURE 2–41 SS block contents.

The pdcch_ConfigSIB1 IE further identifies the time/frequency resources (Control Resource Set or CORESET) that tell the client where to identify the downlink control channel (PDCCH, Fig. 2–43) that contains the signaling information that points to the System Information Block 1 (SIB1).

The SIB1 contains a much larger information set, including more details on cell identifiers, UL/DL configuration, bandwidth partitioning, and so on, providing sufficient information for the random-access procedure to start.

Random-access procedure

After synchronization, the client can track the base station for time and frequency synchronization purposes. It also is aware of allocated preamble format, transmit power, time/

PBCH Payload

MIB					Addit'nal timing related PBCH payload bits		
systemFrameNumber	subCarrierSpacingCommon	ssb-SubCarrierOffset	dmrs_TypeA_Position	pdcch_ConfigSIB1	systemFrameNumber (4 LSB)	Half Radio Frame (HRF)	SSB index MSB

a_0 a_1 a_2 a_3 a_4 a_5 a_6 a_7 a_8 a_9 a_{10} a_{11} a_{12} a_{13} a_{14} a_{15} a_{16} a_{17} a_{18} a_{19} a_{20} a_{21} a_{22} a_{23} a_{24} a_{25} a_{26} a_{27} a_{28} a_{29} a_{30}

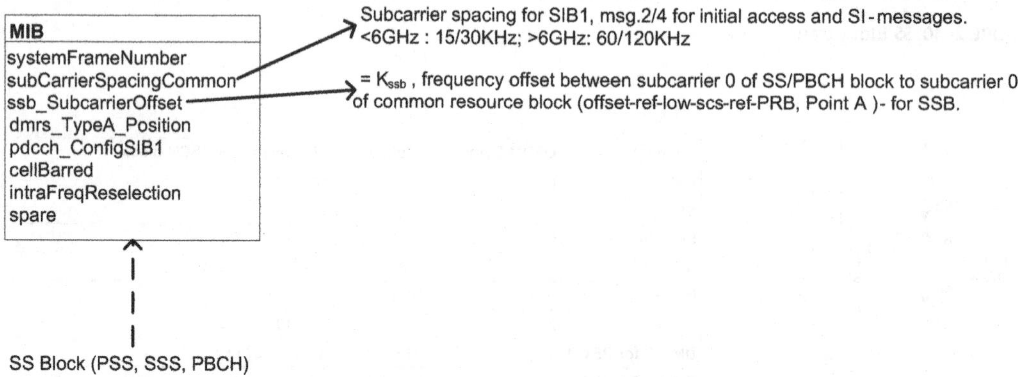

FIGURE 2–42 MIB content. *MIB*, Master Information Block.

frequency resources, and other relevant parameters, through the RACH-ConfigCommon parameter that is signaled either through SIB1 or through an RRC reconfiguration messages with dedicated signaling (NSA operation).

The client can now send a random-access preamble in the uplink (message 1), which triggers a random-access response (RAR) by the MAC layer of the base station (message 2). The RAR is transmitted in the DLSCH and includes a time advance command to communicate the base station estimated timing advance to be applied to compensate for over-the-air latency. It also includes an uplink grant that allocates time/frequency resources to the client and a temporary identifier (TC-RNTI) that (during the random-access procedure) identifies the client.

The first uplink transmission (message 3) from the client to the base station includes this TC-RNTI, a contention resolution identifier and a RRC connection request. The contention resolution identifier is used to resolve any potential contention associated with two clients picking the same preamble. This triggers the last message in the RACH procedure (message

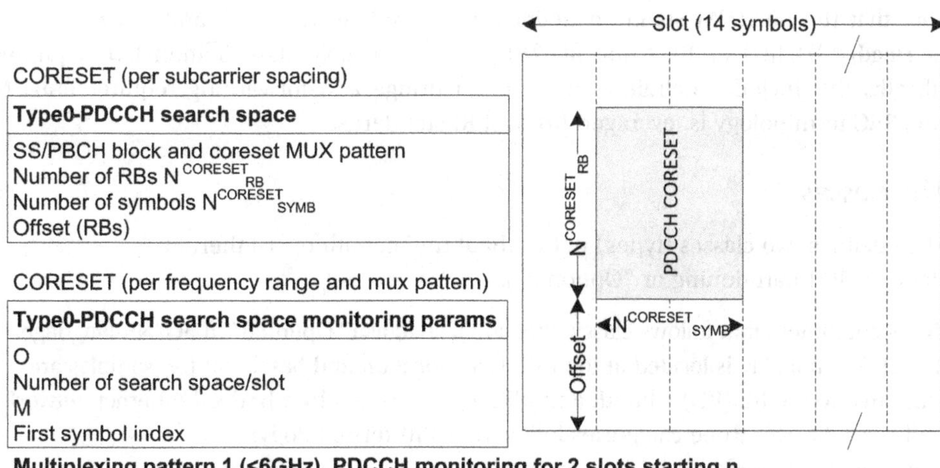

FIGURE 2–43 PDCCH Control Resource Set.

4, from base station to client) which is the contention resolution message that should contain the same contention resolution identifier that the client transmitted to the base station in message 3.

802.1CM

The 802.1CM (Time-Sensitive Networking for Fronthaul) standard defines Ethernet profiles necessary to build networks capable of transporting fronthaul streams. Fronthaul streams are defined as flows between a base station and a radio head.

802.1CM profiles include definition of:

- VLAN
- MAC Service Specifications (802.1AC)
- MAC/PHY Specifications
- Express Traffic Interspersing
- Frame Preemption
- Time Synchronization and PTP (IEEE1588)
- Telecom Profile Specification (IEEE1588)
- Synchronous Ethernet Specification

Note that the 802.1CM standard defines the complete fronthaul, and not only RE (aka Radio Head/RRH/RU) or RE Controller/REC (aka eNB/gNB/Base Station/DU) equipment. Specifically, this includes definition of Ethernet bridge and forwarding requirements. Note that RE/REC terminology is leveraged from CPRI standards.

Traffic classes

802.1CM defines two classes (types) of fronthaul traffic, both over Ethernet:

Class 1 (CPRI partitioning or "Option 8"):

- This is equipment that follows classic functional split, aka "Option 8" in 5G/NR language. The FFT/CP functionality is located at the REC (DU) location and baseband I/Q samples are transferred to the RE (RU). The RE and REC are connected by a bridged Ethernet network.
- Traffic is split over three categories following CPRI terminology:
  - IQ data (samples)
    - IQ data is exchanged regardless of whether end-user data is exchanged (i.e., also when the radio is technically idle) and can be represented as constant bit rate flow. Latency is defined as $<100\ \mu s$ end-to-end REC < - > RE, including propagation delay and network internal bridge delay. Frame Loss Ratio (FLR) must be bounded to $<10^{-7}$.
  - Control and Management (C&M) (exchanged between control and management entities on REC/RE)
    - There is no explicit latency requirement for this traffic. FLR must be bounded to $<10^{-7}$.
  - Synchronization (frequency as well as time/phase) is done using existing solutions and protocols. Synchronization can be done using PTP (IEEE1588), SynchE, GNSS (GPS,) or other (proprietary) methods.

Class 2 (eCPRI partitioning or "Option 7"):

- This is equipment that follows eCPRI split options and as such a flexible split functional split of the 3GPP Physical Layer between REC and RE.
- Traffic is split over three categories following CPRI terminology:
  - IQ data (no longer necessarily samples), with subcategories
    - User data. Latency is defined as $<100\ \mu s$ end-to-end eREC < - > eRE. FLR must be bounded to $<10^{-7}$.
    - Real-time control data, associated directly with user data.
    - Data for other CPRI services (UP support, remote reset, ...).
  - Control and Management (C&M)
    - C&M information is exchanged over commonly used transport protocols such as UDP and TCP. There is no specific eCPRI protocol.
  - Synchronization
    - Synchronization (frequency as well as time/phase) is done using existing solutions and protocols as outlined for class 1.

Timing and frequency synchronization requirements are defined in 802.1CM, based on (and aligned to) base station requirements as defined in 3GPP.

Bridging functions

The 802.1CM standard defines bridging functions that are important to the fronthaul bridged network. Note that these may not apply to the REC/RE nodes.

- Latency components
  - input queueing delay (likely not present)
  - interference delay between time that a frame was selected for transmission and actual transmission—caused by queued-up frame of higher or same priority as the chosen frame
  - frame transmission delay (time to transmit one frame at the transmission rate of the port)
  - LAN propagation delay—physical delay due to LAN link length
  - store-and-forward delay associated with the bridge function
- Frame preemption which is the suspension of the transmission of a "current" frame by a higher priority frame. This feature is optionally supported by fronthaul bridges. Its value diminishes as port speeds increase.
- Network synchronization (PTP, SynchE, GPS, proprietary as described earlier).
- Flow control. Flow control support can invalidate latency guarantees and may have to be disabled on fronthaul bridges.
- Energy-efficient Ethernet. Like flow control, Energy-efficient Ethernet may be prohibitive to fronthaul bridges and needs to be disabled.

Fronthaul profiles

The 802.1CM standard defines two feature profiles for fronthaul that both apply to Class 1 and Class 2 traffic (note, Class 1 and Class 2 traffic requirements are common). The profiles define three types of fronthaul traffic:

- high-priority fronthaul (HPF) with end-to-end latency <100 μs associated with Class 1 and 2 fast user plane (IQ)
- medium-priority fronthaul (MPF) with end-to-end latency <1 ms associated with Class 2 slow user plane and Class 2 fast C&M
- low-priority fronthaul (LPF) with end-to-end latency <100 ms associated with C&M

Profile A shall be supported by the bridges in a fronthaul network:

- It supports maximum frame size of up to 2000 B, with actual applied frame size potentially smaller (e.g., 802.1Q frames of 1522 B).
- It has three traffic priorities supporting HPF, MPF, and LPF in strict order.
- It maintains <100 μs delay through the complete network with all hops included.

- It defines no fronthaul packet loss through network configuration (so no oversubscription), thus maintaining packet loss the 10^{-7} FLR requirement.

Profile B may be supported by the bridges in a fronthaul network and adds frame-preemption feature. Given that nonfronthaul traffic is preemptable, no maximum frame size is defined for it in Profile B.

Fronthaul gateway

In an indoor "small-cell" aggregation use case, multiple (say, 8...16) RUs are connected to a single *logical* DU. All RUs combined support a single logical cell across a larger geographical area. The definition of a logical cell here is that it is a single set of time and frequency resources that can be shared across multiple physical cells. The same signal is broadcasted in downlink (transmit) direction across multiple cells and the receive direction is implemented by Cooperative Multipoint (CoMP)-like combining (IQ summing). In addition, the fronthaul gateway (FHGW) (Fig. 2–44) reduces wiring as it combines the connectivity from multiple RUs (shown on the right-hand side in the figure) to a single interface toward the DU.

In this (enterprise networking) example, the physical connection to the DU is 25 GbE or 2 × 10 GbE (typically, optical), which implies that (given a 2R2T or 4R4T/100 MHz deployment) IQ samples can be transported in uncompressed (16b I/Q) form. The connectivity to the RU is 10 GbE (optical or PoE) and as such would be required to support compression in the case of 4R4T/100 MHz deployment. Connectivity (Fig. 2–45), functionality, and timing requirement at "box" level are summarized as:

- 1 × 10/25 Gbps toward DU
  - Optical
- 1 × 10/25 Gbps toward second FHGW/AU
  - Optical
- ≤ 8 × 10 Gbps toward RU
  - Electrical, PoE support
  - Optional optical

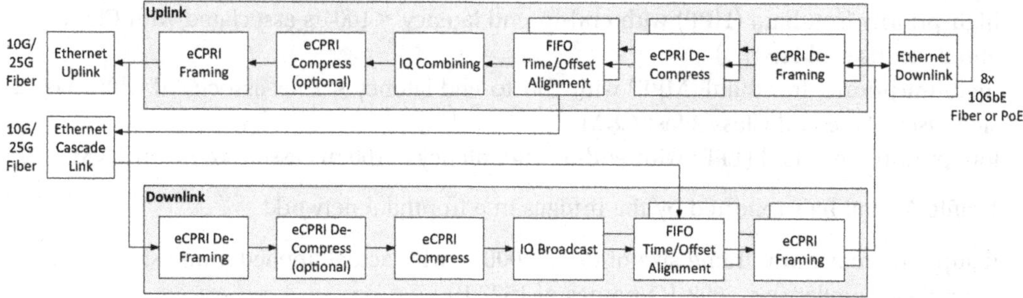

FIGURE 2–44 FHGW/AU U-Plane functionality. *FHGW*, fronthaul gateway.

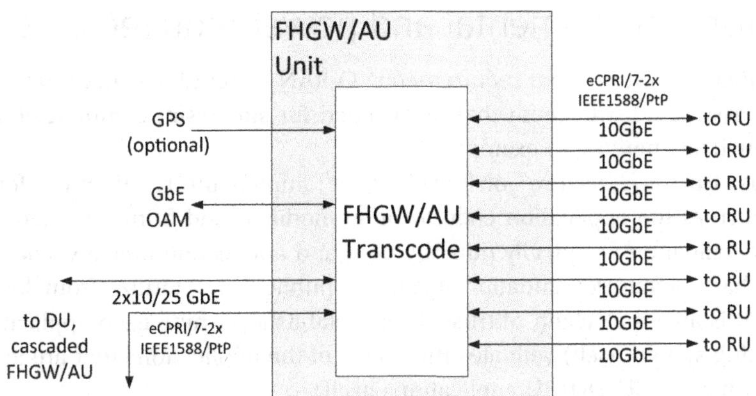

FIGURE 2–45 FHGW/AU baseline connectivity requirements. *FHGW*, fronthaul gateway.

Cell site router/gateway

The cell site router provides access from the CU and DU to the mid- and backhaul aggregation network to the CU and the core network. In its simplest form, when it provides connectivity between a (known) CU and (known) DU instantiation, little more than basic connectivity between the two units over an IP interface needs to be provided. In such case, the cell site router functionality can be absorbed in the DU unit. High-capacity 5G cell sited may need more router capacity to bifurcate a single backhaul IP connection to multiple lower capacity interfaces to subunits. Other use cases for cell site routers include separation of traffic between backhaul network types (say, enterprise and core networks).

Cell site router functionality includes the following:

- Time and frequency synchronization—these functionalities are often included in the cell site router given that this is the first entry point for a connection carrying IEEE1588 traffic from the mid-haul link.
- Security—when required for the operator network, the cell site router can provide IPSec for securing the link.
- Traffic switching and routing toward other cell sites that are chained or otherwise connected to the main site.
- QoS—wireline traffic prioritization and shaping. We discuss this in more detail in the chapter 7.

Cell site routers can either be logical (software) functions implemented on the CU and DU itself or implemented as separate hardware entities that are independently designed, built, and deployed.

Form factor, environmental and power requirements

Besides (3GPP centric) functional requirements, O-RAN system have additional system level requirements associated with them that are needed for successful commercial deployment. We cover a few items here as an example.

Referred to "trust architecture" or "secure boot" implies the requirement for the binary image that executes the application code to be unmodified and from a trusted source. This trusted source validation is typically done by boot and application image validation by comparison to a hash/signature validation against a public key. Starting from low-level boot mechanisms, this allows a "chain of trust" to be established where each application (including the operating system itself) validates the image of the applications that are loaded subsequently (including the CU/DU/RU applications itself).

A set of Ingress Protection (IP, not to be confused with the "other IP" (Internet Protocol)) codes are defined in the IEC standard 60529. They classify the physical enclosure with regards to protection from intrusion from accidental damage, dust, and water. IP standards define a two-digit number that defines solid particle (first digit) and liquid (second digit) protection as follows (Table 2–5).

Some equipment is defined by NEMA ratings rather than IP ratings. NEMA ratings expand beyond IP ratings by inclusion of icing conditions, hazardous areas, and more. Initial translation from NEMA to IP can be found using standard tables, for example in Ref. [8].

Fanless operation is required for more stringent IP-rated products. This dictates the use of components (processors, DSPs, FPGAs) with reduced power consumption of 10–30 W instead of the typical ~100 W power consumption associated with desktop and server processors. Both the component-level and board-level power consumption need to be checked, if required with thermal analysis and modeling tools.

Electromagnetic Compatibility (EMC) standards outline the allowed conducted or radiated emissions by the product. Multiple standards are defined for EMC compliance, both in Europe and European Union. The American FCC Part 15 standards define intentional and unintentional radiation limits for various bands and are the go-to standard for wireless LAN (802.11) and other standards.

RoHS stands for Restriction of the Use of certain Hazardous Substances in Electrical and Electronic Equipment and is a product-level compliance requirement as defined in European Union Directive 2002/95/EC. The 2006 directive restricts the use of specific hazardous materials found in electrical and electronic products if that equipment is to be allowed to be sold in the European Union. Use of lead (Pb) is forbidden by this directive and that includes lead historically used for soldering components to PCBs. RoHS 2 expands RoHS to include cables, spare components, etc., whereas RoHS 3 includes additional substances.

ASN.1

Standardized by ITU-T, ASN.1 is an interface description language that defines data structures as used in many applications including 3GPP CP (RRC, S1AP, NGAP, etc.) and other

Table 2–5 Ingress protection definitions.

Value	Solid particle protection	Liquid protection
X	No data available	No data available
0	No protection	No protection
1	50 mm—for example, the back of a hand	Dripping water
2	12.5 mm—for example, a finger	Dripping water when tilted at 15 degrees
3	2.5 mm—tools or wires	Spraying water
4	1 mm—wires, small insects	Splashing of water
5	Dust protected	Water jets
6	Dust tight with applied vacuum	Powerful water jets
6 K		Water jets with increased pressure
7		Immersion, 1 m
8		Immersion, 3 m
9		Powerful high-temperature water jets

protocols. It is a formal notation that is used in 3GPP standards definitions. ASN.1 standard defines multiple mechanisms to convert ASN.1 data structures to serialize data streams that can be encoded and decoded by both the transmitting and receiving entity. The RRC protocol specifies the use for Packed Encoding Rules.

ASN.1 tools include parsers that parse human readable definitions of ASN.1 data structures and convert them in computer readable definitions (e.g., C language data structures) as well as APIs to encode and decode to/from serialized bit streams. ASN.1 tools come in open-source and commercial-grade options.

DiffServ

Since the introduction of the Internet, its use has changed from purely data based to also accommodate voice, video, and time-critical data such as gaming. The early design of the Internet and the TCP/IP protocol did not accommodate the QoS mechanisms needed to support this functionality. Hence, several protocols have been developed to give greater control over QoS in the Internet compared to the original IPv4 (RFC791) and IPv6 (RFC8200) standards. Such mechanisms include Integrated Services, Multiprotocol Label Switching (MPLS), and the Differentiated Services (DiffServ) model. The DiffServ model is the most popular one.

The word "services" defines several characteristics of packet or cell transmission in one direction in a network. Such characteristics can be throughput, delay, jitter, and packet loss.

Why do we need differentiated services? Apart from what is explained above (accommodation of heterogenous application requirements and user expectations), it also gives the possibility to apply differentiated pricing, which is of course a very interesting feature for an operator.

In general, for describing QoS, the following functional blocks are used:

- Classification/marking. This is the selection of a packet based on the contents of one or more fields in the header.
- Policing/shaping. A method of making sure that a traffic source does not transmit on a higher speed than allowed. Shaping the traffic means modifying it in such a way that the source—after shaping—complies with the preagreed traffic requirements.
- Queueing/scheduling. A mechanism to prioritize the transmission of certain traffic aggregates above others.
- Congestion avoidance mechanisms. A mechanism to lower the transmit speed at the sender, before the router is fully congested; for example, Random Early Discard (RED), discussed later.

There are different ways of providing QoS to a network using the functional blocks described earlier. Examples are as follows:

- IPv4 precedence marking. This is the use of the IPv4 Type of Service (ToS) field as originally described in RFC791. This RFC described a simple priority dequeue for different types of traffic.
- Label switching. For example: MPLS (or Asynchronous Transfer Mode, back in the day). A traffic stream gets a certain label and is switched according to that. On a management level, the characteristics of the traffic stream are known so that it can be routed through a part of the network that is not congested. This keeps the total bandwidth use of the network below maximum, thus guaranteeing the resources to the users. This goes at the cost of higher management overhead.
- Integrated Services (IntServ)/RSVP. In this system, each flow makes a bandwidth reservation in each node of the network, thus guaranteeing the end-to-end QoS. IntServ is not widely used.

The problems with Label Switching and Integrated Services are that they require a large amount of overhead at the central nodes of the network, as every flow needs to be treated individually. This requirement makes these schemes not well scalable. Simple IPv4 ToS-field-based precedence marking on the other hand does not give enough flexibility to provide good QoS. Combined, this explains the need for a different system: Differentiated Services or DiffServ.

DiffServ reaches scalability by limiting the application of classification and conditioning to the boundaries of the network. Inside the network, the Per Hop Behavior (PHB) works on aggregates of flows which have been marked using the DiffServ (DS) field in IPv4 and IPv6 headers. This means that there is no need to keep a per-flow or per-customer state table inside the network.

Main concepts of DiffServ are as follows:

- Classification. As explained before, this means the selection of a packet based on the contents of one or more fields in the header.

- Conditioning. Metering, marking, dropping, shaping, or a subset of these. Metering means the measurement of the characteristics of an incoming traffic flow, by using, for example, a token bucket (Single-Rate Tri Color Marking or srTCM). Marking means modification of a packet to indicate conformance with a preagreed traffic contract. If packets do not comply with the traffic contract, they can be discarded (dropped) or shaped to comply with the contract, for example, by delaying them in time.

A DiffServ network is built around a domain, which is a contiguous network under the same administrative ownership. Inside this domain we differentiate between edge nodes and core nodes. An edge (or boundary) node provides an interface to a non-DiffServ network or a different DiffServ domain, while the core (or interior) nodes have no interaction outside the DiffServ domain.

As the traffic enters the DiffServ domain, it is classified and possibly conditioned (conditioning can also take place at the egress though). After this, enough information is available to assign the traffic to a DiffServ Behavior Aggregate using a DSCP. Inside the network, traffic is routed according to the specified PHB for this DSCP. The PHB is defined as the "externally observable behavior of a DiffServ node to a particular aggregate." This means that the PHB is the means by which a node allocates its egress bandwidth to an aggregate. A simple PHB would be, for example, a percentage bandwidth allocation. It is implemented by means of buffer management and packet scheduling mechanisms.

The use of the DSCP is defined in RFC2474. As it is impossible to add an extra field to a currently existing IP header, the DSCP is carried in the existing ToS field from IPv4 or the Traffic Class Field in the case of IPv6. Both fields are 8 bits in size. Of these 8 bits, the last 2 bits are used for end-to-end congestion notification (ECN), leaving 6 bits for the effective coding of the DSCP field.

The default PHB is Best Effort (BE), meaning that no bandwidth guarantees are given to the incoming traffic. This PHB is coded in the DSCP as "000000." The highest priority traffic on the other hand is classed as Expedited Forwarding (EF, "101110"). This traffic should be routed with the highest priority. However, the ingress must be policed in such a fashion that the total egress traffic from a node is smaller than the available bandwidth. This implies making sure that no user abuses the EF class. There are four different Assured Forwarding (AF) classes with a dequeue priority between the BE and EF. Each of the four AF classes can have three drop precedence levels (low, medium, and high). The DSCPs for the AF classes are noted in the following table:

	AF1x	AF2x	AF3x	AF4x
Low	001010	010010	011010	100010
Medium	001100	010100	011100	100100
High	001110	010110	011110	100110

The class to which a certain flow must be mapped when in enters a DiffServ domain can be determined during the classification stage. Within the class, the drop precedence is determined using, for example, a single-rate Tri Color Marker (srTCM) or a dual-rate Three Color Marker (see RFCs 2697, 2698). Red normally corresponds to the high drop precedence, yellow to medium drop precedence, and green to low drop precedence. It is also allowed to map the three drop precedence levels to two different levels of drop probability. In this case, both red and yellow are given the same drop probability.

The AF class does not allow any packets within a microflow to be reordered. This almost always dictates that it is not possible to use three different queues to transmit the traffic (and does a priority dequeue). Instead, a mechanism such as RED is to be used to discard the packets, which is described next.

One issue for implementing a DiffServ domain is the compatibility with the original IP RFC791, in which the first three bits are used as a precedence indicator. This is only partially compatible with RFC2474. In order to maintain compatibility between DiffServ and older IPv4 network, it is assumed that all traffic coming from a non-DiffServ aware network are remarked at the edge nodes.

There are multiple ways to discard packets from the output queues. The best known algorithms are RED (as developed by Floyd and Van Jacobson) and Weighted RED (WRED). RED is a smart mechanism to discard packets from the tail of a queue which is designed to interact well with the TCP protocol functioning on higher layers. Hence, in order to understand why we need a discard mechanism such as RED, it is necessary to see how the TCP protocol functions.

When a TCP connection between two end nodes is set up, both exchange information about the available buffer space at the two ends of the connection, but no knowledge is available on the speed of the link and the level of congestion. Hence, some form of congestion control is needed between the two nodes. For this reason, both hosts keep a so-called "congestion window" which accounts for data that may be "in flight." At no point in time is the sender allowed to put more data on the line than the size of this congestion window. This keeps the sender from initially overloading the link. When a connection is set up, the sender starts with a congestion window of 1. After the first sent segment is acknowledged to have been received by the other side, the window is increased by 1 segment. Thus, after one Round Trip Time (RTT), two segments may be sent. Whenever these two segments are acknowledged, the window is increased by 2, etc. This procedure is called slow start (even though it is not very slow) and effectively doubles the congestion window every RTT.

The exponential growth of the window stops when its value becomes larger than a certain fixed value. After this, every RTT the window size is increased by only 1 segment per RTT, placing the sender in "congestion avoidance" mode instead "slow start." As the congestion window and hence the number of segments in flight starts to grow, the load on the link does as well. Hence, at a certain point in time, we can expect that the network will become overloaded. This is the congestion phase. What happens is that somewhere in the network (the first overloaded router), a packet will get dropped. This means that the sent segment will not get acknowledged within the expected time. The transmitting TCP node will react to this by slowing down the congestion window to a lower value (originally, back to 1 segment size, but newer TCP versions such as TCP Reno are less defensive), effectively lowering the transmission speed.

The point of explaining this is to show that TCP protocol behavior is that a single lost packet is enough an indication of a congested network to drastically lower its transmission speed. This means that we have no reason to drop multiple packets belonging to one connection if a router gets congested. Another behavior is that when a router carrying multiple TCP connection starts tail-dropping packets, all TCP connections carried by the router will see lost packets, hence all lowering their transmission speed. This causes a phenomenon called known as "global synchronization," where each connection goes through the slow-start mechanism at the same time, causing very ineffective use of the available bandwidth.

Both these issues with TCP as the most used higher layer transport protocol (between the UE/client and the server on the other side of the core network) tell us that we need a smarter mechanism at the routers to deal with a congested link. As said before, RED is an example of this. RED assumes that we can calculate the average length of the output queues in a router. When the average length of a queue increases above a certain level, RED does not simply start tail-dropping all new packets. Instead, with a probability that increases with the queue size, packets are dropped randomly. In other words, when a link is lightly congested, there is a very small chance that packets will get dropped, causing only few TCP links to lower their transmission speed. As the link gets more congested though, more packets will get dropped, causing more TCP connections to back off. The extreme measure of tail-dropping all packets from a queue is only taken when the queue size gets above a certain top-threshold.

The DiffServ RFC specifies that an AF implementation must attempt to minimize long-term congestion in each class while allowing short-term congestion caused by bursty traffic. RED is an example of an algorithm that does exactly this. The same DiffServ RFC also specifies that the dropping probability for low drop precedence packets should be lower than for high drop precedence packets. It specifies that at least two different levels of drop probability must be implemented for each class. In this case, AFx1 is given one level and AFx2 and AFx3 are given the other (higher) level. Specifying different thresholds for both different IP precedence levels and different traffic classes can be done using an algorithm called WRED. This allows standard traffic to be dropped more frequently than premium traffic during periods of congestion.

Multiprotocol Label Switching support for DiffServ

When DiffServ traffic must be carried over an MPLS network, the DSCP needs to be mapped to accommodate MPLS level QoS. For this purpose, the EXP bits in the MPLS header are used. Inside the MPLS network, each node only looks at the MPLS label when forwarding the packet. There are two ways of transporting the MPLS packets over the network. The first is called E-LSP. Here, the PHB for the MPLS node is completely defined by the EXP field inside the MPLS label. If the network needs either more than 8 PHBs or the EXP field cannot be used for some reason, the behavior is specifically set up per MPLS link. In this case, the PHB can be decided either from the MPLS label, by a combination of the MPLS label and the EXP field or from a combination of the MPLS label and some link layer encapsulation. This is called L-LSP.

References

1. Diffserv to QCI Mapping, https://tools.ietf.org/id/draft-henry-tsvwg-diffserv-to-qci-03.html, 2020.
2. Telecom Infra Project, Learnings from virtualized RAN technology trials over non-ideal fronthaul, 2019.
3. Zynq UltraScale + RFSoC, https://www.xilinx.com/products/silicon-devices/soc/rfsoc.html.
4. Layerscape Access LA1200 Programmable Baseband Processor. https://www.nxp.com/products/processors-and-microcontrollers/arm-processors/layerscape-multicore-processors/layerscape-access-la1200-programmable-baseband-processor:LA1200.
5. Crest Factor Reduction (CFR) Concept, https://rfmw.em.keysight.com//wireless/helpfiles/n7614/Content/Main/Crest_Factor_Reduction_Concept.htm.
6. Crest Factor Reduction, https://zone.ni.com/reference/en-XX/help/374264L-01/rfmxspecan/crest_factor_-reduction/.
7. eCPRI Specification v2.0, "*Common Public Radio Interface: eCPRI Interface Specification*," May 2 2019.
8. Ingress Protection Code, https://en.wikipedia.org/wiki/IP_Code.

3 Hardware system dimensioning

This chapter discusses dimensioning aspects. Dimensioning is tightly coupled with requirements and often an iterative analysis between dimensioning and peak/average requirements is needed to come up with a system that is both cost and power efficient as well as high (enough) performance. Key topics that are relevant include

- Use-case dimensioning for throughput, user count, and latency topics including oversubscription
- Derive eCPRI bandwidth for dimensioning of fronthaul interfaces
- Mid/backhaul dimensioning
- Hybrid Automatic Repeat ReQuest (HARQ) context store and restore bandwidth and sizing
- Double Data Rate (DDR) memory size and bandwidth dimensioning
- Radio unit (RU) analog transmit and receive chain dimensioning

System dimensioning involves mapping the application (say, a centralized unit (CU), a distributed unit (DU) or a radio unit (RU)) to the available hardware, or other way around, defining the required hardware to satisfy application requirements. Application mapping is done for multiple reasons. First, it builds confidence that the use-case can be met by determining the technical feasibility of the use-case to within certain margins of uncertainty or error. Second, it defines early in the process where focus areas of performance are. Identifying potential performance bottlenecks early on allows us to dedicate internal or supplier resources to these areas and to track performance for key topics during the implementation cycle so that support can be brought in on-time, or hardware modifications are identified early as possible in the project.

System mapping is about identifying which tasks need to be executed on each available hardware resource. Once we establish this mapping, we can define communication requirements (message rates, sizes, etc.) between components, utilization of each component, and so on. Some level of basic automation can be useful, this can be as simple as a spreadsheet calculator or as complex as a fully automated system model in a dedicated tool.

Centralized/distributed unit use-case dimensioning for throughput

Per the Federal Communications Commission (FCC), an information transfer rate (in either uplink or downlink) of 768Kbps or more defines a connection as "broadband" in the United States. Even though the definition is somewhat ambiguous (is this a theoretical or

Open Radio Access Network (O-RAN) Systems Architecture and Design. DOI: https://doi.org/10.1016/B978-0-323-91923-4.00011-2

measurable throughput?), it sets the first definition of what the expected consumer behavior of a broadband connection is. Most broadband systems—be it wired or wireless—make use of a shared medium (wireline: consider a DOCSIS cable modem sharing coax access or wireless: consider the air/wireless interface as the shared medium). This opens the question of how many users should be allowed to share this (fronthaul) medium (Fig. 3–1). Similarly, the backhaul medium (from the wireline cable head-end or the wireless base station) is limited as compared to the peak throughput sustained by the system it is providing access to.

This topic is referred to as oversubscription or statistical multiplexing of the access medium, using the fact that even during periods of heavy usage, not all subscribers/systems will be trying to use the system at peak throughput and we can thus dimension for a (lower) average rate, with a percentage likelihood that consumer satisfaction is achieved.

As said, oversubscription is not unique to any unit or interface in the system. It applies to the Core Network, CU, and DU components as well as the mid- and backhaul interfaces. We assume—for now—that the fronthaul interface to the RU is dimensioned for peak capacity rather than average, even though even this is not per se the case: techniques such as Bandwidth Partitioning in 5G allow us to implement statistical multiplexing over the fronthaul interface.

Traffic requirements based performance analysis

One way to analyze requirements is to ignore the advertised peak and average consumer rates completely, and instead look at what we know about the overall consumer pool requirements. This averages out *all* users over time and as such applies to the core network, CU, and potentially DU pieces of the overall solution.

Consumer Internet Traffic is sourced from Cisco,[1] shown below graphically below in Fig. 3–2.

The split between mobile and fixed network traffic is shown below in Fig. 3–3.

The overall internet traffic, taking in to account the split over fixed versus mobile network, it is possible to project the amount of internet traffic per household. Note that per the

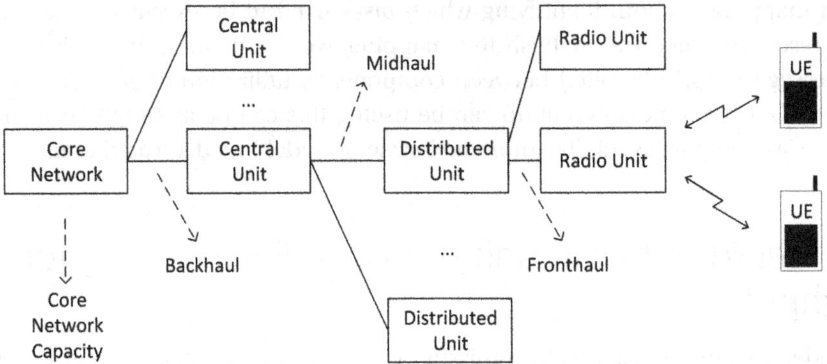

FIGURE 3–1 Abstract view of O-RAN deployment showing points of congestion and/or oversubscription.

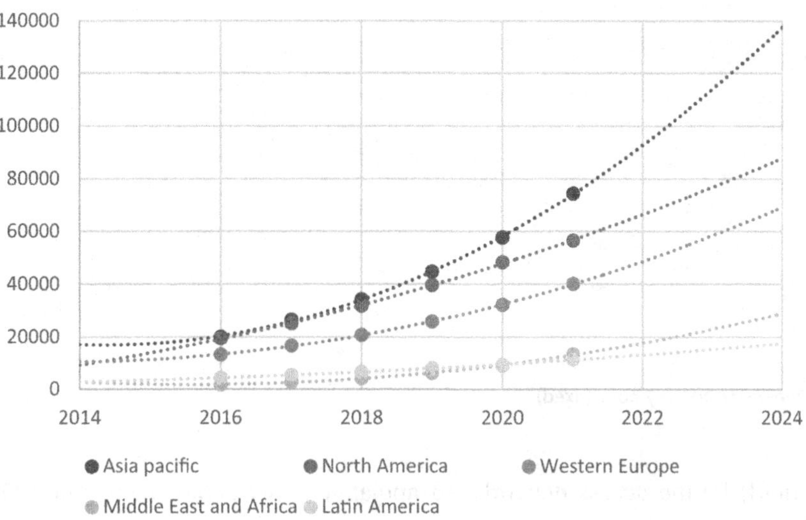

FIGURE 3–2 Consumer internet traffic aggregate across fixed and mobile. *Cisco.*

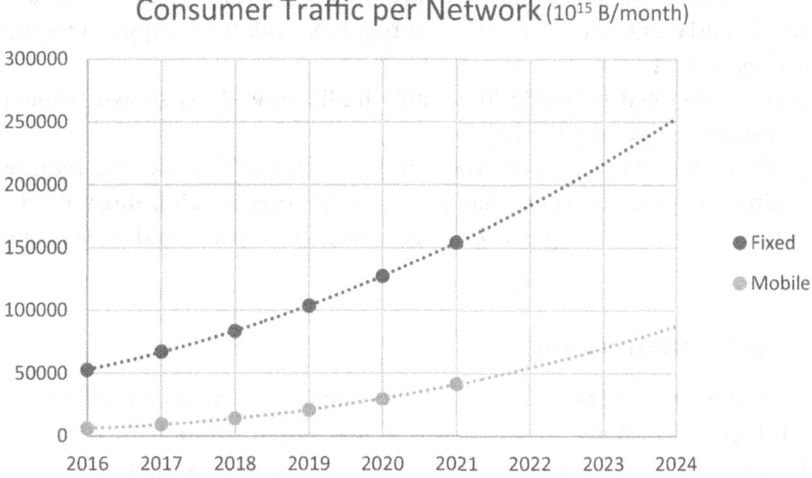

FIGURE 3–3 Split fixed and mobile traffic.

United States government,[2] the average household size in North America (USA) is roughly 2.5 persons. Assuming the household internet traffic is consumed during a certain amount of active "Busy Hours" per day, we can project the required data rate to be supported

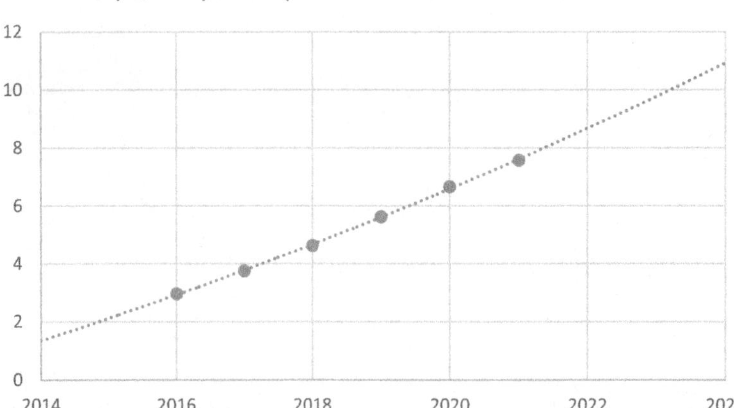

FIGURE 3–4 Dimension to busy hour (fixed).

(per household) by the access network, extrapolated to 5G deployment time (2020–24) as shown in Fig. 3–4:

This graph assumes that for consumer/household use, the complete volume of Internet traffic is consumed during 3 "busy hours" per day. This rate projects the average required throughput during these hours. Projections are only made for North America, which is a well-understood market. Quality of Experience requires peak throughput to be much higher —assume 1 Gbps at least.

A similar projection can be made for mobile traffic as well, as shown below (assuming the same 3 "busy hours" per day) in Fig. 3–5.

This gives us all the data we need. We can now project dimensioning requirements for those networking components where statistical multiplexing is valid, those locations where we are dealing with traffic averages over large amounts of users and a wide geographical area.

User data rate dimensioning

In the previous chapter, we showed how to use a top-down approach to dimension Radio Access Network (RAN) requirements, where the inidividual component throughput targets are derived from overall (expected) Internet use across a large geogrphical area. Another, and very different, more bottom-up way of looking at specifically DU dimensioning is to look at the consumer individual experience as set (for example) by the measured download rate, and to extrapolate from there. We can aim, for example, at a minimum download rate of 10 Mbps that a consumer expects to achieve. A so-called oversubscription ratio is applied to account for the fact that it's highly unlikely that all users, at the same time, will be looking to achieve this download rate. It's a quasi arbitrary number, but typical rates range in the 10–50 range. Assume for our DU example to have roughly 1000 connected users per Cell

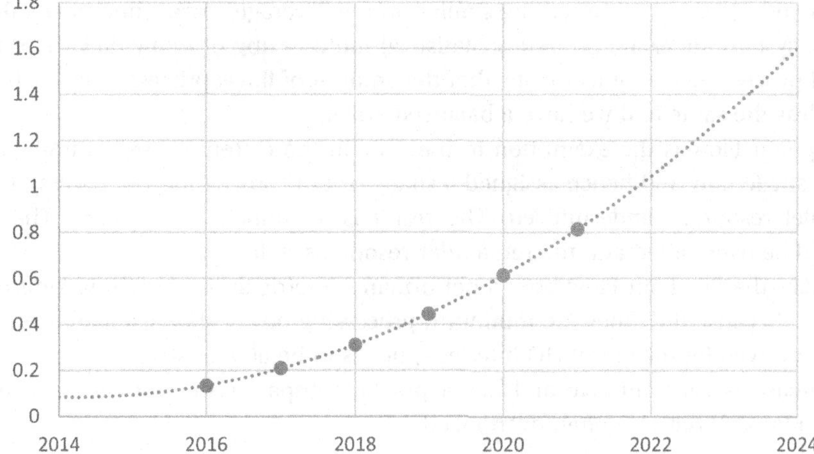

FIGURE 3–5 Dimension to busy hour (mobile).

Site DU (a rule-of-thumb global average from the 4G (LTE) days), we can now calculate the required Cell Site throughput as: 10 Mbps × 1000 users/20 (chosen oversubscription ratio) = 500 Mbps. This would be the capacity that the Cell Site base station needs to support towards the CU link and can be used to dimension specifically Layer 2 stack performance needs, for the Physical Layer components, dimensioning to average is much more complex given the real time constraints that the physical layer is under.

Note how the oversubscription ratio can be correlated to the previously discussed Mbps/ "busy hour" value. If we take (2020) a Mbps/"busy hour" of 0.6 Mbps and a desired consumer "experienced" data rate of 10Mbps, the oversubscription rate is 10/0.6 ∼ = 17x. This shows how the two techniques (top-down and bottom-up) end up matching.

Elephant flows

An elephant flow is defined as a high bandwidth connection that is typically transmission control protocol based. This type of flow could correspond to a user downloading a huge file or similar operation. The characteristics of this flow is that it is a very high bandwidth, over a short amount of physical time.

What makes an Elephant Flow unique is that it poses specific challenges on the underlying hardware implementation. Let's explain this in a bit more detail. Modern hardware implementations for high-bandwidth telecommunications products rely on parallelization to implement high-performance systems where there is more than a single processing element doing each function (say, encryption, classification, and so on). In this type of implementation, the incoming traffic is directed to one of the multiple cores, accelerators, or other processing elements to spread the processing load. Traffic directing is mostly done based on some classification rule that directs all traffic associated with a single flow to a single (set of) processing resource(s). An example of such a classification rule could be the IP Source/ Destination address tuple or something similar. In the "average" case where many flows pass through the system, there are no issues; statistical multiplexing of *many* flows over relatively *few* parallel processing elements means that the amount of flows processed by each resource is more or less the same and we have a balanced system.

The Elephant Flow is the exemption to the rule, in the extreme case, all traffic is associated with a single flow and hence assigned a single (set of) processing resource(s), leaving all other parallel resources underutilized. The result is an unbalanced system: The assigned resource will be overloaded and the remainder resources stale.

This makes the Elephant Flow both a performance metric and a benchmarking/optimization use-case because it defines the minimum processing power that a single resource (core, accelerator, or even input/output (IO interface) needs to be able to sustain.

The use-case is very intuitive and has a practical impact (who likes to wait long for a download to finish?) but also often overlooked.

Use-case dimensioning for latency

Overall application latency is composed of the various components shown in Fig. 3–6.

The picture includes the Radio Access Network (RAN) latency (User Equipment (UE) to RU/DU and CU), but also the latency from the Communication Service Provider (CSP or RAN/ Operator network), as well as the latency from the internet itself. Note that the CSP and Internet latencies can be eliminated by use of Metro Edge Computing where the application server is colocated with the CU/DU components.

- CSP and Internet latencies. This is the combined delay incurred by the operator network as well as the access network connection to the operator network. As an example, a typical network provider guarantees [3] in the order of 20–50 milliseconds one-way latency, depending on the physical location of source and destination. Note, as a comparison for wireline

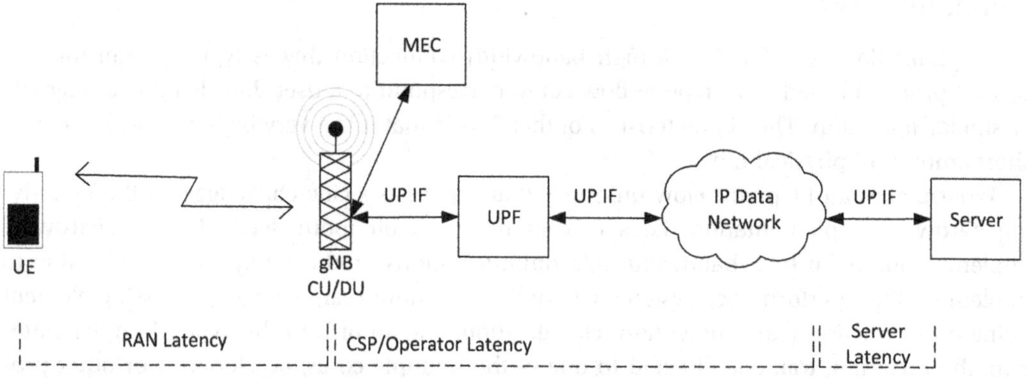

FIGURE 3–6 Network component latencies.

networks, typical digital subscriber line latency is also reported as 20–50 milliseconds. These latencies constitute the bulk of the overall UE to Server latency and have been subject of recent operator focus—and driving data center localization. We expect these numbers to be aggressively reduced in the next years.

- CU latency. The performance of the CU UP depends much on the choice of implementation, which can be Network Processor, Field Programmable Gate Array (FPGA), Application Specific Integrated Circuit (ASIC) or software centric on multicore central processing units (CPUs). Additional latencies can be added by switches or routers that are placed in between the CU UP function and the connectivity to/from the DU and the User Plane Function (UPF or core network). As a rule of thumb and omitting queueing delays caused by congestion, consider the following latencies:
  - <1 microseconds latency for a latency-optimized (cut-through) Ethernet switch
  - <10 microseconds latency for a store-and-forward Ethernet switch
  - 10–100 microseconds latency for a software implemented switch or router function running on high-performance CPU core (example: Data Plane Development Kit (DPDK) software-based implementations). This latency includes IPSec or other tunneling cost.
- RAN latency. This latency consists of 3 main components that can be analyzed in more depth independently.
  - Protocol processing latency associated with L2 protocol stack handling that is outside of the slot-based 3rd generation partnership project (3GPP) timing domain. We assume this component (as per software implementation for switch/router functions in the CU) to be relatively small as <100 microseconds.
  - Queueing delay incurred through the slot-based timing architecture of the air interface as well as the latency incurred by scheduler prioritization of other users. This topic is discussed in more detail below when we discuss the Transmission Time Interval. Assuming prioritization of low-latency users and light system loading with regards to these low-latency users, queueing delay should be limited to below 10 milliseconds.
  - Air interface latency associated with the MAC/PHY processing and HARQ retransmission. This latency is discussed in detail in 3GPP 36.912 for LTE where similar math can be applied for 5G/NR. For Frequency Division Duplexed (FDD) operation in LTE, the one-way latency through the RAN is calculated as D_{UP} [milliseconds] $= 1.5 + 1 + 1.5 + n \times 8 = 4 + n \times 8$, where n is the number of HARQ retransmissions. Considering a typical case where there would be 0 or 1 retransmission, the approximate average U-plane latency is given by $D_{UP,\ typical}$ [milliseconds] $= 4 + p \times 8$, where p is the error probability of the first HARQ retransmission. The minimum latency is achieved for a 0% HARQ Block Error Rate (BLER), but a more reasonable setting is 10% HARQ BLER. $D_{UP,0\%HARQ\_BLER}$[milliseconds] = 4 (0% HARQ BLER) $D_{UP,10\%HARQ\_BLER}$[milliseconds] = 4.8 (10% HARQ BLER). Time Division Duplexed (TDD) numbers are slightly worse. Assuming a 30KHz subcarrier spacing and 500 milliseconds slot time in 5G/NR (as opposed to 1 millisecond slot time in LTE), these numbers can be divided by two to achieve an estimated <3 millisecond RAN latency.

Aggregate latency through the RAN is estimated as 5–10 milliseconds in a realistic one-way latency through the RAN before optimizations such as discussed in Ultra Reliable Low Latency Communication (URLLC) protocols come into play.

Users/transmission time interval

The "UE/TTI" metric relates to the definition of how many users (User Equipment (UE)) should be scheduled each 3GPP slot (Transmission Time Interval or TTI). This metric is important because it impacts the Medium Access Control (MAC) and Physical (PHY) layer implementation/test complexity as well as testability. To deduce what we think is a reasonable number of UE/TTI, we show a couple of strategies, with a few Small Cell use-cases as practical examples. There are three strategies we can take:

Option 1: Engineering deduction

Step 1, assume a M/D/1 queueing model [2] with a single server representing a single scheduled UE/TTI, a 2000 slot/second (FR1) service rate μ and a ($\rho = 0.95$, so $0.95 \times 2000 =$) 1900 "virtual Transport Block"/second arrival rate, where we assume a 95% loaded system
($\rho = 0.95$)

Plugging this into the equation for average waiting time:

$$\omega = \frac{1}{\mu} + \frac{\rho}{2\mu(1-\rho)}$$

we get $\omega = 1/2000 + 0.95/(2 \times 2000 \times (1-0.95)) = \sim 5$ microseconds.

Step 2, given that M/D/C queues are tough to evaluate with a lot of complex math we would like to avoid, we are going to make the gross simplification that we can simply multiply this waiting time ω by the ratio of active/scheduled UE to approximate a (say) round robin scheduler that reduces the effective service rate. So, in case of 64 active UE and 8 scheduled UE we get to $(64/8) \times 5$ milliseconds = 40 milliseconds average waiting time in the system.

Now, we can plot all of this in a graph to play with numbers (Fig. 3–7).

Note that the waiting time for low system loading here is unreasonably high because we assume a very basic round robin scheduler that doesn't skip users for which no data is available. As such, we focus on the right-hand side of the chart. What it shows is that when the system gets highly loaded, queueing time can get reasonably high. Let's define "high" as 20 milliseconds—an arbitrary but reasonable choice for an Enhanced mobile broadband (eMBB) use-case.

This would suggest a ratio of 8 between active UE and Scheduled UE to be an "pretty good" choice for a highly loaded system (90%). Assuming an Integrated Small Cell with 64 Connected UEs, and a 5x ratio between Connected and Active UEs means $64/5 \sim = 13$ Active UEs and ($13/8 \sim =$) 2 (rounded to an integer number) scheduled UE/TTI are enough.

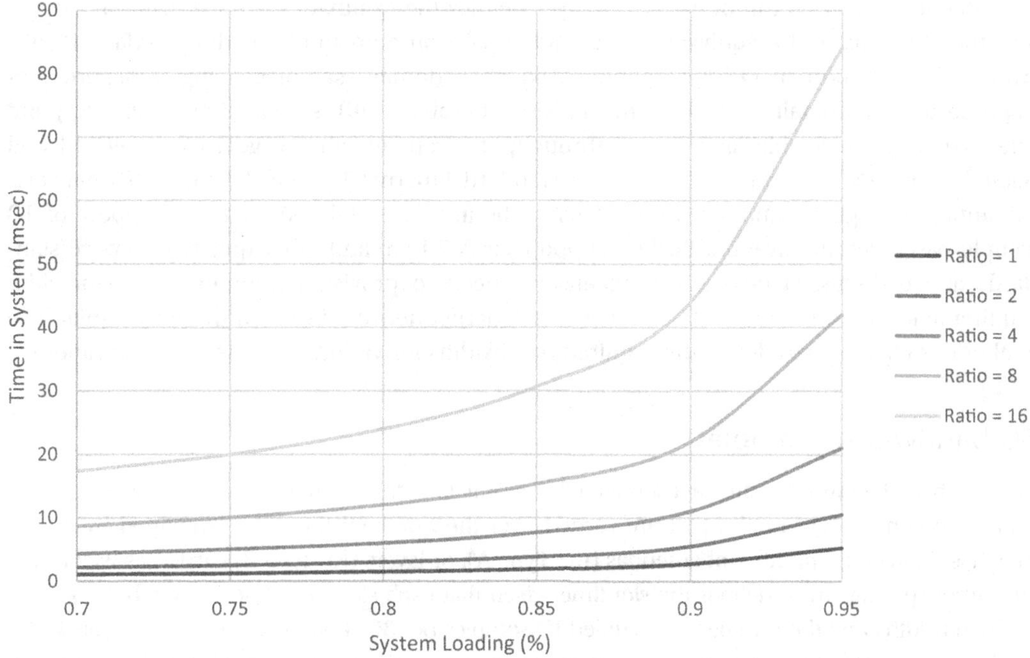

FIGURE 3–7 System waiting time evaluation to evaluate scheduled UE/TTI.

Option 2: Common Sense

Above approach is complex even when we take many simplifications in the process. A different model would be to look at the usage profile and see if this can help us provide a scheduling guideline. Consider a mmWave (3GPP: Frequency Range FR2) system with beamforming employed.

If we have 64 Connected Users evenly spread over (90-degree radius, 10-degree beam width) approximately 10 beam directions, each beam would be able to direct traffic at 8 users maximum. Given that users tend not to be perfectly evenly spread, we add some safety margin, let's double this to 16 users/beam.

Assuming (per "Option 1") a ratio of 5 for Connected to Active UE count, we come to 3 Scheduled UE/TTI.

Option 3: The Marketing Way

"Do what the customer asks" or "do same or better than the competition" may not be the perfect engineering answer but often defines reality. Some rule-of-thumb numbers that can be used are a Scheduled: Connected UE ratio of 1:10 and Scheduled: Active UE ratio of 1:5 in the small cell application space. This would put our example requirements to 6 UE/TTI.

eCPRI, fronthaul bandwidth and latency

The eCPRI standard and the control user and synchronization (CUS) plane interface that O-RAN defined to run on top of the eCPRI layer defines the fronthaul interface between the DU and the RU. This is the equivalent of a traditional Common Public Radio Interface (CPRI) (with the exemption of O-RAN defining frequency domain samples being transported as opposed to time domain), which is the challenge because CPRI is defined as a point-to-point interface with predictable latency and throughput. Even though the goal of the eCPRI and related standards is to run IQ traffic on a standard Ethernet/Internet Protocol (IP) network, minimum throughput and maximum latency bounds are still required for support of IQ sample-based communication. Besides support for 802.1 cm and other quality of service standards as we discussed before, this means we need to provision minimum IO bandwidth requirements at the DU and RU, as well as the Ethernet network between these two units. The goal of this chapter is to define the required bandwidths for various use-cases/configurations.

5G bandwidth examples

5G fronthaul bandwidth can be estimated by calculating the aggregate number of (frequency domain occupied) IQ samples that are carried over the fronthaul link. This is defined by the air interface bandwidth, number of antennas (RU Type A), or layers (RU type B) carried as well as the subcarrier spacing which defines the slot time, given that each slot constitutes 14 symbols. 5G carrier bandwidths and the number of occupied IQ samples per 3GPP are copied below (Table 3–1):

As an example, assuming 30KHz subcarrier spacing, per baseline unit of 100 MHz occupied bandwidth, a single layer corresponds to 3276 (occupied subcarriers) × 14 (symbols/slot) × 2000 (slot/second) × 32b (uncompressed IQ data) = 2.93 Gbps. We multiply this by the number of layers to define the transmit or receive bandwidth requirements.

The fronthaul bandwidth can optionally be reduced by using IQ compression techniques such as modulation, Block Float, a-law/μ-law and other compression techniques as defined in the O-RAN CUS Plane fronthaul specification. 16 to 9- or 12-bit compression are often used in O-RAN implementations, depending on the use-case defined.

Scaling up to the defined use-cases, (peak) bandwidths are summarized in below tables for some low-end and higher-end examples (Tables 3–2 and 3–3).

Table 3–1 Number of 5G/NR occupied subcarriers for 30KHz subcarrier spacing for different RF bandwidths.

5G b/w	5	10	15	20	25	30	40	50	60	80	90	100
Carriers	132	288	456	612	780	936	1272	1596	1944	2604	2940	3276

Table 3–2 Example of 5G O-RAN fronthaul eCPRI IQ bandwidths—low-end.

	#Antenna layer	RF Occupied bandwidth (OBW)	Uncompressed IQ bandwidth	9b/16b compressed	Modulation compressed
Consumer indoor	2-layer DL, 1-layer UL	40 MHz	2.3 Gbps DL 1.2 Gbps UL	1.32 Gbps DL 0.66 Gbps UL	427 Mbps DL
Consumer	4-layer DL, 2…4-layer UL	100 MHz	11.72 Gbps DL 5.86–11.72 Gbps UL	6.6 Gbps DL 3.29–6.6 Gbps UL	2.2 Gbps DL
Outdoor	8-layer DL, 4-layer UL	100 MHz	23.44 Gbps DL 11.72 Gbps UL	13.2 Gbps DL 6.6 Gbps UL	4.4 Gbps DL

OBW, Occupied bandwidth; DL, Downlink; UL; Uplink.

Table 3–3 Example of 5G O-RAN fronthaul eCPRI IQ bandwidths—high-end.

	MU-MIMO configuration	RF Bandwidth (IBW)	Uncompressed IQ bandwidth	9b/16b Compressed	12b/16b compressed
16RT	8 layer DL, 4 layer UL	100 MHz	23.44 Gbps DL 11.72 Gbps UL	13.7 Gbps DL 6.8 Gbps UL	18 Gbps DL 9 Gbps DL
32RT	8 layer DL, 8 layer UL	100 MHz	23.44 Gbps DL 23.44 Gbps UL	13.7 Gbps DL 13.7 Gbps UL	18 Gbps DL 18 Gbps DL
32RT	8 layer DL, 8 layer UL	200 MHz	46.88 Gbps DL 46.88 Gbps UL	27.4 Gbps DL 27.4 Gbps UL	36 Gbps DL 36 Gbps DL
64RT	8 layer DL, 8 layer UL	200 MHz	46.88 Gbps DL 46.88 Gbps UL	27.4 Gbps DL 27.4 Gbps UL	36 Gbps DL 36 Gbps DL

DL, Downlink; IBW, Instantaneous bandwidth; RT, real time; UL; Uplink.

Note:

- These tables assume 14 occupied symbols (all layers occupied) which is unrealistic typically, a subset of the symbols is used for Physical Downlink Control Channel (PDCCH) and other "overhead."
- These tables exclude overheads associated with (Ethernet) packetization of imaginary and real (IQ) sample data as well as signaling overhead of beamforming information. A 10% margin is typically assumed to accommodate these components.

4G bandwidth examples

Like 5G, 4G fronthaul bandwidth can be estimated by calculating the aggregate number of (frequency domain occupied) IQ samples that are carried over the fronthaul link. This is defined by the air interface bandwidth and number of antennas (RU Type A) or layers (RU type B) carried. 4G subcarrier bandwidth is defined to 15KHz which corresponds to a 1 millisecond slot time at 14 symbols/slot. 4G carrier bandwidths and the number of occupied IQ samples per 3GPP are copied below (Tables 3–4 and 3–5):

Fronthaul latency

Like the end-to-end system latency that we calculated previously, we can estimate the fronthaul latency by accounting for each component in the system.

- DU termination latency is the cost of terminating the eCPRI frames in the DU. This cost includes the hardware overhead through Ethernet PHY and MAC components (assumed less than 1 microsecond), the (optional) latency associated with an Intelligent Network Interface Card (NIC) implements the fronthaul functionality in an offloaded manner (assumed around 10 microseconds or less).
- When termination is implemented as a software function (say, in DPDK), Virtual Machine or application wakeup latency needs to be accounted for in the receiver—this is the latency between arrival of an interrupt that indicates packet arrival and the start of the (high priority) application that processes the packet. This latency can be anywhere between 10 microseconds (optimized PREEMPT_RT application with no virtualization supported) and approximately 50 microseconds with virtualization worst-case. In addition, the software function to process the packet takes compute clock cycles that translate to about 1–5 microseconds depending on software optimization level.
- Ethernet Serialization and Physical Transmission Latency. Physical Ethernet transmission latency is calculated based on Ethernet speed and packet size. For example, the serialization delay of an 800-byte packet (as used by the end-to-end latency benchmark) on a 1 Gbps link is (800 bytes $\times$ 8 bits)/1 $\times$ 10^9 bps = 6 microseconds. Note that this

Table 3–4 Number of 4G/NR occupied subcarriers for different RF bandwidths.

4G b/w	1.4	3	5	10	15	20
Carriers	72	180	300	600	900	1200

Table 3–5 Example of 4G O-RAN fronthaul eCPRI IQ bandwidths—low-end.

	#Antenna layer	RF bandwidth (OBW)	Uncompressed IQ bandwidth	9b/16b compressed	12b/16b compressed
Consumer indoor	2-layer DL, 1-layer UL	10 MHz	538 Mbps DL 269 Mbps UL	302 Mbps DL 151 Mbps UL	404 Mbps DL 202 Mbps UL
Consumer	4-layer DL, 2-layer UL	20 MHz	2.15 Gbps DL 1.07 Gbps UL	1.21 Gbps DL 0.61 Gbps UL	1.61 Gbps DL 0.81 Gbps UL
Outdoor	4-layer DL, 4-layer UL	20 MHz	2.15 Gbps DL 2.15 Gbps UL	1.21 Gbps DL 1.21 Gbps UL	1.61 Gbps DL 1.61 Gbps UL

latency increases significantly when accounting for large (sets of) packets, for example as needed for L2/L1 interfacing. A high-performance deployment is expected to use 10GbE + link speeds. In addition to the serialization latency, we need to account for the physical transmission latency through the medium. The speed of light is approximately 3.34 microseconds/km but due to refraction in the fiber, fiber speeds are slightly lower at approximately 4.9 microseconds/km. Given a 'typical maximum' distance between DU and RU of 10s of km, the physical transmission latency through the medium can dominate the overall system latency.
- Cost of Ethernet switching/routing in the fronthaul. As we indicated earlier, as a rule of thumb and omitting queueing delays caused by congestion, consider the following latencies
    - < 1 microsecond latency for a latency-optimized (cut-through) Ethernet switch—the most likely option for fronthaul applications.
    - < 10 microseconds latency for a store-and-forward Ethernet switch.
    - 10–100 microseconds latency for a software implemented switch or router function running on high-performance CPU core (example: DPDK based implementations). This latency includes IPSec or other tunneling cost.
- RU termination latency which is expected to be like the DU termination latency.

End-to-end latencies over Ethernet in a virtualized, software centric world is measured below 100 microseconds but this can be optimized in a nonvirtualized environment or more hardware centric implementations to 10 or lower microseconds.

Distributed unit internal IO

Bandwidths between various components inside the DU need to be understood well when the implementation of the DU is split over multiple physical components with limited bandwidth interconnect. Refer to Fig. 3–8 below showing the Physical Layer flow which is the key component for discussion of bandwidths internal to the system given that we're dealing with IQ samples:

Various Physical Layer internal bandwidths can be calculated in a similar fashion as the eCPRI bandwidth. In fact, we can use the eCPRI calculations as a starting point for bandwidth allocation. We assume a FR1/ < 6 GHz (30 KHz Subcarrier Spacing (SCS)) 4-antenna (2-layer) system with 100 MHz carrier bandwidth (single carrier) and 64QAM support. Use the uplink processing chain as an example, where

- Baseband time domain sample bandwidth: 4 (carriers) × 4096 (subcarrier/Fast Fourier Transform (FFT)) × 14 (symbol/slot) × 2000 (slots/second) × 32 (16b I + 16b Q) = 14.7 Gbps.
- Baseband frequency domain sample bandwidth: 4 (carriers) × 3276 (subcarrier/FFT) × 14 (symbol/slot) × 2000 (slots/second) × 32 (16b I + 16b Q) = 11.7 Gbps.
    - This is the eCPRI bandwidth discussed previously (2.93 Gbps/carrier).
- Beamforming and precoding reduce the decoded bandwidth from 4 to 2 layers or 5.9 Gbps.

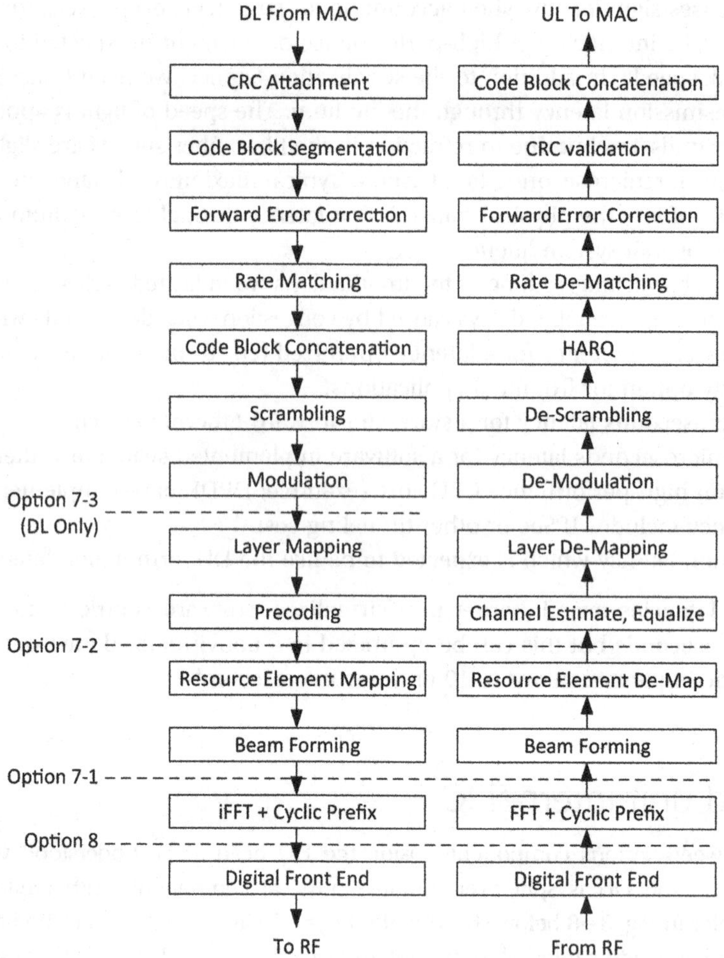

FIGURE 3–8 Physical layer functional components.

- De-modulation generates an 8-bit Log Likelihood Ratio (LLR) for each bit that is input to the decoder. We calculate this bandwidth as 2 (layers) × 3276 (subcarrier/FFT) × 14 (symbol/slot) × 2000 (slots/second) × 8 (bit in 256QAM) × 8 (bit/LLR) = 11.7 Gbps. Note that this bandwidth is higher than the bandwidth between the DU and the RU.
- This LLR bandwidth is input to the HARQ combining and Forward Error Correction decoding stages. We will discuss HARQ in more detail later to explain how the peak bandwidth input to the Forward Error Correction decoder can be up to 3x the LLR bandwidth of 35 Gbps.

- Output from the LLR decoder can be calculated like the LLR rate but taking into account the Code Rate as 2 (layers) × 3276 (subcarrier/FFT) × 14 (symbol/slot) × 2000 (slots/second) × 8 (bit in 256QAM) × 0.9 (Peak Code Rate approximation) = 1.3 Gbps.
- Above rate is assuming 14 occupied symbols/slot which ignores Physical Downlink and Uplink Control Channels (PDCCH/PUCCH), reference symbols and other overheads. Assume approximately 20% reduction to define the MAC/PHY throughput at approximately 1 Gbps.

Memory dimensioning

Memory dimensioning needs to be done for both memory size, enough to accommodate the Operating System (OS), packet buffers, user contexts etc., as well as memory bandwidth to ensure that the system can sustain the Read and Write bandwidth demands of the application at peak throughput.

Memory sizing

DDR sizing is relevant for defining memory components to be used in the system. As an example of a low-end DU, we work through a numerical:

- OS image—an embedded Linux image can most often be accommodated in approximately 0.5GB.
- We assume a similar memory requirement for the application images, approximately 0.5GB.
- Bulk U-Plane packet buffers can be implemented in the CU (dual/multiconnectivity) or DU component, depending on how the system is deployed. Buffer memory estimation can be done from the average (or peak) user data rate to be supported together with an indication of required buffer memory. As an example, assume the 500Mbps DU site throughput we estimated before with approximately 1 second of buffer memory of this throughput. This equates to approximately 500Mbit of buffer memory 62Mbyte. Assuming static buffer sizing (say, 2KB buffer storing a 500 byte packet), this equates to approximately 0.25GB.
- HARQ buffer memory is dimensioned as (more detail later) as 8 (maximum HARQ processes) × 64 (active users) × 2MB (TB HARQ buffer) or approximately 1GB.

Aggregate memory requirements are approximately 1GB for the OS, approximately 0.25GB for packet buffering and 1GB for HARQ buffering. Rounding up to power-of-two numbers, 4GB of DDR memory is chosen.

Life-of-a-packet memory bandwidth analysis

An often-overlooked aspect of performance is DDR bandwidth consumed by the system. All practical Linux based systems as used in CU and DU applications use an external DDR memory for storing both instruction and data. Most modern processors have a cache-based memory hierarchy where both CPU access to external memory and often also accelerator and IO traffic access to external memory is shielded by on-chip cache memory that reduces the bandwidth to external memory and provides a lower (average) latency to instruction and data access, thus improving performance. A generic System-on-Chip (SoC) architecture is shown in Fig. 3–9 below. We identify three levels of cache:

- L1 cache, separate for Instruction and Data storage. This cache memory is local to the CPU only and characterized by very low access latency (single digit clock cycles from the CPU). Typically dimensioned to small values such as 32 or 64KB.
- L2 cache, which is typically shared by a group of CPUs (cluster) with slightly higher latency of 10 seconds of clock cycles for CPUs within the cluster (or more for access by a CPU from a different cluster). Typically dimensioned to few 100 KB or up to several MB across multiple CPU clusters.
- L3 or platform cache which is either integrated into the SoC interconnect or implemented as part of the memory controller to offload DDR traffic explicitly. This cache has higher latency and is dimensioned to multiple MBs in higher-end chips. Lower-end devices can exclude the L3 cache altogether. Note that L3 cache memory is (optionally) accessible by accelerators and/or IO interfaces—consider the use-case where the content of an

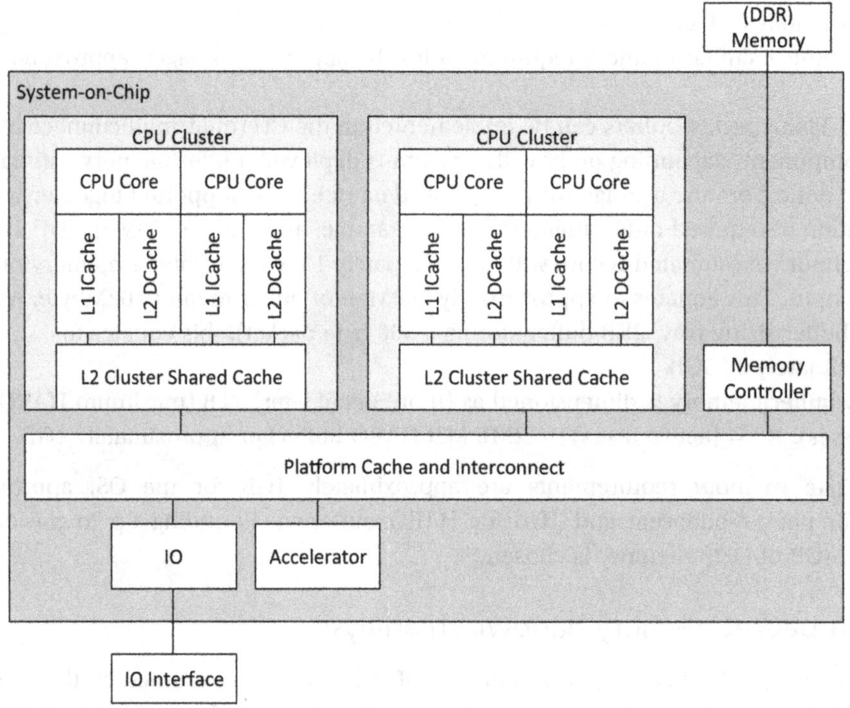

FIGURE 3–9 Generic (modern) system-on-chip (SoC) architecture showing cache hierarchy.

Ethernet frame coming in on an IO interface is read by a CPU immediately on reception and thus the path to and from DDR is not needed.

Each of the (SoC internal) interfaces has an associated bandwidth and latency, but these interfaces are assumed dimensioned for enough performance by the SoC manufacturer for most, if not all, use-cases. The typical bottleneck in the system as such becomes external (DDR) memory bandwidth.

DDR interfaces are defined by their interface speed (counted in (giga) transactions/second or GTPS) and the transaction width which is typically 64 (or with Error Correcting Codes (ECC) enabled, 72) bits. Combined, these two values give a raw throughput of the external memory, calculated from the multiplication of those two values. Consider a modern multicore CPU with dual 72-DDR4 interfaces running at 2.4 GHz which gives a raw throughput of 2 × 64bit (ECC bits are not used by the application) × 2.4 GTPS = 307.2 Gbps (note, unlike peripheral component interconnect express (PCIe) and Ethernet, this bandwidth is shared across Read and Write operations). Latency to DDR memory is above 100 (if not 100 seconds) CPU clock cycles.

For various well-documented reasons, this raw performance is not achieved in real-world applications that access memory in a near-random fashion. We identify two challenges:

- Latency to DDR memory tends to increase as the loading to the memory increases. This behavior is different from on-chip memories, latency to which tends to be bounded.
- The theoretical throughput number to the DDR memories is not achieved in practice.

Both aspects can be made visible by benchmarking the DDR performance as described elsewhere. We show an example below that shows DDR benchmarking for random memory accesses with small (single cache line) access width as shown in Fig. 3–10:

As the benchmark shows, when DDR loading exceeds approximately 50%, latency starts increasing exponentially, negatively impacting system performance. This "50%" number is used as a rule-of-thumb and can be relaxed in specific use-cases, for example, with large sequential accesses. Innovative new external memory techniques (Low Power DDR (LPDDR), High Bandwidth Memory (HBM), etc.) can have very different performance characteristics that need to be assessed independently. Note how external memory accesses are done in discrete quantities of cache lines sizes, typically 64b (8byte).

Besides establishing the peak or maximum practical loading of the memory interface, we need to find the application required throughput. Given the nondeterministic nature of cache-based systems, this is done by estimation. We track what is referred to as "life of a packet" (PDCP/RLC/MAC) or "life of a sample" (Physical Layer) processing through the applicable SoCs and memory subsystems to determine the memory transactions associated with the processing steps. Let's look at an example for a CU/DU combined flow associated with PDCP/RLC/MAC U-Plane processing below in Fig. 3–11.

The description is done for the downlink direction. The uplink direction is assumed reverse from downlink or can be left as an exercise for the reader. Key components are described below:

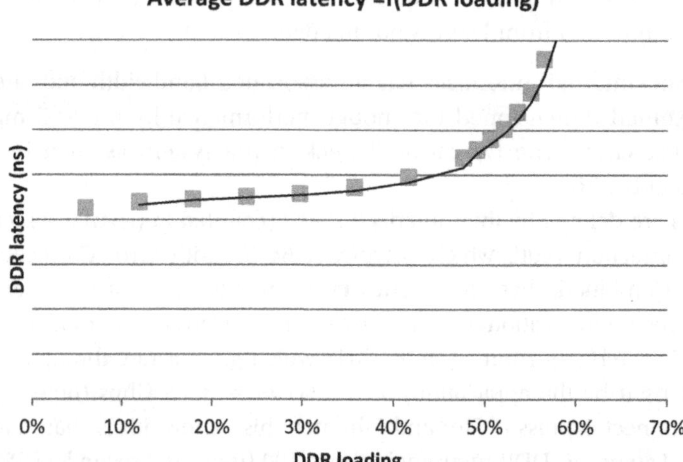

FIGURE 3–10 DDR latency as function of bus loading.

- Ethernet Rx—Pre-IPSec
  - The packet Rx function receives an Ethernet frame from the Ethernet interface, either on-chip from an embedded controller or off-chip through an Intelligent NIC. The Ethernet interface writes some form of frame/buffer descriptor to(wards) DDR memory that contains at a minimum a pointer and length field with some "valid" bit. The Ethernet controller also writes bulk payload (Ethernet frame) to the DDR memory and signals the GPP core through an interrupt that (a) frame(s) are available for processing.
  - The GPP core reads the frame/buffer descriptor from memory and parses it to conclude this is a U-Plane IPSec frame that needs to be de-ciphered
- IPSec de-ciphering
  - IPSec de-ciphering can be implemented as a software function on a GPP core or in an accelerator, for the purpose of memory dimensioning there is little difference. For both cases, there is a memory read operation, followed by a de-cipher stage within the SoC and a memory-write operation.
- IPSec postprocessing and PDCP preprocessing
  - The de-ciphered (now plaintext) frame is parsed to process the GTP-U header and classified to identify the bearer or PDCP context as required for the PDCP cipher operation.
- PDCP ciphering
  - The main memory-intensive operation in PDCP is ciphering which (like for IPSec) involves a read operation followed by a cipher stage and a write operation.

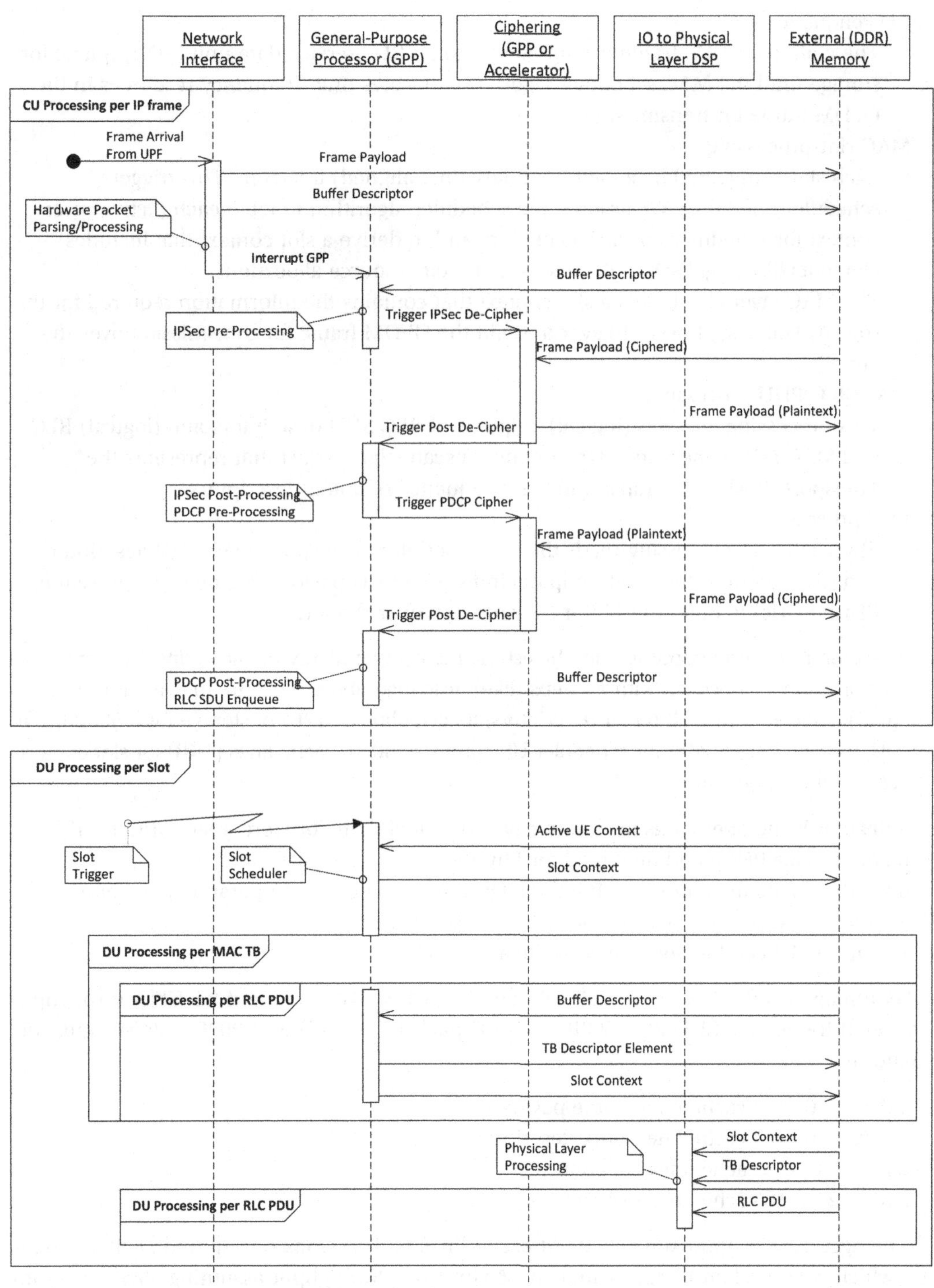

FIGURE 3–11 Tracking DDR operations in PDCP/RLC/MAC U-plane.

- RLC enqueue
  - The ciphered PDCP frame is enqueued into an RLC service data unit (SDU) queue for storage until the MAC scheduler algorithm allocates time/frequency resources in the OFDM frame for transmission.
- MAC slot-processing
  - Each slot is triggered by a (slot boundary time aligned) interrupt. This triggers scheduler operation. We assume the scheduler algorithm to fetch each (active) user context for scheduler metric calculation, and to derive a slot context that includes elements like target scheduled users and their resource allocations.
  - The MAC layer also builds a slot context that contains the information required for the (mostly stateless) Physical Layer to build the OFDM frame for transmission over-the-air.
- MAC/RLC PDU processing
  - This processing includes (logical) dequeue of RLC SDU descriptors and (logical) RLC and MAC PDU generation steps to build a scatter/gather list that represents the Transport Block to be transmitted as a sequence of buffer descriptors.
- PHY processing
  - Physical Layer processing reads the slot descriptor, Transport Block (TB) descriptor and (derived from the TB descriptor) individual PDU payload segments (representing PDCP SDUs) to build the TB for transmission over-the-air.

Assessing DDR throughput is done by estimating (in granularity of cache lines) the memory transaction counts associated with all steps (like) described above. For example, assume 1 cache line (64B) associated with all buffer descriptors, 8 cache lines (512B) of storage for a 500B frame or PDU, 2 cache lines (128B) for scheduler UE contexts and 16 cache lines (1KB) for slot context.

We count an aggregate:

- 6 Read or Write operations for frame payload (initial write for store, read/write for IPSec, read/write for PDCP and final read for Physical Layer transmit)
- 6 Read or Write operations for frame or TB descriptors that are operating per-frame
- <# active users> × 2 cache lines UE context Read
- 1 Write + 1 Read for slot context Write and Read

Assuming an FR1/<6 GHz small cell targeting 64 Active Users and 2.5 Gbps throughput with 500B frame size (2.5 Gbps/500B = 625 K packets/second) and 2000 slots/second, this equates to

- 625 K × 6 × 8 cache lines frame payload
- 625 K × 6 × 1 cache line frame descriptor
- 2 K × 64 × 2 cache lines UE context
- 2 K × 2 × 10 cache lines slot context

The aggregate is approximately 34 M cache lines or 17.5 Gbps of required DDR throughput, which is equivalent to approximately 35 Gbps raw throughput assuming 50% maximum

DDR utilization. A 2.4GTPS, 72-bit DDR controller provides 153.6 Gbps, leaving (more than) enough headroom for this application.

Above analysis is highly theoretical and optimistic in that it assumes very few overheads or software inefficiencies. For that reason, typical implementations add significant safety margin (2x) on top of this number. At the same time, above example doesn't consider L3 cache offload of DDR bandwidth which typically reduces the DDR load by up to half, when present and not implemented as write-through cache.

HARQ memory and bandwidth dimensioning

One of the largest bandwidth consumers in the DU is the Physical Layer HARQ operation and its associated context store/re-store operation.

Transport blocks, code blocks, and HARQ

3GPP defines a MAC PDU as the unit that is transferred between MAC and Physical Layers, or transmitted/received by the MAC layer. In the Physical Layer, the MAC PDU is transformed to a TB by:

- Segmenting the TB into a set of Code Blocks (CB) that are smaller (and bounded) in size
- Adding a cyclic redundancy check (CRC) to each CB
- Adding a CRC across all CBs to have a TB level CRC

as shown below in Fig. 3–12.

HARQ operation between base station (gNB) and client (UE) includes HARQ Ack/Nack operation which is acknowledgment of (in)correct reception of the TB by the other side. In addition, CB level CRC pass/fail information is transmitted, at groups of multiple CBs (Code Block Groups). This is a difference from LTE where HARQ information was only transmitted on a TB basis.

For determining code block size, refer to 3GPP 38.212, assuming Low Density Parallel Codes (LDPC) base graph 1 to find a nominal CB size of 8448 bits. The maximum transport block size is calculated from the frame configuration as follows:

- FR2/mmWave
  - Assume (120KHz SCS, 125 microseconds slot time) 273 Resource Elements (RE) allocated, 12 SC/RE,
    6 bit/subcarrier (64QAM), 14 symbol/slot, 2 layers/TB and 948/1024 code rate
  - $273 \times 12 \times 6 \times 14 \times 2 \times 948/1024 = 63690$ Byte
- FR1/ < 6 GHz

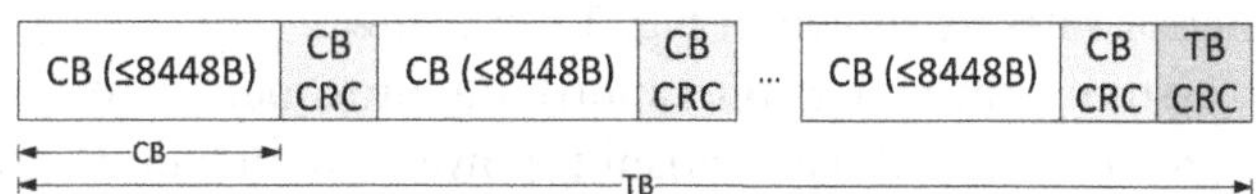

FIGURE 3–12 Transport block segmentation to code blocks.

  - Assume (30KHz SCS, 500 microseconds slot time) 273 RE allocated, 12 SC/RE, 8 bit/subcarrier (256QAM), 14 symbol/slot, 2 layers/TB and 948/1024 code rate
  - 273 × 12 × 8 × 14 × 2 × 948/1024 = 84920 Byte

Note how in the receive side, each physical bit at the decoder output is represented by a so-called "soft bit" or LLR which is a 6- or 8-bit representation of each bit. In addition, given HARQ operation, the 3x datarate that is generated from a 3x LDPC decoder equates that the maximum uncoded data block going into the LDPC decoder can be approximately 3 times the size of the Transport Block. In aggregate, taking a 3x code rate and 8-bit LLRs, each TB has a peak memory footprint sized as:

- FR2/mmWave
  - 273 × 12 × 6 × 14 × 2 × 1 Byte/LLR × 3 = 1.65 Mbyte
- FR1/ < 6 GHz
  - 273 × 12 × 8 × 14 × 2 × 1 Byte/LLR × 3 = 2.2 Mbyte

For (worst-case) retransmissions, these buffers need to be stored and re-stored every decode opportunity (slot). This implies large buffers are typically too large to be stored in dedicated on-chip static random access memory and hence target external DDR memory. The bandwidth associated with this operation is similarly high: up to 24x (3x code rate, 8x LLR size) the MAC/PHY bit rate, driving high memory speed.

Limited buffer rate matching

Limited Buffer Rate Matching (LBRM) is a technique to reduce HARQ buffer size (and/or associated PCIe bandwidth for save/restore operation). For large transport blocks, HARQ buffer is about 25KB per CB. Since this is a large amount of RAM, the 5G standard has a feature that allows for reduced HARQ RAM usage. This feature is called LBRM and is defined in 3GPP TS 38.212 Section 5.4.2.1. Using LBRM the HARQ RAM usage goes down to about 12.5KB per CB.

On code block versus transport block level CRC

On the first transmission the forward error correction acceleration (FECA) engine verifies both the CB and TB level CRCs. However, in retransmit scenarios, the TB CRC is not explicitly (re-)verified by the FECA engine. This opens the question of (error) performance rate—what is the chance that a corrupted TB is passed as correct to the MAC layer?

To analyze this scenario, start by plotting Bit Error Rate (BER) (wrongly decoded bit) versus both CB and TB error rates as shown in Fig. 3–13 below:

HARQ operation assumes (in typical operation) a approximately 10% TB error rate. This is the target TB error rate to which transmit power, modulation and Forward Error Correction rates are tuned by the scheduler algorithm.

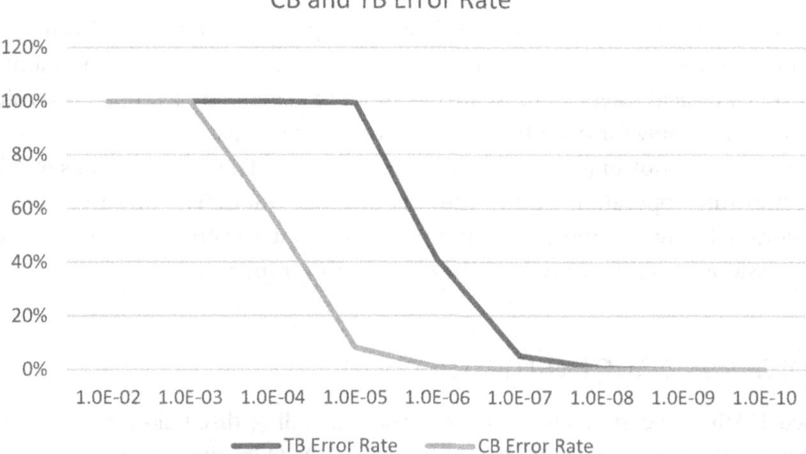

FIGURE 3–13 Code block (CB) and transport block (TB) error rate as a function of bit error rate (BER).

The above chart implies a BER of somewhere between 10^{-6} and 10^{-7}. We assume 10^{-6} as the nominal operating point.

In this scenario:

- The CB Error Rate (the probability that a CB has at least 1bit error) is calculated as:
  $CB\_error\_rate = 1-(1-BER)^{CB\_size} = 1-(1-10^{-6})^{8448} = 8.4 \times 10^{-3}$.
- The TB Error Rate (the probability that a CB has at least 1bit error) is approximated as:
  $CB\_error\_rate \times num\_CBs\_in\_TB = 8.4 \times 10^{-3} \times 55 = 0.46$.
- CRC24B = 0xC00031 is used for CB CRC which has a Hamming distance of 4 over Payload length up to 8M bits. Therefore, all combinations of 3-bit errors will be detected but some combinations of 4-bit errors may not be detected within the CB.
- Probability of 4-bits in error within a CB can be estimated assuming Binomial distribution and using probability mass function:
  Probability of 4-bits in error = nchoosek(CB_size, 4) × (BER)^4 × (1−BER)^(CB_size − 4) = 2.1×10^{-10}.
  A CB CRC *may* not detect this error combination, this is the probability that the CB may "falsely" pass the CB.
- The probability that the TB has at least 1 CB with 4 bits in error = (approx.) $2.1 \times 10^{-10} \times$ num_CBs in_TB = 1.15×10^{-8}.
  If TB CRC was checked, then the above scenario could "likely" be detected. In the absence of TB CRC validation, 1.15×10^{-8} is the probability that a "bad" TB could be passed up to L2 as a "good" TB.

As such, this probability is low enough without considering the Hamming Weights, which will make it even smaller. All 4-bit error combinations are not undetected by CB CRC, only some of them. Also, the BER was chosen conservatively in the above example.

Transmit operation

In the transmit operation, the Transport Block is typically re-encoded (from scratch) for every transmission and/or retransmission. This means that there is no requirement for maintaining a HARQ Physical Layer context in the modem. Each TB is retrieved (typically from the MAC layer entity) new for each transmission and retransmission.

Note that this does not impose any bandwidth penalty. The system is designed for 100% successful transmit operation and the associated memory or I/O bandwidth(s). Retransmissions take away time/frequency resources that would otherwise be allocated to "new" transmissions. Overall, there is no I/O bandwidth impact.

HARQ process count dimensioning

We discussed HARQ process definitions previously, including discussion on synchronous and asynchronous HARQ options. Even with asynchronous HARQ though, the maximum throughput that the communication path between base station and client can achieve is limited by the number of active HARQ processes, given that a single HARQ process can't be re-used until the acknowledgment for that process comes in (first time or after retransmission). This defines that the round-trip time between gNB (DU + RU) and UE needs to be matched to peak throughput and maximum latency of this round-trip time. HARQ timing definitions are shown in Fig. 3–14:

Key processing times in the HARQ Round Trip Time (RTT) chain are:

1. T1—Required time to transmit one TB, i.e., the physical time the TB transmission consumes—compare it to the serialization/de-serialization delay of a wireline interface.
2. T2—The time required at the UE/client between the end of the TB and the time that the acknowledgment/negative acknowledgment (ACK/NACK) for this TB is sent. It includes UE/Customer Premises Equipment (CPE) PDSCH Rx processing latency, UE/CPE UCI (PUCCH or PUSCH) Tx processing, time until next ACK/NACK transmission is available. For synchronous HARQ in LTE Uplink, this was defined to 4Ms by 3GPP standards, but 5G/NR allows more flexible timing for this time.
3. T3—Physical time to transmit one ACK/NACK. Typically, 1 (short PUCCH) to 14 (long PUCCH, PUSCH) symbols.
4. T4—The transmission delay between gNB BaseBand Unit (DU) and UE/client antenna ports. It includes physical over-the-air latency, gNB RU to DU delay (if Baseband & Analog are on different locations).

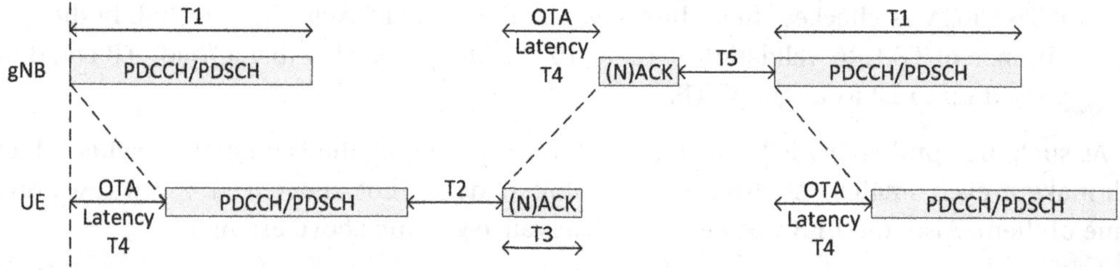

FIGURE 3–14 Critical timing in HARQ.

5. T5—The time required at the gNB to (re)transmit data following ACK/NACK reception. It includes ACK/NACK reception at gNB PHY, retransmission scheduling at gNB MAC and of course RU delay.

Overall HARQ latency is defined as $T_{HARQ} = T1 + T2 + T3 + T4 + T5 + T4$. T_{HARQ} drives the minimum number of required HARQ processes to satisfy the "capability to transmit a continuous stream." It can be found using the equation NMIN, HARQ $\times$ TTI $\geq T_{HARQ}$.

Minimizing T_{HARQ} implies reducing each T value. From the UE side, the only capability driven is T2. T5 is about gNB capability, T4 depends on cell size, T1 and T3 are NR standard driven, and gNB implementation dependent. Lower T_{HARQ} improves performance—low latency is a key 5G target—as well as reducing system (memory) resources. This drives reduction of T1 and T5. Key aspects of this reduction include

- Linux scheduling performance. The MAC/RLC portion of the application assumes Linux as an OS, and the Linux scheduler as thread scheduling mechanism. To maintain hard-Real Time (RT) operation, the PREEMPT_RT patch is applied. The maximum RT response time is defined as the worst-case response time to a trigger such as a frame timer or Interrupt Request (IRQ). When applying the PREEMPT_RT patch, the latency bound is to <20 microseconds (see section PREEMPT_RT).
- Memory and interface throughput. The latency of a data transfer through a link such as PCIe SerDes is inverse proportional to the utilization of the link. For example, a PCIe gen3 dual lane controller supports a goodput of approximately 14 Gbps. When connecting to a 5 Gbps 5G/NR modem with a 125microseconds TTI (~80KByte/TTI), the Transport Block transfer time consumes approximately 45 microseconds or 36% of the frame latency.
- DU/RU latency which is accounted for multiple times—both during transmit and receive operation.

With reduced frame times in 5G/NR, the impact of PREEMPT_RT performance becomes more important. Mitigation options include

- "Early wakeup" where the trigger to the core running MAC layer is offset by the expected PREEMPT_RT latency. This comes at the cost of potential underutilization of the core (e.g., a wait-loop for ACK/NACK reception).
- Minimize the overhead of interacting between MAC and PHY. For example, move to a push model where the MAC host uses a Direct Memory Access (DMA) write (as opposed to PHY DMA read) to transfer the data to the PHY.
- Increase T_{HARQ} (and accept increased HARQ process count).
- Increase the link capacity of the MAC/PHY physical interface.

Radio unit

RU implementation is about dimensioning of the analog processing chain and the digital algorithms required to improve the performance of this chain, including Crest Factor Reduction (CFR) and Digital Pre Distortion (DPD) algorithm selection and implementation.

Instantaneous bandwidth and occupied bandwidth

After RF operating frequency and output power, Instantaneous BandWidth (IBW) is probably one of the most important RU parameters. It defines the frequency boundaries of the allocated bands—from the lowest to the highest frequency occupied. Occupied BandWidth (OBW) defines the sum of allocated/occupied bandwidths in the frequency occupied (Fig. 3–15). In case of contiguous spectrum, IBW and OBW are the same, but in case of multicarrier deployment, IBW and OBW are different—consider a deployment example in the United States CBRS bands where (worst case), IBW can be 145 MHz (3.55–3.7 GHz) but OBW can be 20 MHz (one 10 MHz carrier in 3.55–3.65 GHz and one 10 MHz carrier in 3.69–3.7 GHz). OBW is relevant for low-PHY components such as beamforming and FFT complexity whereas IBW is relevant for Digital Front End (DFE) functions that are implemented across the whole IBW. This includes the data converters and related components.

Receiver chain analysis

Some back-of-the-envelope calculations can help us understand the performance of the receiver chain, to understand the receiver sensitivity and dynamic range metrics. This also helps explain the bypass option for the secondary Low Noise Amplifier (LNA) and the requirement to be able to configure the demodulator gain and/or baseband Variable Gain Amplifier (VGA) gain.

Consider a general Radio Frequency (RF) transmitter & receiver diagram with a baseband I/Q interface, as shown below (Fig. 3–16):

Without dictating physical chip boundaries, we can analyze the sensitivity and dynamic range by assessing the component performance of the various components in the (bottom) receive chain. Table 3–6 shows some examples with typical values.

In this example, we assume that the Analog-to-Digital (ADC) maximum input of 532 mV is at or near the ADC clipping level. Because the minimum Rx sensitivity is associated with a quadrature phase-shift keying (QPSK) modulated signal that requires less Signal to Noise Ratio (SNR) performance as compared to the higher-order modulated signal for the "near" scenario with maximum Rx input levels, we can get away with not using the full dynamic range of the ADC by not achieving the maximum ADC input level.

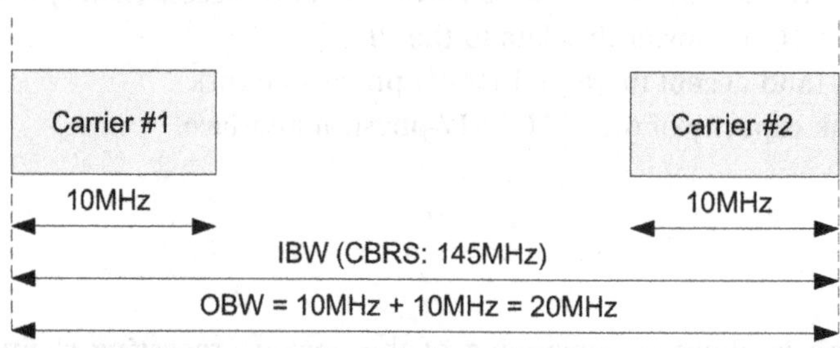

FIGURE 3–15 Instantaneous bandwidth (IBW) and occupied bandwidth (OBW) examples.

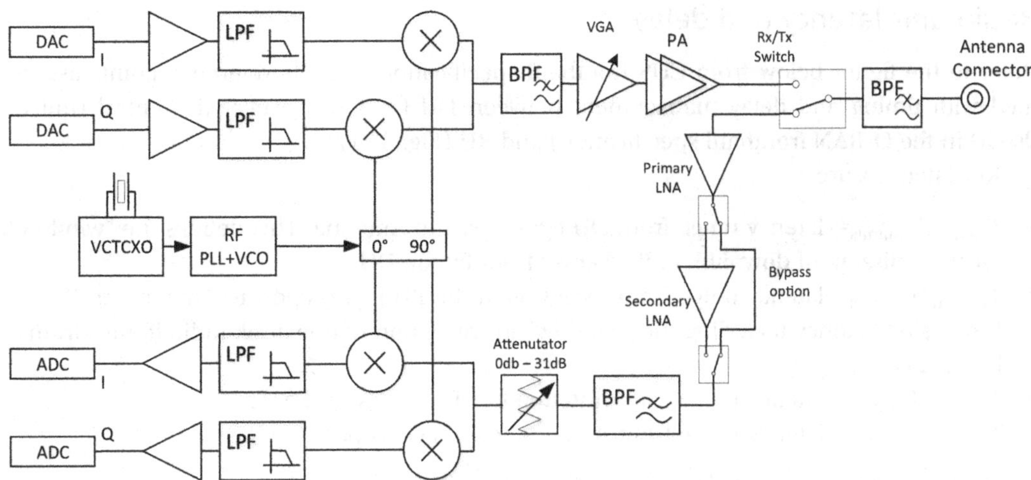

FIGURE 3–16 Generic RF receiver block diagram with zero-Intermediate Frequency (IF) interface.

Table 3–6 Example receiver sensitivity analysis.

	Minimum Rx sensitivity ("far")	Maximum Rx input level ("near")
Antenna Rx Sensitivity	−90dBm (min)	−25dBm (max)
BPF Filter Insertion Loss	−1.5 dB	−1.5
Primary LNA gain	20 dB	20 dB
Secondary LNA gain	20 dB	0 dB (bypassed)
De-modulator gain/attenuation	8 dB	−25dB
De-modulator output (dBm/Vpp)	= −43.5 dBm/6.7 mV	= −31.5 dBm/27 mV
Baseband gain	20x (13 dB)	20x (13 dB)
ADC input (Vpp)	134 mV	532 mV

We estimate the receiver SNR for the by looking at the overall received signal level, noise level and noise that is introduced by the receiver as characterized by the Noise Figure (NF). The NF is an indication of the performance reduction in the receiver chain caused by non-ideal components that introduce noise and is measured in dB. Together with gain, it is the key performance number in LNAs. Lower numbers imply better performance, typical values for LNAs being around 1–2 dB. The first LNA in the chain dominates as we know from Friis' equation for noise. As we showed in Chapter 1, O-RAN Overview, we count received SNR as:

SNR = S (Signal) – NF –174dBm (Thermal Noise) + 10*LOG(used bandwidth)

Assuming −90dB received signal (minimum sensitivity), 2 dB NF and 20 MHz bandwidth, we have a signal level of −90dB and a Noise level of −174 + 73−2 = −103dB or an SNR of 13 dB which should be enough to sustain QPSK operation or better. This 13 dB (with margins added, see below) also defines the required ADC performance.

Radio unit latency and delay

Refer to the figure below from CUS fronthaul specifications for key reference points associated with timing and delay management between DU (= Low Level Split Central Unit or lls-CU in the O-RAN fronthaul specification) and RU (Fig. 3–17):

Key latencies are

- $T_{1a,min}/T_{1a,max}$—latency range from DU egress port to Antenna. This defines the "window" for transmission of downlink radio frames from by the DU.
- $T_{2a,min}/T_{2a,max}$—latency range from reception at the RU ingress port to Antenna or RU Processing Latency including the "window" for reception of downlink radio frames from by the RU.
- $T_{a4,min}/T_{a4,max}$—latency range from Antenna to DU ingress port.
- $T_{a3,min}/T_{a3,max}$—latency range from Antenna to RU egress port.

Or summarized as shown in Fig. 3–18 below.

In this figure, the DU Processing Latency represents the time spent in the Physical Layer of the DU, from MAC layer Transport Block generation to eCPRI I/Q sample generation. During the DU Transmit Window, the I/Q samples are transferred from the DU to Ethernet using eCPRI encapsulation. The RU Receive Window defines a buffer time that compensates for any jitter in the fronthaul interface. Remainder time is spent in the RU Processing stage which are the low-PHY and DFE components which are typically highly deterministic.

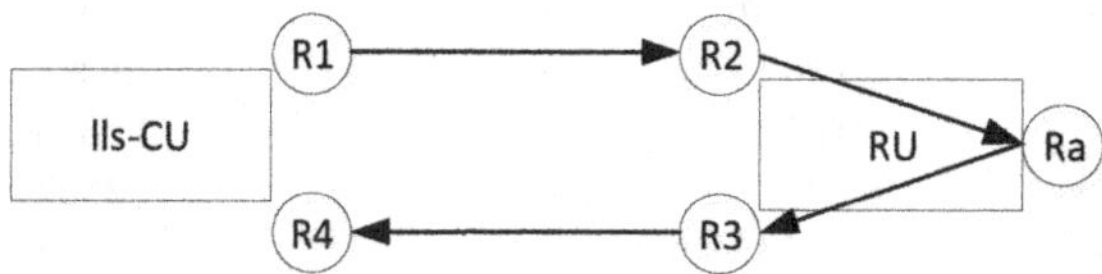

FIGURE 3–17 Fronthaul and radio unit latencies per O-RAN specifications.

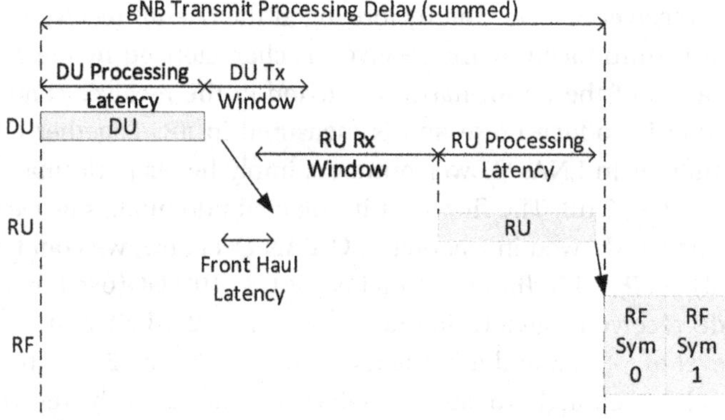

FIGURE 3–18 Transmit fronthaul and radio unit internal latencies.

Analog latency is defined as the time it takes for the analog signal to transfer from the RF antenna to the data converter interface and is typically counted in nanoseconds or ignored altogether.

Digital predistortion

Digital Predistortion is one of the more complex topics to define in the RF system as it combines aspects of regulatory and 3GPP standards compliance (Adjacent Channel Leakage Ration (ACLR, also known as Adjacent Channel Power Ratio or ACPR) limits) with efficiency (DC power consumption) and compute complexity (clock cycles, FPGA Lookup Tables (LUTs), or gates spent on implementing the DPD algorithms). Striking a balanced tradeoff between these aspects can become an iterative and complex process.

Characterization of the Power Amplifier (PA) is typically done in a dedicated testbench environment, often operated by the manufacturer of the PA. These testbed environments characterize the output power, efficiency, gain, and ACLR of the device under varying conditions such as temperature, applied bias, input signal conditions (signal type, spectral width,...), frequency band of operations. Analysis results are shared in component datasheets or characterization reports that are unique to the PA and depend on implementation technology, operating frequency, bandwidth, and so on and can greatly range over cost.

An example of such characterization is shown below in Fig. 3–19.

Performance of a PA with DPD applied is analyzed subsequently to determine best possible performance targets after linearization and the impact of different DPD algorithms. An example is shown below in Fig. 3–20.

Although the absolute numbers of this benchmark are not necessarily relevant, we can validate two intuitive characteristics of the PA. First, as the bandwidth of the transmitted

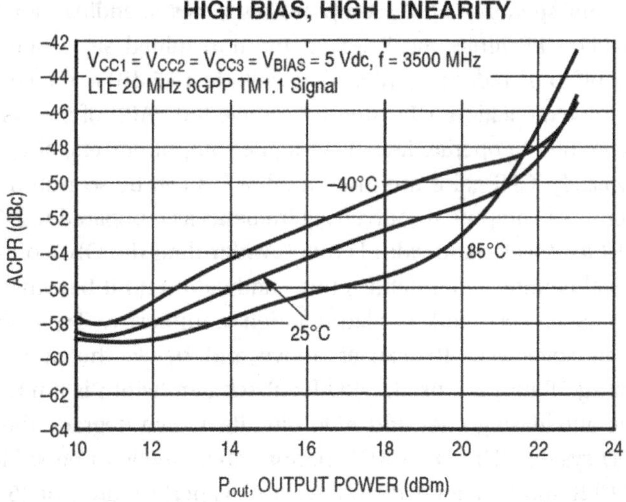

FIGURE 3–19 ACPR/ACLR characterization example.

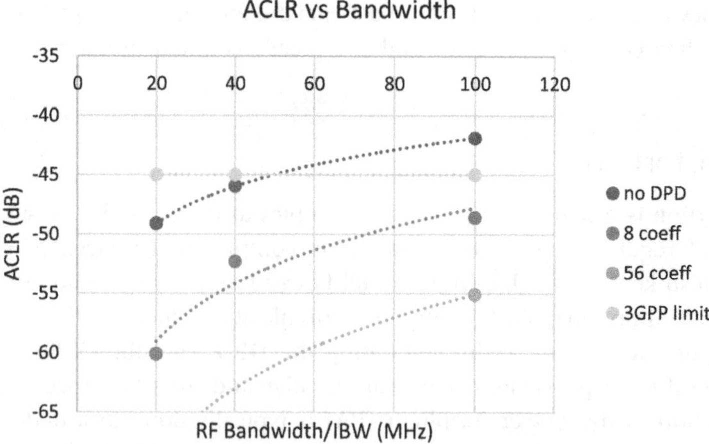

FIGURE 3–20 ACLR versus bandwidth for a typical low-power power amplifier (PA).

signal becomes wider, the ACLR performance drops—for the example here, a drop of 7 to more than 10 dB depending on whether DPD algorithms are applied. Second, the complexity of the DPD algorithm defines the level of improvement of the ACLR, with incrementally worse value. At some point, the PA simply cannot be corrected beyond its inherent capabilities. The question is whether (from a system point of view) it is worthwhile investing in complex DPD algorithms with maximum performance or live with a "good enough is good enough" system. As we discussed before, for all but indoor applications, the value of investing in compute resources for advanced DPD algorithms is worth it with regards to DC power consumption metrics but the engineering effort or financial cost may be considered too high.

Note the trade-off on spending DFE (compute) resources spending can become a tradeoff between CFR and DPD. CFR limits the PAPR of the transmitted signal and hence allows the Power Amplifier back-off to be reduced (order of magnitude 3–5 dB) at the cost of compute clock cycles spent on CFR implementation. DPD doesn't change the PAPR of the system but allows the Power Amplifier to continue to operate linearly at higher output powers, also at the cost of compute clock cycles. Typically, CFR gains are more modest (as can be seen in the chart above) but also lower effort in terms of compute resource requirements as compared to DPD gains.

The required DPD feedback bandwidth is much wider than the OBW of the signal. There is no single mathematical equation that defines this bandwidth and it becomes part of the trade-off between performance, cost and implementation complexity. Typically, 3–5x OBW or enough bandwidth to cover one RF carrier above and below the target band (cover the *Adjacent Carrier* Leakage Ratio) are used. DPD feedback bandwidth is limited by the quality of components, like the antialiasing filter and ADC circuits, which degrade the linearization performance of the DPD system. The DPD ADC performance needs to be sufficient to support a system with 45 dB ACLR and hence needs to have a ADC performance of 45 dB or better.

Data converter and phase locked loop (PLL)

ADC performance requirements are dominated by the dynamic range which is defined as the difference between the maximum allowed input signal level and the noise plus interference floor level. Thus we need to consider the maximum modulation and coding scheme which requires the maximum possible SNR in order to decode the signal with the desired rate. These are depending on the (algorithmic) modem implementation but we gave some values before (−5.5 dB for QPSK, 5−15 dB for 16QAM, 15−20 dB for 64QAM and 25 dB or better for 256QAM). Let's look at an example ADC analysis for a simple FR1/ < 6 GHz baseband I/Q use-case, ignoring DPD feedback path support which requires to capture a wide RF bandwidth:

- Sample rate F_{sample} is defined from the signal bandwidth which is 100 MHz in the RF domain or 50 MHz for I and Q each. Given 3GPP sample rates in multiples of 30.72MSPS, we assume 122.88MSPS as a minimum sample rate above Nyquist criterium or 245MSPS with 2x oversampling to simplify aliasing filter implementation.
- SNR requirements are defined as
  - Assume 25 dB required SNR for 256QAM decoding.
  - 12 dB required PAPR to be supported, assuming the transmitting side did not reduce PAPR with CFR.
  - Gain errors (4 dB) consist of gain and offset errors as caused by inaccuracies in internal reference and amplifier offset voltages reduce the usable dynamic range. Assuming gain error and offset errors to be 10% of full scale, then each contributes a 1 dB reduction in dynamic range. The ADC must be backed off 1 dB to prevent input clipping and another 1 dB to account for limited dynamic range. Add 2 dB dynamic range margin.
  - Analog Gain Control (AGC) error. Assume + /− 10% (i.e., 20%) error. Add 2 dB dynamic range margin.
  - 10 dB rule-of-thumb fade margin which is the assumed channel fading over (short term) time and frequency in the target spectrum. If not taken as a rule-of-thumb, this number can be derived from channel measurements or similar.
  - 10 dB blocker margin. Receiver blocking occurs when the signal that is intended to be received is "blocked" because unintended signals (typically from nearby sources) radiate in or near the passband of the RF receiver and reduce the sensitivity to the intended signal. Specifically, in use-cases where IBW and OBW are not the same, analog (baseband and RF) filters do not match exactly with the signals intended to be received and blocker signals should be taken into consideration. Note that blocker margin is a key design consideration that impacts the analog (filter) and ADC performance. Consider a 2G (GSM) example where the required receiver sensitivity is −102 dBm with an adjacent (3 MHz carrier offset) specified at up to −23 dBm. The approximately 80 dB difference in signal strength needs to be overcome by a combination of (RF but mainly baseband) analog filtering combined with blocker margin in the ADC specification.

- −3dB processing gain from oversampling (245.76 MHz over 122.88 MHz). The Processing Gain is defined as the gain had from oversampling the desired signal with a rate above the Nyquist frequency.

$$PG = 10\log_{10}\frac{F_{sample}}{2F_{max,signal}}.$$

- In aggregate: 25 + 12 + 4 + 2 + 10 + 10−3 = *60 dB*

Note, this does not include any additional safety margin. The analysis is shown graphically in Fig. 3–21.

Digital to Analog Converter (DAC) performance is assessed in a similar fashion to ADC performance:

- Sample rate F_{sample} is defined from the signal bandwidth which is 100 MHz in the RF domain or 50 MHz for I and Q each. Given 3GPP sample rates in multiples of 30.72MSPS, we assume 122.88MSPS as a minimum sample rate above Nyquist criterium or 245MSPS with 2x oversampling to simplify aliasing filter implementation or 491.76MSPS if we add enough bandwidth for DPD correction.
- SNR requirements are defined as:
  - Assume 30 dB required for 256QAM support as derived from the 3GPP specified 3.5% Error Vector Magnitude (EVM) ($EVM_{RMS} \sim = \sqrt{\frac{1}{SNR}}$) per [4].
  - Add SNR margin—the EVM shall be dominated by the PA, after all. To estimate marging, we assume the transmitter DAC cannot degrade EVM by more than 0.25% (0.6 dB impact). Assuming the following equation for system SNR < System SNR = $-20\log\left(10^{-SNRa/10} + 10^{-SNRb/10} + \ldots 0.10^{-SNRn/10}\right)^{1/2}$, this requires adding 8.86 dB SNR margin, or rounded to 9dB.

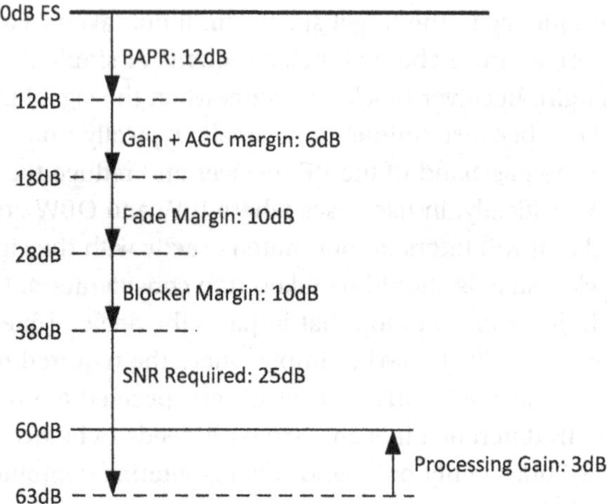

FIGURE 3–21 ADC Dynamic Range Analysis.

- 12 dB required PAPR to be supported.
- Gain errors (4 dB) consist of Gain and offset errors as caused by inaccuracies in internal reference and amplifier offset voltages reduce the usable dynamic range. Assuming gain error and offset errors to be 10% of full scale, then each contributes a 1 dB reduction in dynamic range. The DAC must be backed off 1 dB to prevent output clipping and another 1 dB to account for limited dynamic range. Add 2 dB dynamic range margin.
- In aggregate: 30 + 9 + 12 + 4 = *55 dB*.

Note, this does not include any additional safety margin. Again, a graphical representation below in Fig. 3–22.

PLL performance is defined by the SNR degradation for the ADC/DAC caused by clock jitter from the PLL. This degradation is defined by:

$$SNR_{jitter} = -20\log_{10}\left(2\pi f_{in} T_{jitter}\right)$$

Phase noise and clock jitter are closely related—phase noise is the instability of a frequency expressed in the frequency domain and clock jitter is fluctuation of the signal waveform in the time domain. Hence, the clock jitter in time domain can be calculated by integrating the phase noise of the clock signal over a specific section of the frequency domain, where most phase noise definitions are stated as the ratio between carrier power and noise power as a function of the offset frequency from the carrier as shown in Fig. 3–23.

The integration can be estimated by adding the individual section areas (A1, …, A6 in the plot above), from typically a few KHz in the low side up to (rule of thumb) 2x F_{sample} in the high side (say, 245.76 MHz for the 122.88 MHz VCXO taken in the example above):

- A7 ~ $= -161$ dB + $10\log_{10}(245\times10^6 - 10\times10^6) = -77$ dBc

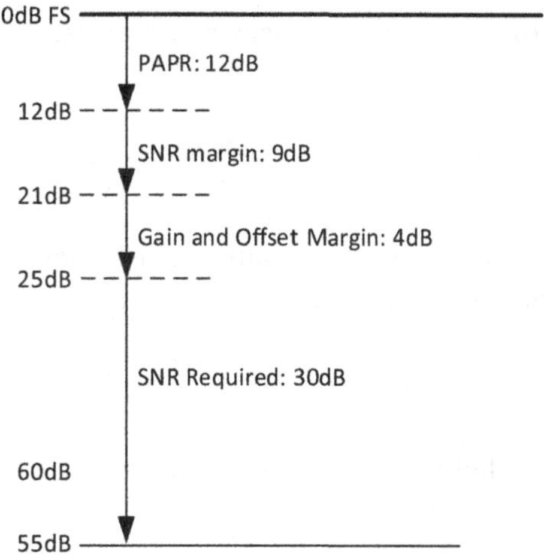

FIGURE 3–22 DAC dynamic range analysis.

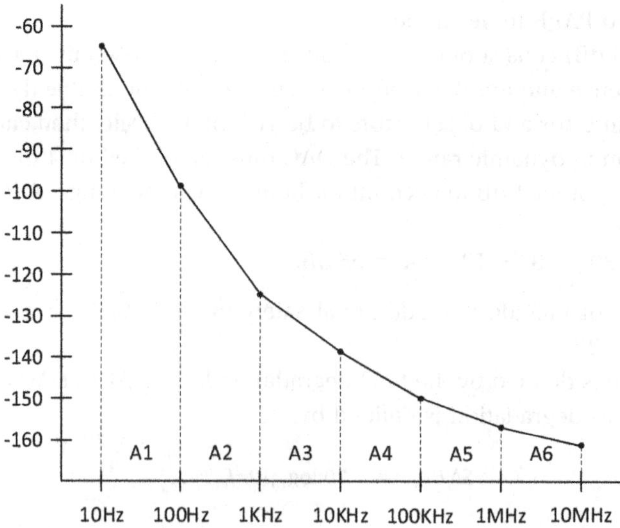

FIGURE 3–23 Phase noise plot for typical Voltae Controlled Crystal Oscillator (VCXO) (example: Microchip/Vectron MX-805).

- A6 $\sim = -160\text{ dB} + 10\log_{10}(10 \times 10^6 - 1 \times 10^6) = -90\text{ dBc}$
- ...
- A3 $\sim = -140\text{ dB} + 10\log_{10}(10 \times 10^3 - 1 \times 10^3) = -100\text{ dBc}$

Assuming the broadband noise (A7) is dominating the performance, we calculate the Phase Jitter in radians [root mean square (RMS)] as:

$$Phase_{jitter} = \sqrt{2\, x\, 10^{\frac{A}{10}}}$$

or (A = −77 dBc) $\sim 2 \times 10^{-4}$ for our example, which translates to an RMS jitter time value in seconds using:

$$T_{jitter} = \frac{Phase_{jitter}}{2\,\pi\, f}$$

or ($f = 245 \times 10^6$) ~130fs. Note that when multiple clock components (say, oscillator and a clock generator/PLL) are combined, we can calculate combined T_{jitter} as:

$$T_{jitter,combined} = \sqrt{T_{jitter,A}{}^2 + T_{jitter,B}{}^2}$$

Online tools such as [5] can help in deriving the clock jitter from the phase noise plots.

References

1. World Broadband Statistics, http://point-topic.com/wp-content/uploads/2013/02/Point-Topic-Global-Broadband-Statistics-Q1-2013.pdf, 2013.
2. Wikipedia, M/D/1 queue, https://en.wikipedia.org/wiki/M/D/1_queue.
3. Verizon, *Verizon IP Latency Statistics*, http://www.verizonbusiness.com/about/network/latency/.
4. Mahmoud, H.; Arslan, H. Error vector magnitude to SNR conversion for nondata-aided receivers. *IEEE Transactions on Wireless Communications* **2009,** *8* (5), 2694–2704 May.
5. Silicon Labs, *Phase Noise to Jitter Calculator*, https://www.silabs.com/jittercalculator/phase-noise-jitter-calculator.aspx.

Further reading

FCC Form 477. *Federal Communications Commission.* https://www.fcc.gov/economics-analytics/industry-analysis-division/form-477-resources.

Silicon Labs AN739, *Estimating Clock Tree Jitter*, November 2012.

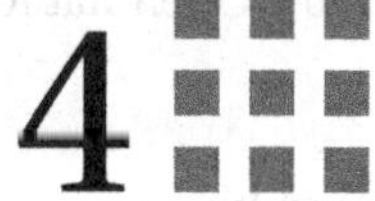

Hardware architecture choices

This chapter covers the different hardware architectural choices for implementing O-RAN systems, from the very high-end (data center scale-out) to the low-end (Integrated Small Cell). We discuss typical scale/dimensioning for each of these solutions, as well as some of the implementation choices that the hardware/systems architect has.

Implementation options evolve over time, as silicon technology improves. This makes way for the possibility that any device performance claims made in this chapter are likely to be outdated as soon as they are written down. So we aim to stay away from calling out any specific devices or device vendors but instead stay in generic vendor-independent terminology.

Scalability

Depending on the scale of the deployment, we consider the following architectural options (in order from large-scale to small-scale). These four examples will be covered in more depth later.

- *Data Center*—connecting tens of cell sites and potentially 100 s of radios. Fronthaul bandwidth is in the order of Tbps maximum and aggregate MAC/PHY throughput is 10 Gbps.
- *Cell Site Solution*—connecting single cell tower with multiple (3, 9, 6, 12) sectors. Fronthaul bandwidth is in the order of 100 Gbps maximum and aggregate MAC/PHY throughput is in the order of 20 Gbps.
- *Central Unit/Distributed Unit (CU/DU) based small cell solution*—distributed small cells, for example, for indoor deployment as picked for the example in Chapter 1, O-RAN Overview. Fronthaul bandwidth is in the order of 25–50 Gbps and aggregate MAC/PHY throughput is in the order of 5–10 Gbps.
- *Integrated Small Cell*—all-in-one solution with optimized power/cost. Aggregate MAC/PHY throughput is <1–5 Gbps.

The scalability aspect is important to take note of here. Overall fronthaul bandwidth and MAC/PHY throughput scale is up to one hundred times! Clearly, a single solution does not match all deployment options. Scalability was a key target of 5 G standards development, and we can see some of this at work here. Now, on the other hand, let's look at how scalability works from a component/system point of view. Let's look at some

Open Radio Access Network (O-RAN) Systems Architecture and Design. DOI: https://doi.org/10.1016/B978-0-323-91923-4.00014-8

possible scaling options, ordered from simple and low-impact to complex and high-impact:

- Scale the performance of an existing device by changing core frequency. This is what Personal Computers do when they change to a "turbo" frequency for peak performance requirements—you don't change the system at all, but just run it at a different frequency to match performance requirements. Obviously the software impact is pretty much nonexistent, there doesn't need to be any re-partitioning of software. Also, the hardware changes that are required are limited. The hardware system may want to support dynamic voltage and frequency scaling to incrementally reduce power consumption. Overall, this approach can save tens of percentage of consumed power, with minimum impact.
- Scale the number of cores in a device, without changing device families. Silicon devices from (almost) all vendors come in device families that are pin-compatible but support different numbers of CPU cores, programmable gates, memory interfaces, and so on. This allows a single hardware design to support multiple devices with associated price and power targets, typically with approximately two times overall performance range from low-end to high-end. There is a software impact, however, where software needs to either be modified to scale "automatically" between number of deployed cores (Linux as an Operating System does this, obviously, but a application level thread scheduling algorithm could be built similarly) or the software needs to have multiple instantiations, each hard-configured to the number of CPU cores deployed in the specific system.
- Scale by adding or removing devices in the system. Consider building a board design that allows to de-populate some of the components for a single hardware design to be able to support multiple product categories. This strategy doesn't save in form factor but given that silicon constitutes the bulk of the hardware cost, the cost savings may well be worth the effort. Approximately two times scaling factor can be achieved through this strategy.
- Scale by adding or removing hardware acceleration. Quite intrusive to software so unlikely to be done as a quick derivative to build a lower- or higher-end product within the same family, but high-impact with regards to the potential performance increase that can be achieved. As an O-RAN specific example, consider a DU implementation that implements Enhanced Common Public Radio Interface (eCPRI) processing as a CPU centric software component in a low-end system that uses embedded Ethernet for fronthaul processing in the low-end versus using a dedicated Field Programmable Gate Array (FPGA) or Application Specific Integrated Circuit (ASIC) for fronthaul processing in a higher-end system. The rule of thumb is that when a single component occupies tens of percentage of the overall system load in a software centric implementation, it becomes worthwhile offloading this component to a dedicated accelerator. Overall performance increase potential can be quite large: about two times scaling factor can be achieved.
- Scale by adding or removing the number of blades in a chassis or the number of chassis in a deployed system. This is all about large scale-out scenarios as well-documented for datacenter applications where scale-out is the primary purpose of system architecture. Scale-out scenarios require the support of a backplane of sorts in the system (typically

Ethernet based) that allows the blades or systems to connect to each other and to a common fronthaul and backhaul transport component. Potential scaling factors can be huge—this is the whole point of scale-out deployments after all—but it comes at the cost of disaggregation: scaling to smaller sizes doesn't work well because of component count. This scaled-out architecture applies to the "data center" scale that we discuss in more detail below.

Development cycle

Besides the scale of the product, we should consider the cost associated with the hardware and software development cycles. We consider three product development options:

- Commercial-off-the-shelf (COTS) based systems where the board is a "white box" hardware system that is commercially and readily available. In theory, hardware design can be eliminated from the development process, saving time and R&D cost (hardware headcount). The point of the COTS system is that it can be procured instantly, and therefore there are no risks and costs associated with hardware development. The cost is instead paid for in the deployment phase, the COTS hardware, is not optimized to the target use-case and hence costly in terms of financials, power consumption, form factor, or other aspects.
- Reference Design is based where the hardware is largely copied from a manufacturer or ODM provided 3rd party implementation and hence both low-effort and low risk in terms of potentially needing a re-spin of the hardware. In addition, board bring-up and Board Support Package (BSP) software components are largely available. We assume less than 3 months for any board customizations and/or the time required to procure the development systems.
- Bespoke and highly customized solution to the use-case. Components can be selected in a mix-and-match fashion, but we need to accommodate for the time and effort spent on the system design. It is obvious that this approach is time consuming and potentially risky—hardware teams are often squeezed for time and headcount and don't have the engineering or procurement scale to make choices that 3rd party (reference design or COTS) systems have. But in a world where product differentiation and final product cost are important, bespoke hardware is obviously the most optimized path.

Note that the concept of O-RAN is made to imply "software defined" and "COTS," but this is *not* actually the case: when we look at the O-RAN standards (fronthaul and other interface specifications, use-case definitions and others), they don't define the use of any specific hardware or software. The other implementation options are perfectly valid for consideration.

Also, consider the overall development timeline associated with the product, as shown in Fig. 4–1. In the end, the overhead associated with custom hardware development is dwarfed by software development and deployment timelines. Moving to a COTS based system saves time but is not a silver bullet.

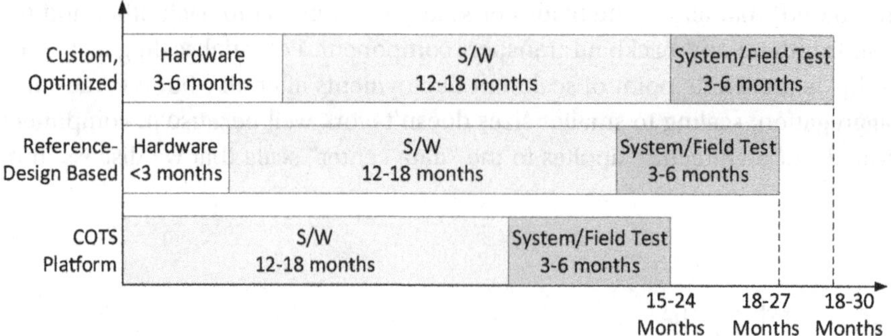

FIGURE 4–1 Development timelines for a typical product based on commercial-off-the-shelf (COTS), reference design and customized systems (indicative only).

The tradeoffs here are important are in the end driven by commercials where balance needs to be struck between:

- Target volume of deployment
- Unit cost overhead associated with COTS platform
- Development overheads associated with a more custom platform
- Development timeline and product lifecycle: how long until the next revision of the product comes out and a more optimized COTS platform has overtaken the optimized but now outdated custom unit.

Data center architecture

In a disaggregated environment, resources are de-coupled, and each resource can be implemented (and scaled) independently from each other. In traditional data center implementations, elements of disaggregation include compute, storage and networking that can be scaled independently to be used for on-demand allocation by any (cloud) application.

In a 5G centric data center, the application is better known (much less generic). This allows for optimization: choice between general purpose (GPP) hardware or application optimized (accelerated) implementations to be chosen depending on requirements with regards to cost, performance, power and flexibility/time to market.

A generalized data center architecture for a CU/DU system is shown below in Fig. 4–2: We'll describe individual components here:

Fronthaul/Top of Rack (TOR) switch. Provides the fronthaul eCPRI/Ethernet connectivity between the RUs and DUs together with 1588 timestamping and eCPRI (de)compression. This functionality is implemented in high-end FPGAs or switch-like ASICs with Terabit per second (Tbps) performance classes.[1] Connectivity to the DU blades is provided over Ethernet links carrying uncompressed eCPRI IQ samples for low intensity CPU processing in the DU processor.

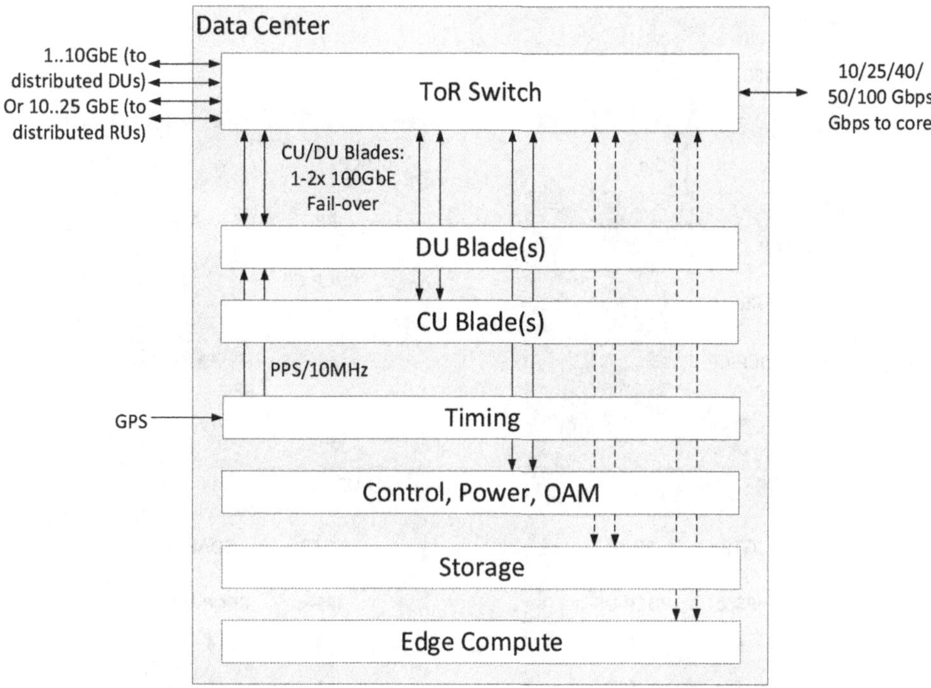

FIGURE 4–2 Data center architecture.

Timing is provided by a separate time server solution that is optimized for performance and provided by a specialized solutions vendor. The time server acts as a timing source towards many RUs at once through the TOR switch as well as providing a Pulse Per Second (PPS) and/or frequency accurate source towards the DU blades.
CU blades perform CU processing, including (optional) support for IP Security (IPSec) for backhaul security. Key interfaces are shown in the figure below (Fig. 4–3), with an integrated (traditional macro base station) option shown on the left side and the disaggregated solution on the right side. The 5G New Radio (NR) gNB (Base Station) is connected to Access and Mobility Management Function and User Plane Function in the 5G Core Network (5GC).

Key interfaces, all defined by 3GPP and shown below in Fig. 4–4, include:

- NG-C: Control Plane (CP) interface between NG-RAN and 5GC
- NG-U: User Plane (UP) interface between NG-RAN and 5GC
- E1: CP/UP interface within the (logical) PDCP component and provides transport of signaling between Control Plane and User Plane
- F1-C: CP interface between CU and DU
- F1-U: UP interface between CU and DU.

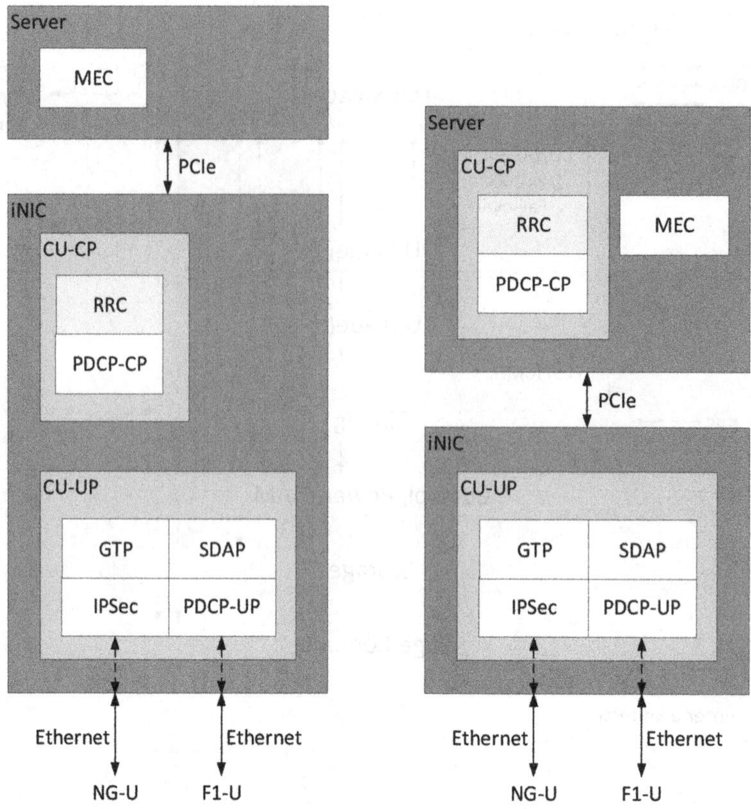

FIGURE 4–3 iNIC solutions for disaggregated CU.

Combined, the F1 interfaces separate transport network layer and radio layers.

Depending on required balance between performance, power and scalability, this CU unit can be implemented on a single device or split across multiple devices and even Instruction Set Architectures, balancing capacity/performance, cost, and power. Let's look at a few options from high levels of integration with relatively low performance to low integration but highly scalable solutions.

At the highest level of integration, a single Intelligent Network Interface Card (iNIC) card can support combined CP and UP functions or be plugged in to a server where UP is implemented in the iNIC and CP in the server environment. These options are shown below (left and right) in Fig. 4–3:

An obvious next level of scale-out can be supported by utilizing multiple iNICs, separating UP functions (as shown below in Fig. 4–5):

An active Ethernet backplane is provided by a TOR switch becomes the method to scale-out as performance requirements increase (Fig. 4–6):

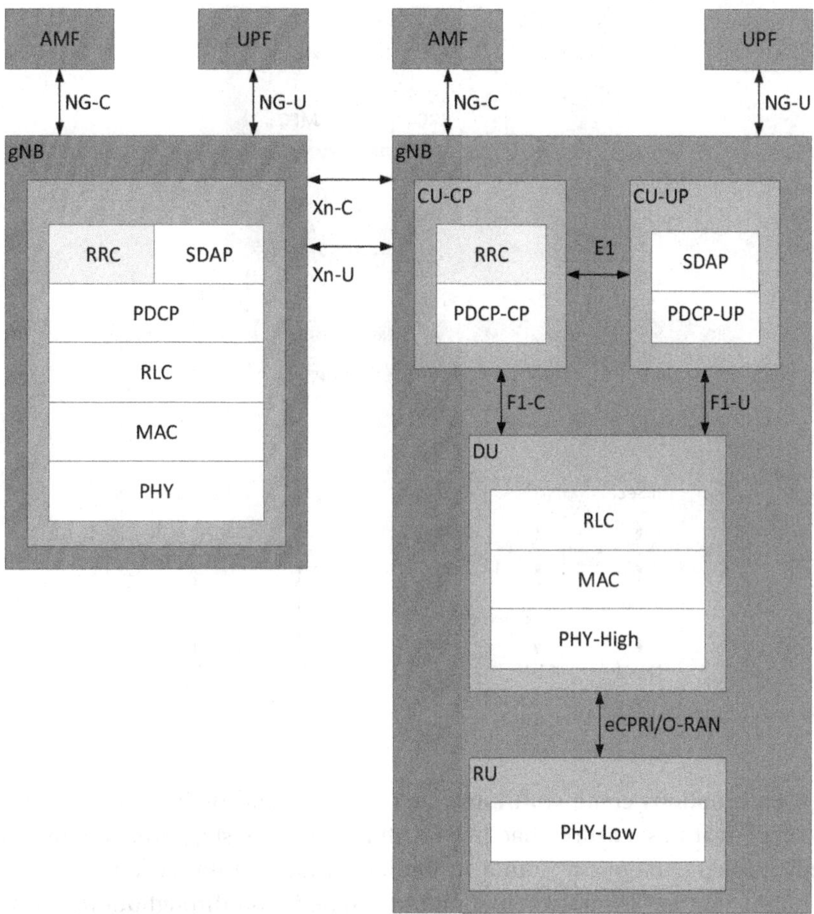

FIGURE 4–4 Key CU interfaces as defined by 3GPP.

In scale-out option, multiple accelerators devices are placed on a processing blade that supports increments of 100 Gbps or more of packet processing per function with an Ethernet switch to provide connectivity between devices. Multiple blades are placed into a 100GbE backplane to provide close to infinite scaling. To support this deployment model, the S1 interface can no longer be implemented as a single (IPSec) tunnel but needs to be split into multiple tunnels that can each be routed to a unique target accelerator. We assume that the statistical traffic properties are such that issues around Elephant Flow handling are minimized (see discussion elsewhere). Traffic distribution between transport and GTP/PDCP modules can be done, for example, based on the GRPS Tunneling Protocol Tunnel Endoint Identification (GTP TEID) field.

The hardware implementation options for the accelerator are typically either FPGA-based or multicore networking processors (historically PowerPC or MIPS ISA but more recently

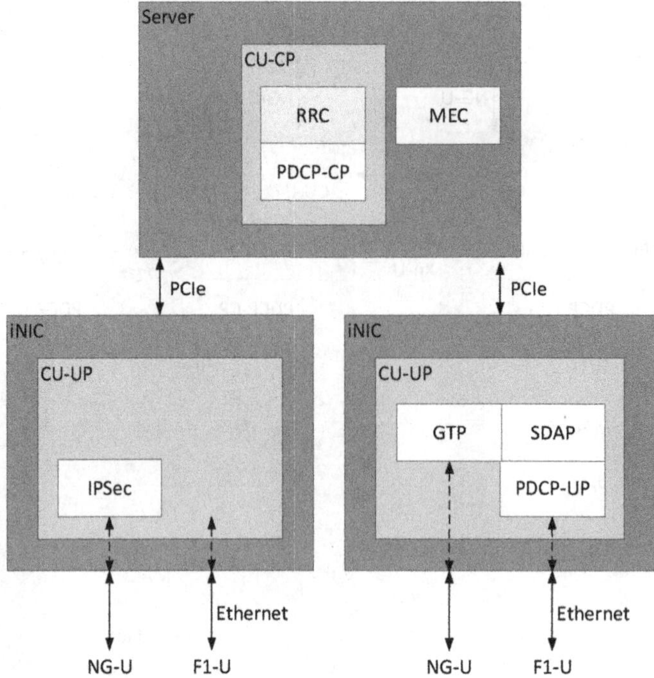

FIGURE 4–5 Scale-out by multiple iNICs.

Arm based). Both options come with integrated Ethernet and dedicated logic for ciphering protocols. These features ensure that power efficiency is a step function improved over software-only implementations as found in traditional (x86-based) servers. This type of offload is not unique to the CU implementation but can be found throughout data center applications where FPGAs and Network Processor Units (NPUs) are often used to offload IO from server cores.

DU blades implement Layer 2/Physical layer stacks, either as a software-only function (for legacy 3GPP networks such as 2G and 3G) or with dedicated hardware acceleration (5G and in limited fashion for 4G). An example implementation is shown below in Fig. 4–7:

Cell site integrated (CU/DU) solutions

These solutions are more similar to the traditional (macro) base station as they provide integrated baseband functionality (the DU, optionally including CU functions) into a form factor that is reminiscent to macro base stations: typically, a 19-inch rack with 2U-4U height. We discuss two implementation options: server-based where the DU is built around an off-the-shelf server with acceleration card to turn it into a DU and optimized hardware that is tuned towards the use-case.

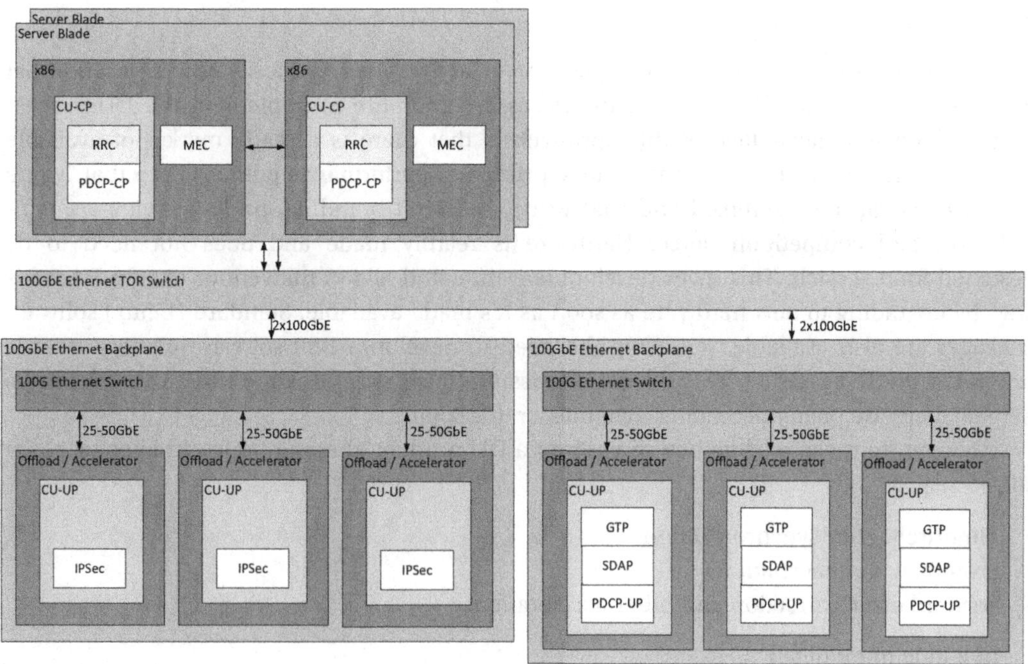

FIGURE 4–6 Scale-out to multiple blades.

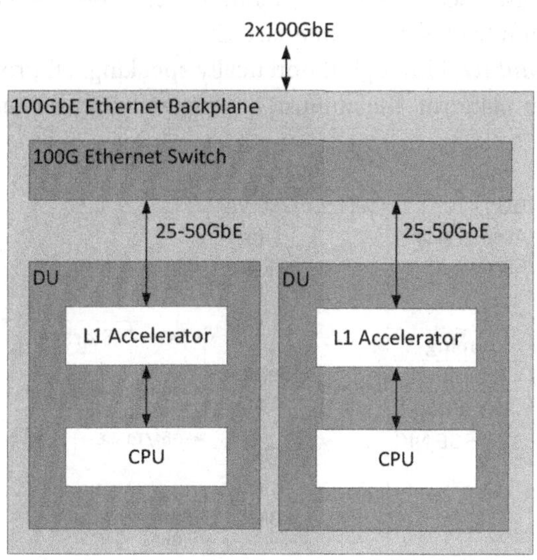

FIGURE 4–7 DU acceleration blade.

Server-based solutions

Server-based systems use a standard edge server architecture (typically x86 Instruction Set Architecture (ISA) based) to provide the necessary compute to implement the DU processing. The obvious advantage of this approach is that there is a ready market of available server platforms in a variety of form factors, price, and performance points. Given that "white box" servers can be assembled and sold with or without branding, profit margins are typically low and competition fierce. Hardware is readily made and does not need to be designed from scratch. This saves development time and allows the vendor to ride "Moore's law" by upgrading to new hardware as soon as it's made available. Standard (Linux) software packages are also available, removing the need to develop a BSP software platform that is unique to the DU product. We will not discuss the intricacies of white box server platforms here as there are plenty of reading materials on the topic.

However, to make a white box server into a DU product, there are a few challenges to be solved (Fig. 4–8):

- Time/frequency synchronization
- Fronthaul and backhaul IO
- Forward error correction and other acceleration.

In a little more detail:

Time/frequency synchronization. Given that standard white box servers do not provision for IEEE1588 or Global Positioning System (GPS) timing, this component is typically provided by an external unit. The external unit can either be a PCIe based plug-in board that incorporates a GPS receiver and Phase Locked Loop (PLL) infrastructure, or a GPS unit that provides a timing source to the fronthaul Network Interface Card (NIC).

Fronthaul and backhaul IO. Although theoretically speaking, all processing can be done in software in a server platform, the amount of processing power required is more than

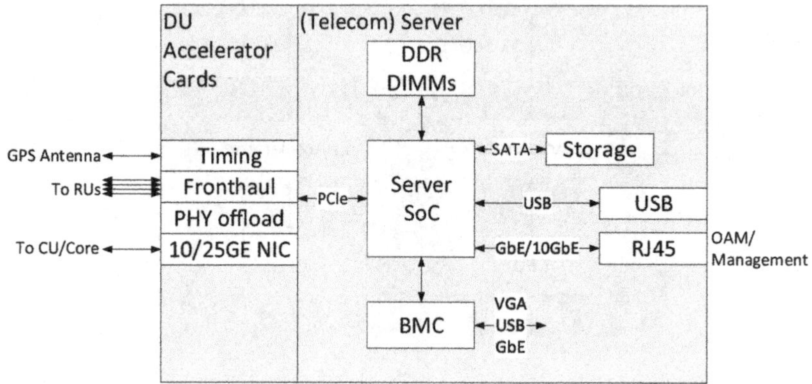

FIGURE 4–8 Server-based cell site DU solution.

supported in a cost-effective manner. One solution is to provide hardware offload for components of the processing chain. For example, a "smart Network Interface Card" (SmartNIC) can offload mid- or backhaul traffic, including the IP/IPSec and GTP encapsulation, offloading the host processor from implementing these functions. Given that the server System-On-Chip (SoC) itself is typically limited on integrated Ethernet interfaces (instead relying on external devices for Ethernet IO), this offload is an obvious candidate. A SmartNIC typically uses a dedicated networking processor (from vendors like Cavium/Marvell, Netronome or NXP) or FPGA (Altera/Intel or Xilinx/AMD) to implement the offloaded protocol stacks on dedicated engines that are more power and cost efficient than the server they are plugged into. The programmability of the SmartNIC allows it to be upgraded to support custom features, at the cost of design, implementation, verification and optimization of this firmware.
High throughput fronthaul traffic can be offloaded through an eCPRI NIC that takes care of the full Control and User plane components of the fronthaul specifications, transferring uncompressed IQ buffers to/from the host processor and offloading the CPU from doing these functions. This NIC can be physically the same smartNIC or one that is optimized for eCPRI.
Forward Error Correction and associated Physical Layer Acceleration targets to offload the host CPU even more by taking the most processor-intensive functions from the processing chain and implementing them in an FPGA or dedicated ASIC/accelerator. Forward Error Correction (FEC) Low Density Parallel Codes (LDPC) decoding is an obvious candidate in 5G/NR. Although the concept of offloading is counter-intuitive for a so-called software-centric solution, performance requirements dictate some form of offload for all but the lowest-end solutions.

Some of the limitations of the standard White Box server architecture are being addressed by evolving standards. Consider the Open Telecom IT Infrastructure (OTII) white-box server project from the Chinese Open Datacenter group.[2] This white box design is driven by China Mobile, China Telecom, China Unicom, China Academy of Information and Communications Technology (CAICT), and Intel. The OTII server is optimized for 5G and Edge Computing both in form factor (physical and environmental) and IO (PCIe connectivity designed for fronthaul IO and Physical layer accelerator). Early versions of the standard call out the need for time/frequency synchronization without specifics on whether this is implemented as a feature on the main board or through a PCIe plugin board.

Some of the offload functions obviously can be combined into one, with an accelerator board like shown here in Fig. 4–9:

Note that this type of accelerator model requires multiple transfers of payload between server and offload engine. The associated bandwidth can become a bottleneck as we will see later.

An alternative accelerator split is shown in Fig. 4–10:

The advantage of this architecture is that the PCIe bandwidth is constrained (there is only a single path to/from the accelerator). In this example, the fronthaul I/O (eCPRI and framing) can reside either in the host processor or in the accelerator card. The latter option is of course most performance optimized and likely to be implemented.

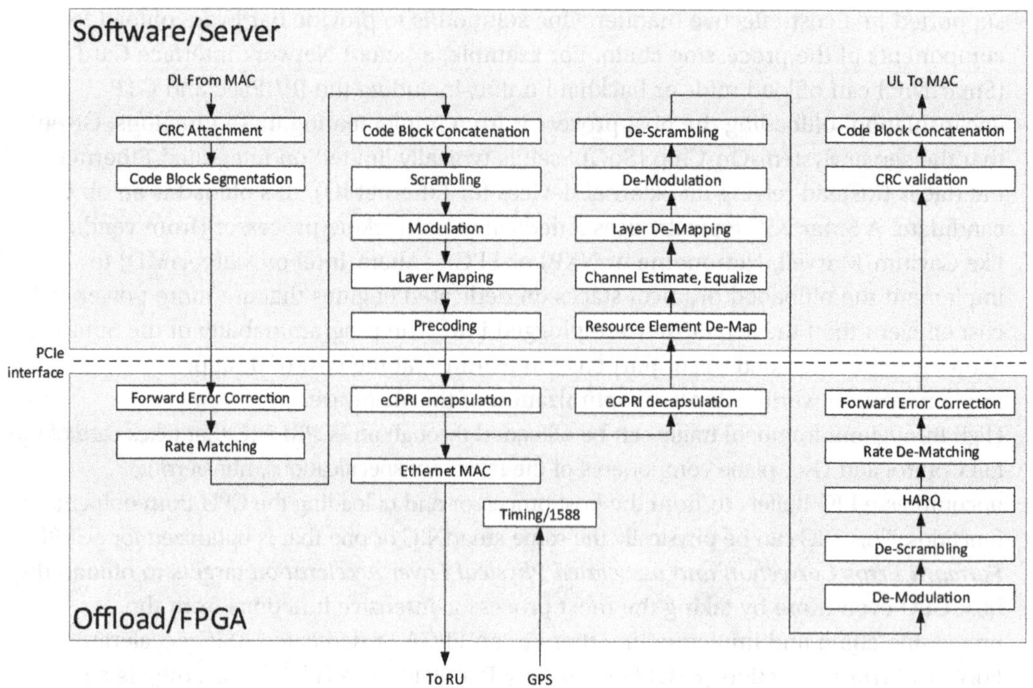

FIGURE 4–9 Typical server PCIe/offload scenario for eCPRI and forward error correction.

PCIe accelerator boards form factors are characterized by physical dimensions and power draw:

- Length. Full-length (long) cards can be up to 312 mm and half-length (short) cards are up to 175 mm.
- Height. A full-height bracket is 111 mm versus a low-profile bracket of 69 mm.
- Width. A dual-width PCIe card requires access to a second exhaust slot on the back even though it does not need to plug into two PCIe connectors.
- Power. Power can be supplied with power through the motherboard for <75 W DC power consumption. If more power is required, an auxiliary power cable needs to be plugged in from the power supply. Typical maximum power is 300 W with an additional 225 W delivered from this auxiliary set of connectors.

Optimized hardware

Server-based solution save time and effort associated with hardware development but are generic in nature which implies that they are not optimized for the bulk of the use-cases—it is unlikely that the combination of compute horse power, IO and power/cost targets are met. So, we end up with a custom hardware design that is either sourced from an Original Device Manufacturer (ODM) that designs and produces the device under the name of the Original Equipment Manufacturer (OEM) or built in-house (by the OEM).

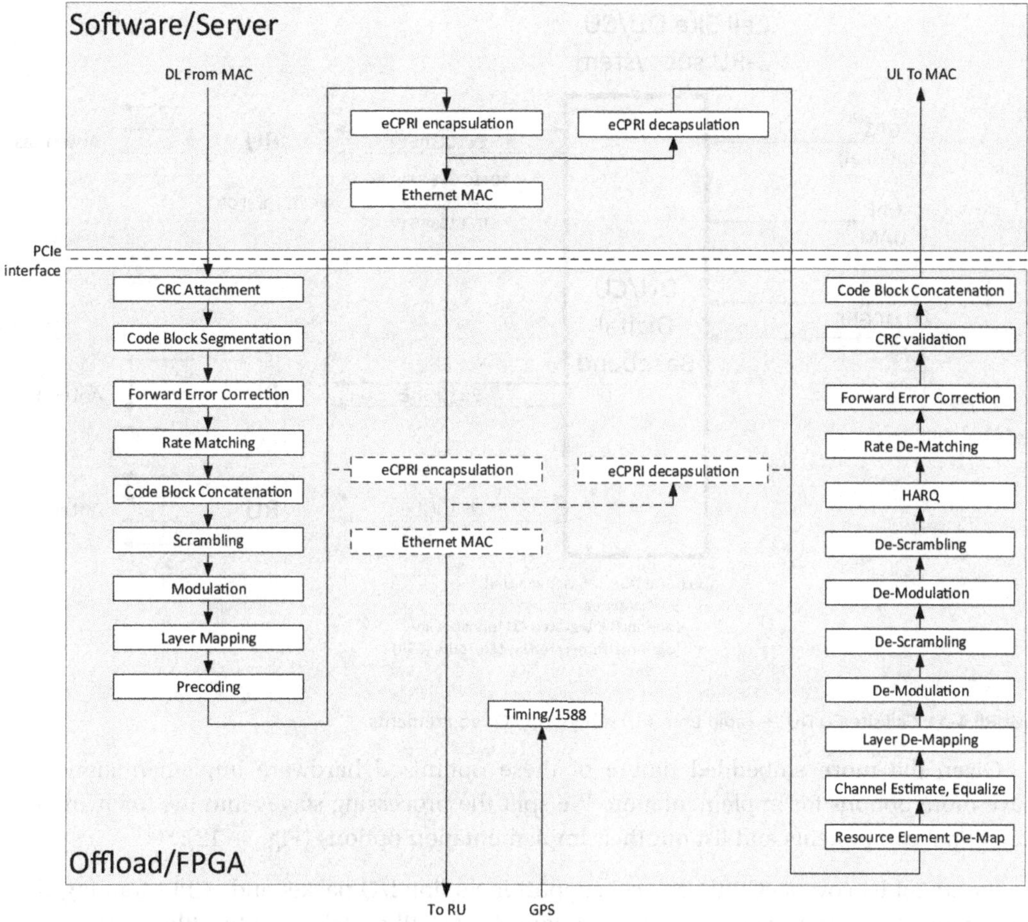

FIGURE 4–10 Alternative server PCIe/offload scenario for eCPRI and forward error correction.

These custom designs often employ the same components/component families as server-based designs but are more flexible on component choice (custom, after all) as well as interconnect/interfacing.

The starting point of design of an optimized hardware solution is to assess the connectivity requirement. We give an example depiction below for a low-end cell-site solution in Fig. 4–11:

Like for the server-based product, the main component in the CU/DU Digital Baseband unit is a programmable (processor) solution. But also, here, for the unit to become a DU, we need to solve some challenges that are unique to the DU application:

- Time/frequency synchronization
- Fronthaul and mid/backhaul IO
- Forward error correction and other acceleration.

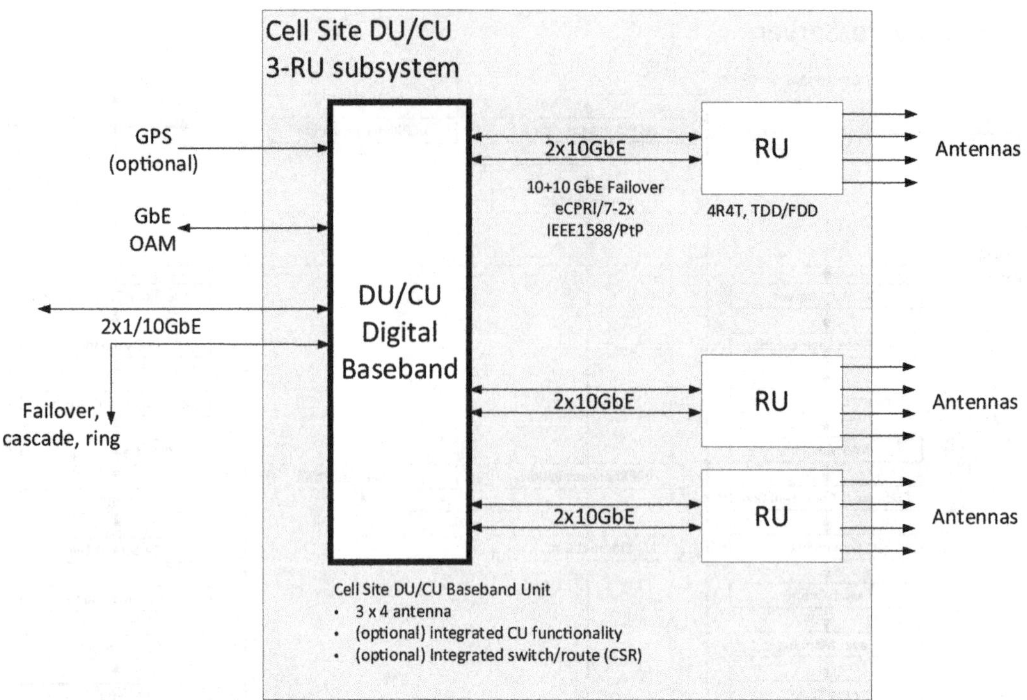

FIGURE 4–11 Cell site CU/DU + radio unit (RU) subsystem, IO requirements.

Given the more embedded nature of these optimized hardware implementations, we have more options for implementation. We split the processing stages into the following key functional components and list out their implementation options (Fig. 4–12):

- Fronthaul IO covers eCPRI protocol support including I/Q packet and eCPRI framing/de-framing. Optionally, this can include support for eCPRI switching to be able to connect multiple baseband devices to each other. Implementation options include:
  - Software implementations on a GPP
  - FPGA based designs
  - Optimized ASIC IP, either implemented in a separate device or as part of a custom SoC.
- Wireless physical layer acceleration
  - FPGA based designs
  - Optimized ASIC IP, either implemented in a separate device or as part of a custom SoC. This ASIC IP is mostly a combination of embedded (programmable) Digital Signal Processing (DSP) cores and hardware accelerators that support the most compute intensive functions.
- Layer 2 stacks are typically implemented as a software function running on embedded processor cores. Acceleration for commonly used functions including PDCP ciphering, wireless scheduling can be integrated.

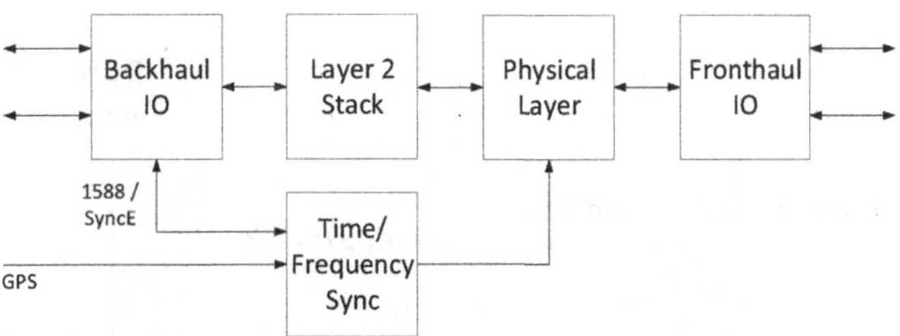

FIGURE 4–12 Functional components of the optimized cell site integrated CU/DU hardware solution.

  - Most embedded implementations are Arm or x86 based, with potential for RISC-V implementations.
- Mid/backhaul IO potentially including a dedicated NPU or similar hardware to offload tunneling and Ethernet/IP switch/routing functions.
  - This functionality is often colocated with the layer-2 stack processor cores
- Time/frequency synchronization

Given the I/O requirements and sector configuration that we previously established, we can calculate all the required I/O bandwidths between the functional components allowing us to define the mix of components that serves the functional purpose whilst hitting cost/performance/development time targets. Note that multiple functional components are likely physically integrated onto a single chip: FPGAs have embedded DSP and Arm cores, integrated ASIC/ASSP solutions can include all functionality into a single device. The question is whether the provided levels of integration make sense for the target application.

We give a few example options for implementation here, in order from deeply integrated/embedded to the disaggregated and software defined.

Integrated Architectures such as[3] for 4G and[4] for 5G combine all functional components associated with a given partitioning/use-case into a single deeply embedded SoC, like shown below in Fig. 4–13:

The level of integration implemented here is both the advantage and the Achilles heel of this approach: either the solution matches the problem-statement very well, or not at all. This goes both for choice of Instruction Set Architecture for the L2 stack, the Physical layer implementation, protocol (4G? 5G?) and IO pieces of the puzzle. Deeply integrated solutions are often seen used in macro base station implementations from the established industry players. A challenge is how to deal with a "scale out" scenario where—for example—we need to support a larger amount of front- and mid/backhaul. Adding an external Ethernet switch to the system for a scale-out purpose defies the target of integration and adds cost to the system that cannot be carried.

Orthogonal to the deeply integrated approach, a more disaggregated option is shown below in Fig. 4–14:

This approach overcomes the challenge of scalability: each of the key components in the system can be scaled independently of the other ones and selected (from different vendors)

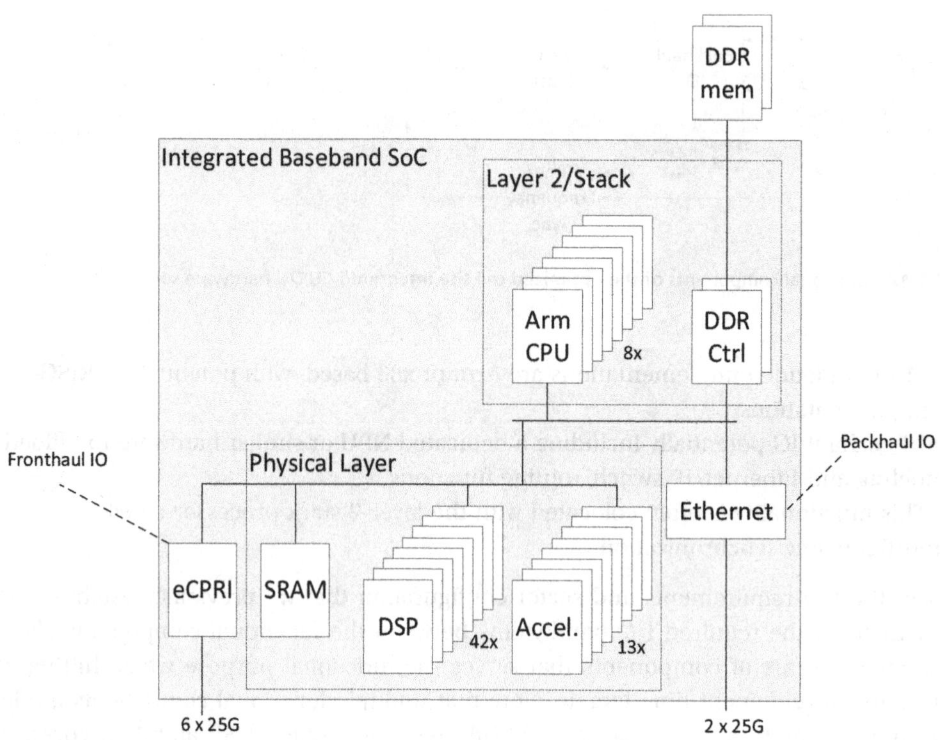

FIGURE 4–13 Deeply integrated/embedded approach.

on their own merits. Typical implementations of the fronthaul functionality would be using an FPGA. The host processor can be x86 or Arm based and Physical Layer accelerations can be FPGA and/or ASSP/ASIC such as.[5]

The *software defined* solution is a simplification from the disaggregated approach for those use-cases where the required system capacity allows us to combine multiple functions into a single device, for example, the front and mid/backhaul functionality that can be combined with the host processor (assuming the host processor has sufficient networking IO) and we only offload limited Physical Layer processing to an accelerator to stick more to the "Software Defined Radio" mantra (Fig. 4–15):

Radio unit

The Radio Unit (RU) performance range runs wide.

- Low-end implementations can have similar radio frequency (RF) specifications as the Integrated Small Cell (ISC) with a 2–4 antenna, <6 GHz, <250–500 mW/output port. These systems are very cost sensitive.

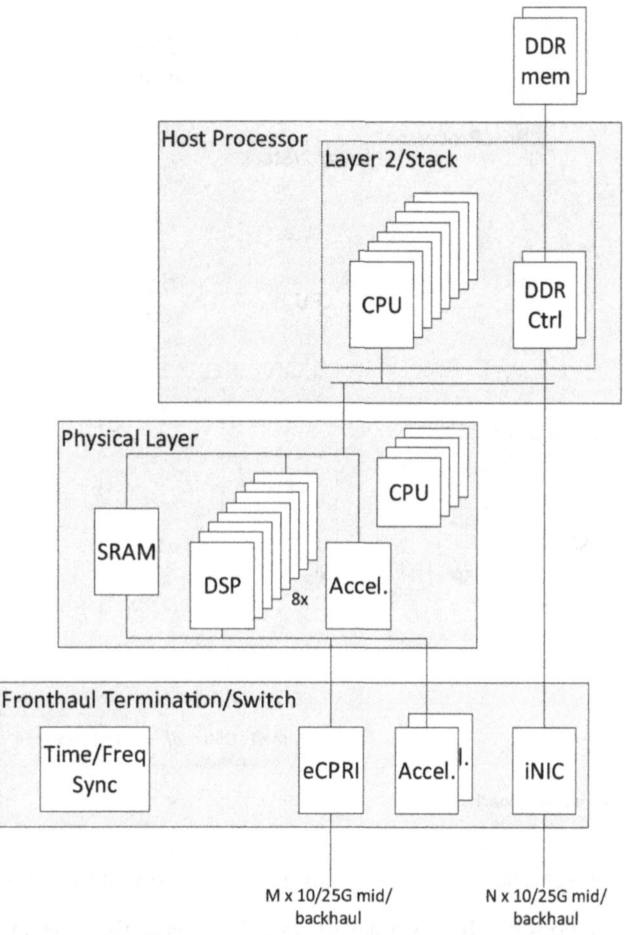

FIGURE 4–14 Disaggregated approach.

- High-end implementations that have limited antenna count high output power of 10s of Watt, such as used in large, outdoor applications. Defined here as a 2. . .8-antenna system targeting the <6 GHz 3GPP Radio Access Network market. These systems typically employ digital beamforming in the RU to direct energy towards the User Equipment (UE). Given the high output power, power amplifier (PA) efficiency is key.
- Massive Multime Input Output (M-MIMO) systems that employ tens of antennas (32. . .64) in the <6 GHz market. The output power from each antenna is in the 5 Watt range. The main challenge in these RUs is scalability and cost: how to efficiently build a system that can scale to "any" antenna count.

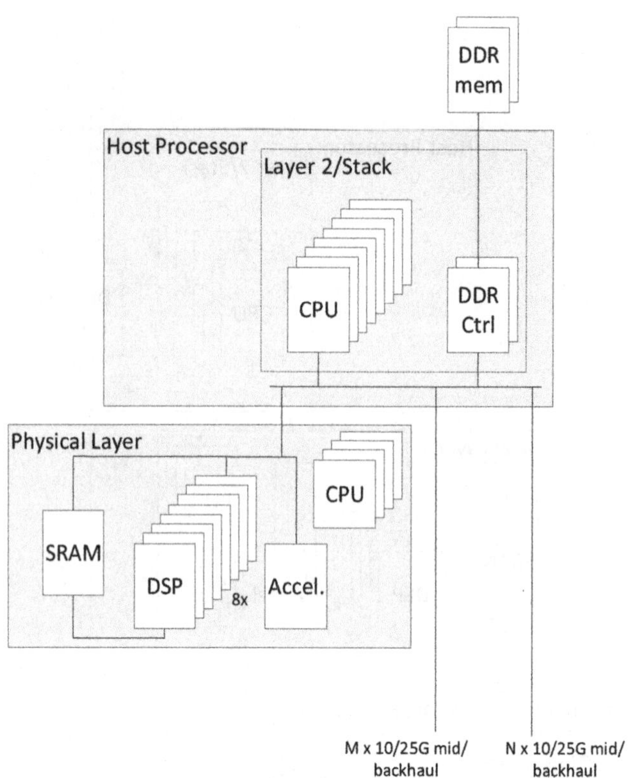

FIGURE 4–15 Software defined approach.

In the mmWave space, scalability can also be from the very low to the very high-end:

- Low-end implementations can have similar RF specifications as the Integrated Small Cell (ISC) with 16 or 32 Antenna elements and output power (EIRP) like that of a handset (30–40 dBm).
- Higher-end implementations can support multiple antenna panels, with 128 or more Antenna Elements each and a large DC power budget (Fig. 4–16).

Key components in the RU include:

1. Fronthaul and eCPRI Termination
   a. This entails termination of the Ethernet/CPRI based fronthaul interface to the Distributed/Baseband Unit and interface conversion from external optical link to board-level interface to Digital Signal Processing components
   b. Includes termination of network-based timing (IEEE1588/Synchronous Ethernet)
   c. Includes Operation and Maintenance (OAM) and related “slow path” functions
   d. We assume a Time Domain, IQ based interface between Baseband/Digital Unit (DU) and RU, where time-to-frequency domain conversion is done inside the RU—corresponding to the O-RAN 7–2x split defined earlier.

2. Lower Physical Layer
   a. Implements the Digital Signal Processing that converts the eCPRI information (frequency domain I/Q data) into time-domain analog I/Q that can be transmitted by the RF subsystem
   b. This typically (for category "B" RUs in the <6 GHz marker) includes the beamforming functionality associated with user-to-antenna mapping
   c. Includes time-to-frequency domain conversion (FFT) functions.
3. Digital Front End
   a. Includes Digital Up/Down Conversion, Crest Factor Reduction and Digital Predistortion functions, specifically for <6 GHz implementations.
4. RF Subsystem
   a. Implements the analog conversion from baseband IQ to (IF and) RF and the antenna Receive (Rx) and Transmit (Tx) functions including analog beamforming.

Note how the picture above shows which components are naturally centralized (specifically beamforming) and which components are explicitly unique to each antenna path. The centralized nature of the beamforming component comes from the fact that in the uplink (receive) direction, this function is implemented as a matrix multiplication between all antenna data and antenna weights to produce the per-Layer (user) samples that are further processed in the DU. This processing stage is difficult to break down into an easily scalable solution involving multiple physical devices. This beamforming aspect can complicate the partitioning, especially when building Massive MIMO systems.

eCPRI termination

This component is very similar to the eCPRI piece of the DU solution and often there is re-use between the DU and the RU, if both are implemented by the same vendor. This saves development time and increases chances of successful (internal) interoperability, after all. eCPRI termination in the RU is implemented (as for the DU) as one of:

- Software implementations on a GPP—typically used in the low- to mid-end

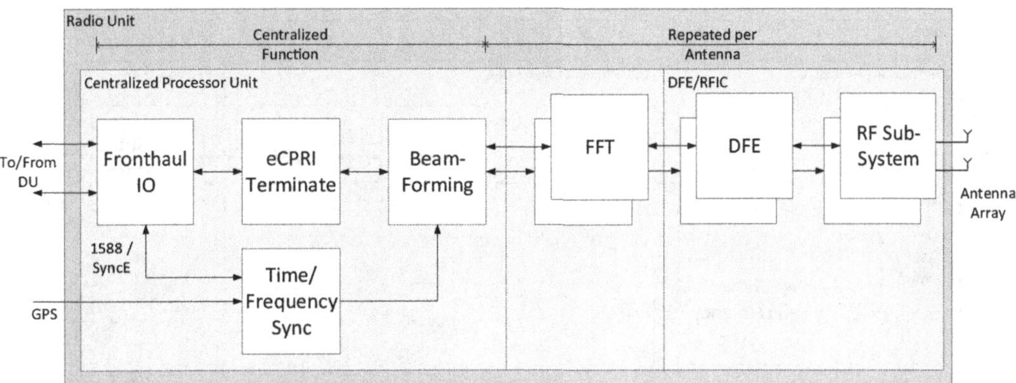

FIGURE 4–16 Key components in the integrated radio unit.

- FPGA based designs
- Optimized ASIC IP, either implemented in a separate device or as part of a custom SoC.

eCPRI termination may optionally push the (de-)compression component to the low-Physical Layer (PHY) processing element, if this element is physically on a different device. This way, the eCPRI termination complexity is reduced, and IO bandwidth between this component and the low Physical layer processor is reduced.

Low physical layer and digital front end

Low Physical Layer and Digital Front End (DFE) functions can be implemented in a centralized or distributed manner, depending on the system scaling requirements. We show a few implementation options here:

The *Software Defined Approach* (Fig. 4–17) uses a general-purpose Networking Processor or alternatively, an FPGA that integrates control plane Arm CPU cores for termination of the eCPRI traffic. The eCPRI buffers are transferred over PCIe to a secondary device that implements the Lower Physical Layer processing for all associated antenna paths, including beamforming and (relevant) DFE components. More about the DFE↔RF Subsystem splits is covered later.

As the antenna count goes up, the Low Physical Layer/DFE functionality doesn't fit into a single device anymore. At this point, we need a scale-out scenario where the beamforming component (which is a naturally centralized function) is separated from the remainder Low Physical Layer implementation, allowing the latter to scale per unit of a single or more antennas as shown in Fig. 4–18.

Baseband to radio frequency conversion

RF to baseband implementation is typically implemented as one out of the three following options, in increasing form of digital (and decreasing form of analog) complexity:

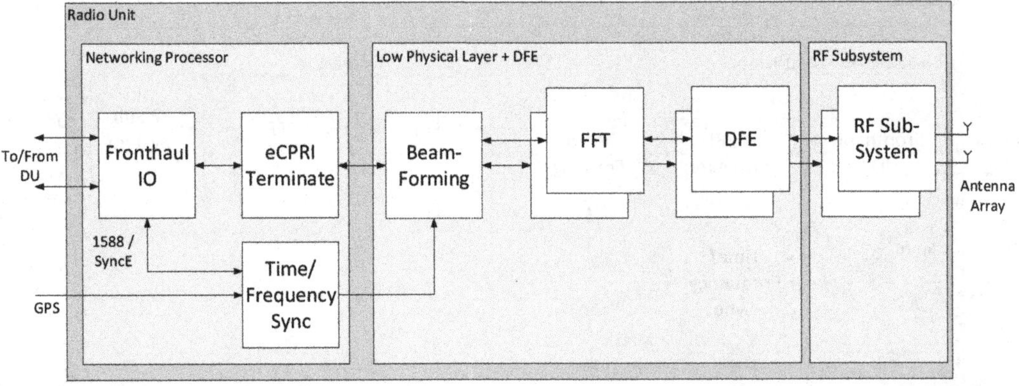

FIGURE 4–17 Software defined approach.

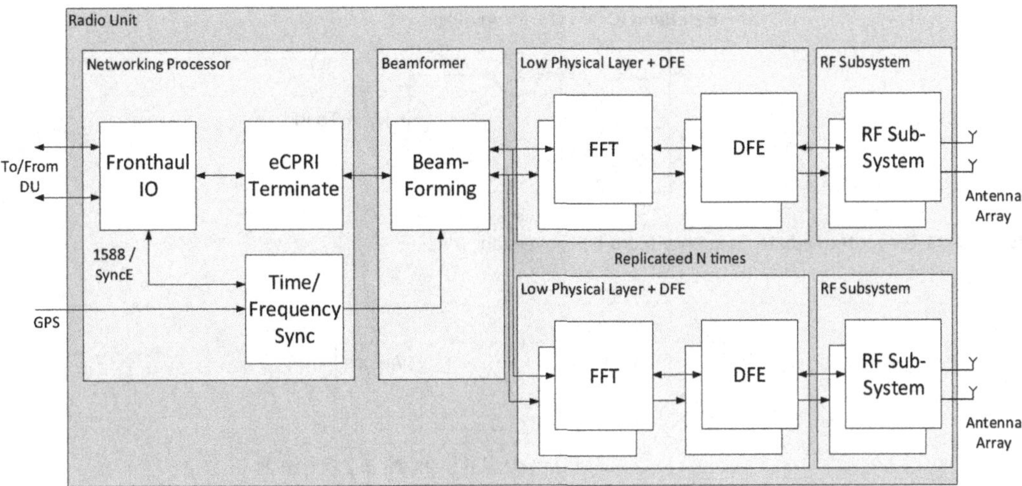

FIGURE 4–18 Disaggregated approach.

Zero-Intermediate Frequency (IF) interface (Fig. 4–19) uses a baseband I/Q interface to/from Analog-to-Digital Conversion (ADC)/Digital-to-Analog Conversion (DAC) with modulation/de-modulation (up/down conversion) in the analog domain:

This traditional zero-IF architecture has the advantage of low-complexity filtering at the baseband (only low-pass filtering is required). There are no relevant considerations with regards to mixing images. However, care needs to be taken to remove sideband image and Local Oscillator (LO) feedthrough through IQ mismatch compensation in the digital baseband domain.

IF interface (Fig. 4–20) uses digital IF signal generation with on-chip ADC/DAC. External up/down conversion to RF frequency:

In this IF architecture, the baseband device implements (QAM) modulation/de-modulation using a digital Numerically Controlled Oscillator (NCO) and digital mixers. The output signal is converted to a final RF frequency with a second (analog) mixer. The advantage of this architecture is that it doesn't require the IQ mismatch compensation from the zero-IF architecture, but this comes at a higher system cost due to increased data converter complexity and more difficult (band pass) filtering at IF/RF stages.

Direct RF conversion (Fig. 4–21) signal generation with on-chip ADC/DAC. External RF only:

In a direct RF sampling architecture, all (QAM) modulation/de-modulation is done in the digital domain, removing the need for a secondary (analog) mixer and associated cost/complexity. This comes at a higher system cost associated with high-speed data conversion and associated digital signal processing.

Digital-to-analog conversion and analog-to-digital conversion

The receiver ADC dynamic range is defined as the difference between the maximum allowed input signal level and the noise plus interference floor level. Thus we need to consider the

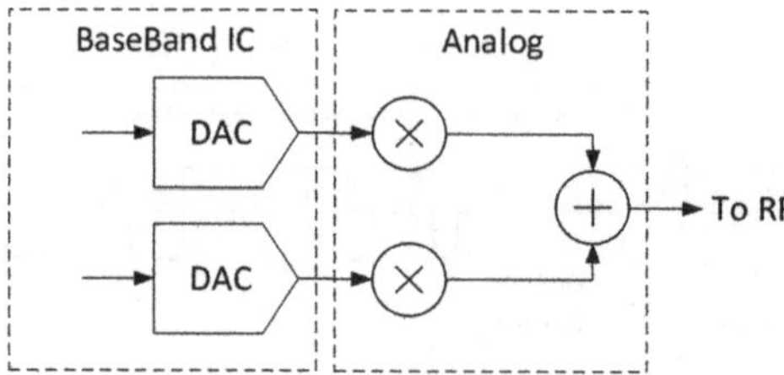

FIGURE 4–19 Zero-intermediate frequency (Zero-IF) conversion.

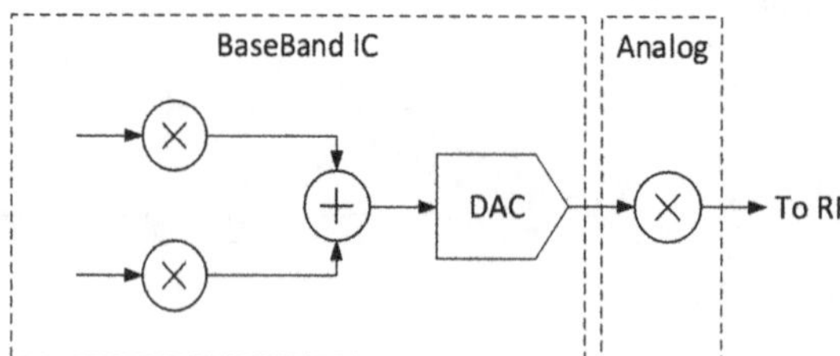

FIGURE 4–20 Intermediate frequency (IF) conversion.

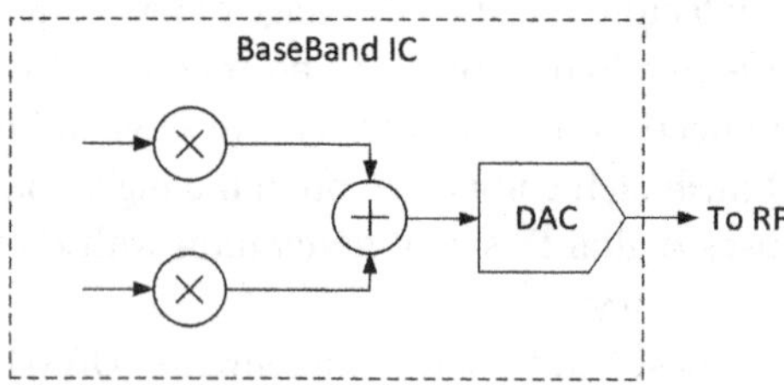

FIGURE 4–21 Direct radio frequency (RF) conversion.

maximum modulation and coding scheme which requires the maximum possible Signal to Noise Ratio (SNR) in order to decode the signal with the desired rate, say, modulation level, layer count, etc. Assuming a (simple to understand) zero-IF interface:

- The Peak to Average power ratio of the OFDM signal as discussed elsewhere, let's assume it to be 12 dB. We don't assume any PAPR optimization techniques implemented in the baseband *receiver*.
- Gain/offset errors. Gain and offset errors as caused by inaccuracies in internal reference and amplifier offset voltages reduce the usable dynamic range. Assuming gain error and offset errors to be 10% of full scale, then each contributes a 1 dB reduction in dynamic range. The ADC must be backed off 1 dB to prevent input clipping and another 1 dB to account for limited dynamic range. Add 2 dB dynamic range margin.

- Fade margin. Fading plays an important role in the dynamic range, causing fluctuations of the signal level. This fading needs to be accounted for in:
  - AGC—Slow fading (rain, night/day, blockage)
  - Dynamic range—Fast fading (moving reflectors, wind, leaves).

  Fade margin requirements obviously depend highly on the RF conditions of the system we're dimensioning for and field test results can be found in literature. As a rule of thumb, assume 15 dB.
- Blocker interference in cochannels or adjacent channels. The industry standard number for this is about 10 dB but this number can be reduced if we're implementing a single-channel receiver or other favorable conditions apply (say, mmWave communications where blocker performance is less relevant due to the highly beamformed nature of the communication path).
- Sensitivity SNR is the required SNR by the (digital) baseband implementation as required to successfully decode the received signal. Performance depends on many factors: Channel model, code block size, Hybrid automatic repeat request (HARQ) retransmissions, MCS and coding rate, MIMO decoder algorithmic performance and so on. Typically, this number is provided by Matlab models of the baseband implementation. We assume 25 dB required for successful decoding of a 64QAM signal.
- Processing gain from oversampling—sampling above the minimum Nyquist requirement. Oversampling is good both because of the processing gain as well as because it simplifies the requirements on the external (analog) baseband reconstruction filter. Assuming a useful RF signal bandwidth of (say) 100 MHz which means a baseband signal rate of 50 MHz I and 50 MHz Q in a baseband ADC/DAC, implying a baseband sample rate of ~100 MHz to adhere to the Nyquist rate. When we sample at 245.76 MHz, the processing gain is calculated as 10*log10(245.75/100) = 4 dB.

Aggregate, it looks like we need a minimum of $12 + 2 + 15 + 10 + 25 - 4 = 60$ dB or 10 Effective Number of Bits (ENOB). This is also enough precision to capture DPD feedback signals as we will be discussing later.

The transmitter DAC dynamic performance requirements are dependent on the signal dynamic range which is calculated by doing an SNR budget analysis:

- Modulation Error Vector Magnitude (EVM). This is the EVM that is targeted by the system given a specific modulation type. EVM is standardized by 3GPP[6,7] to 8% (64QAM) or 3.5% (256QAM). EVM can be converted from % to dB value SNR = -20log (EVM). Assuming 256QAM operation, assume an 29 dB EVM.
- SNR margin. Assume the transmitter DAC cannot degrade EVM by more than 0.25% (0.6 dB impact). Assuming the following equation for system SNR < System SNR = -20log $(10^{-SNRa/10} + 10^{-SNRb/10} + \ldots 10^{-SNRn/10})^{1/2}$, this requires us to add 9 dB SNR margin.
- Headroom for Peak to Average Power Reduction (PAPR). Assume an 8 dB PAPR after applying Crest Factor Reduction (CFR).
- Gain/offset errors. Gain and offset errors as caused by inaccuracies in internal reference and amplifier offset voltages reduce the usable dynamic range. Assuming gain error and

offset errors to be 10% of full scale, then each contributes a 1 dB reduction in dynamic range. The DAC must be backed off 1 dB to prevent input clipping and another 1 dB to account for limited dynamic range. Add 2 dB dynamic range margin.

- Processing gain from oversampling—sampling above the minimum Nyquist requirement. Oversampling is good both because of the processing gain as well as because it simplifies the requirements on the external (analog) baseband reconstruction filter. Assuming a useful RF signal bandwidth of (say) 100 MHz which means a baseband signal rate of 50 MHz I and 50 MHz Q in a baseband ADC/DAC, implying a baseband sample rate of ~100 MHz to adhere to the Nyquist rate. When we sample at 245.76 MHz, the processing gain is calculated as 10*log10(245.75/100) = 4 dB.

Aggregate, it looks like we need a minimum of $29 + 9 + 8 + 2 - 4 = 44$ dB. However, keep in mind that the DAC performance is defined by a secondary factor which is the requirement to provide an output signal with sufficiently low ACLR – per 3GPP specs better than −45dB leakage to the adjacent channel. This -45dB number defines a lower bound to the performance of the DAC , say ~60 dB or 10 ENOB to have implementation margin. ACLR is also impacted by the intermodulation distortion (nonlinearity) of the DAC which causes spectral regrowth.

The above example is a case of a zero-IF interface where the data converter is operating in what is called the 1st Nyquist zone. This zone is defined as the bandwidth between DC and ½ of the sampling frequency F_{sample}. Similarly, the 2nd Nyquist zone is defined as ½ F_{sample} ... F_{sample}, the 3rd Nyquist zone as F_{sample} ... 1½ F_{sample}, and so on. The signal in these other (nonprimary) Nyquist zones can still be captured by the data converter but they are combined with (folded into) the baseband signal. If we filter out all other input signals to make sure that only the signal from the desired Nyquist zone reaches the data converter, we can convert the signal between analog and digital domains. For a data converter to be used for undersampling (or subsampling), its input bandwidth needs to be wide enough to be able to capture the target RF bandwidth. Such data converters are commonly referred to as RF data converters and are used for IF and RF data use-cases as we described earlier.

There is a similar discussion on the digital predistortion (DPD) feedback path that also includes an ADC which we will discuss later.Andersen et al.[8] shows how Zhu's general sampling theorem[9] can be applied to PA characterization, thus removing the requirement of feedback ADC bandwidth to be wide enough to capture all out-of-band distortion.

Radio frequency subsystem

The interface between the modem and the RF subsystem can be either digital through Jedec Standard (JESD) digital sample interfacing or analog through a baseband or IF connection. We discuss both options here.

JESD

The JESD204 standards define high-speed serialized data interfaces used for communication of digital radio samples and control data among the logic devices (the 5G modem) implemented on FPGAs or ASICs or Application Specific Standard Products (ASSPs) and external data converters (DAC/ADC). It is the Serial (SerDes) based evolution of the JESD207 parallel data converter interface that was in use in the early 2000s. Multiple versions of the standard have been developed since 2006 with JESD204b (2011) being the most recent one, supporting up to 12.5 Gbps per serial lane. JESD204 incorporates embedded clocking through use of 8b10b encoding, removing the need for an external clock interface. Synchronization between the baseband ASIC and the data converter is achieved by using an external synchronization signal (SYNC or SYSREF) to achieve deterministic performance (defined and aligned latency). A high level partitioning example is shown in Fig. 4–22 below.

Most standalone data converters from vendors like Analog Devices and Texas Instruments support the JESD interface, but it is also used to interface to integrated RFICs. What we mean with integrated RFICs are devices that include the data converters in addition to baseband ↔ RF conversion as well as (optional) support for some of the digital processing components such as DPD.

The challenge with any digital interface, JESD included, is the required bandwidth as the system scales up with regards to antenna count and RF bandwidth. Consider a 100 MHz 5G/NR system with a 14-bit (nominal) data conversion requirement and 4 antennas. Given a 122.88 MHz sample rate, we have 4 (antenna) × 122.88 (MSPS) × 2 (I, Q) × 14 (bit) ~ = 14 Gbps. Even though JESD204b supports multiple speed grades (grade 1: ≤ 3.125 Gbps; grade 2: ≤ 6.375 Gbps; grade 3: ≤ 12.5 Gbps), multiple lanes are necessary, and the solution can become more complex and costlier than desired. Nonetheless, with bandwidth-limited systems (our example of 100 MHz bandwidth) and lower antenna counts JESD works well. At the same time, once we scale up to mmWave capacities and 5G evolves to 6G, it is clear that the digital interface becomes less and less desirable.

This limited interface capacity also imposes challenges when integrating Digital Front End (DFE) functionality. Digital predistortion requires the data converter subsystem to operate at a higher sample rate than the (native) baseband rate, multiplying the required IO and associated SerDes count to become unsustainable, specifically when power consumption is prime. The answer begins to carve out some of the DFE functionality and migrates this to the RFIC, as pioneered by ADI in 2018[10] but also supported by other RFIC vendors. The computational requirements associated with DPD are high, driving for an aggressive process node roadmap and hence a potentially costly solution. The advantage of this solution is the high level of functional integration: the customer doesn't need to be concerned with DPD algorithmic development, ADC/DAC specs and other items considered complex due to their analog nature.

Analog

The alternative to integration of DPD into the RFIC is to integrate the RFIC functionality into the baseband chip, moving the integration complexity from the digital to the analog domain (Fig. 4–23). In this partitioning option, data converters are integrated into the baseband device, either as a baseband I/Q interface or as an IF or RF data converter, both of which

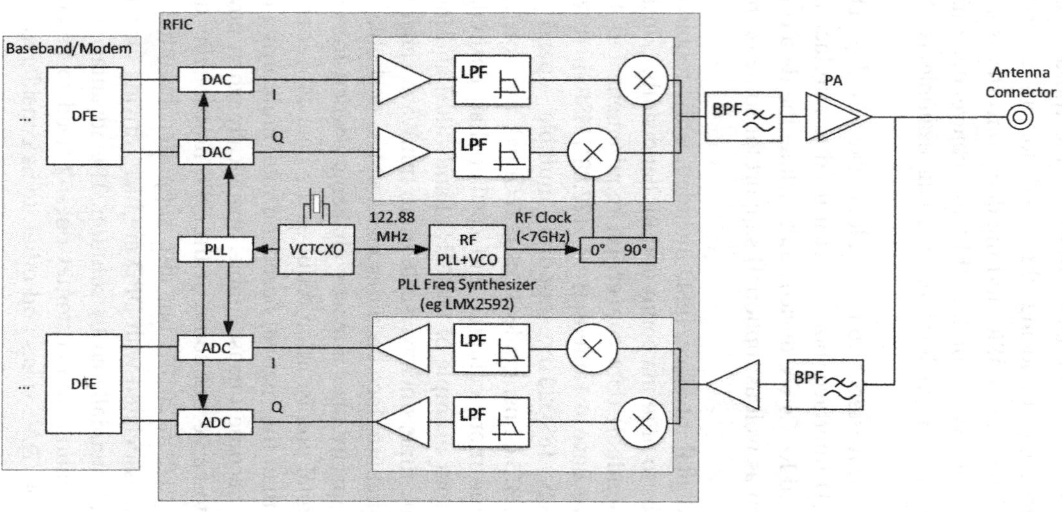

FIGURE 4–22 Digital baseband (JESD) interface between modem and RFIC.

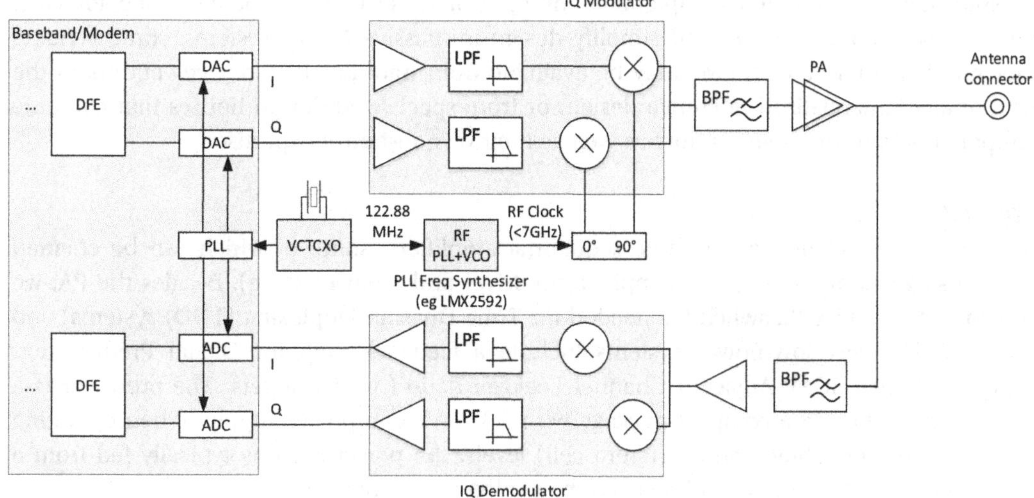

FIGURE 4–23 Analog baseband interface between modem and RFIC.

have been explained in the JESD chapter before. In all cases, the interface is an analog one with associated reduction in pin count.

An IF-based interface is commonly used in the handset world, where the associated pin-count reduction (only a single pair of wires is needed to be routed between baseband chipset and RF subsystem) is key. For the same reason, such IF interfaces can route (modulated) digital control and DC power between the baseband chipset and the RF subsystem.

In a base station the pin-count argument is not very relevant, but cost reduction associated with re-use from a handset solution can be worthwhile. Note that the IF interface is proprietary to each vendor, so expect the baseband and RF solutions to have to be provided by a single party.

In the case of a baseband I/Q interface, the RF subsystem can, if needed, be custom built for any desired band, bandwidth and other requirements. Integrated analog-only RFICs are available from multiple vendors including Analog Devices, Maxim and Texas Instruments at different price and performance targets.

Sub 6 GHz front end module

After modulation, the system needs to amplify the RF signal before it goes to the antenna. The Front-End Module (FEM, Fig. 4–24) includes transmitter, receiver, and controller components into a single subsystem that can be designed to the target use-case or bought as a premade subsystem. Given the wide range of RF bands that the final product may need to support (as we noted in Chapter 1, there is a plethora of RF options in 5G), modularity of the RF front-end can reduce design overhead by moving to a "plug and play" system where new RF bands are enabled by swapping out the FEM.

Note that MIMO systems implement multiple of these paths in parallel, one for each antenna. To reduce cost/size and simplify design in massive MIMO systems, some vendors offer multiantenna FEMs. FEMs may be available from traditional RF device vendors in the form of a commercialized reference design, or from specialized design houses that combine components from multiple vendors as well as their own in-house expertise.

Transmit

The PA transmit chain has multiple sequential amplifiers, some of which can be chained inside a single component (for example, a predriver and a power stage). Besides the PA, we find an isolator, a Tx/Rx switch (as needed for Time Division Duplexing (TDD) systems) and filters. All but very low-power systems include a feedback path for Digital Predistortion required to meet 3GPP Adjacent Channel Leakage Ratio (ACLR) targets. The predriver is a low-power solution in a compact package with 3.3- or 5-Volt power supply. When operating at transmit powers above indoor (femto cell) levels, the power stage is typically fed from a higher voltage (28 V) supply and needs to be cooled appropriately.

Receive

The Rx/Tx switch routes the antenna signal to the Low Noise Amplifier (LNA) during Transmit mode. During Receive mode, the input to the LNA is connected to ground through a 50 Ohm resistor to make sure that the LNA does not pick up any reflected signals. The LNA can be single stage or dual stage, depending on external addition LNA stages and required receiver performance. Some modules include a band-pass filter as part of the design.

Control

The FEM control block includes Transmit/Receive control logic, PA bias control, current and temperature module, and other "low speed" logic. Interfacing to the control logic is done

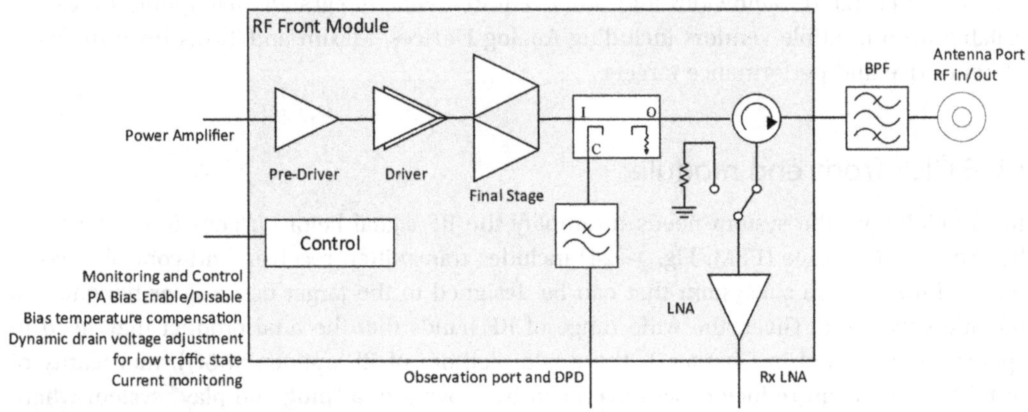

FIGURE 4–24 RF front end module (FEM).

through I2C, Serial Peripheral Interface (SPI), General Purpose Input/Output (GPIO), and similar control interfaces.

Key RF specifications include:

- The *output power* requirements map to the desired EIRP of the final product and is likely defined by FCC or other regulatory limitations.
- *ACLR* targets are a boundary set by 3GPP and other regulatory limitations on the ratio between the transmitted power on the intended channel and the unintentionally transmitted power in adjacent channels. We discuss ACLR more in-depth in discussions on DPD.
- *Power efficiency* is important as for all besides indoor deployments, RU power consumption is likely dominated by Power Amplifier PA efficiency, given that power amplifiers are typically consuming the highest amount of physical power in the system. In outdoor deployments, passive cooling is often a system requirement to be able to adhere to environmental (ingress protection against dust/water) and noise limitations.
- Given their proximity in a massive MIMO system, achieving *channel isolation* across the many antennas is a key performance metric. Channel isolation is a metric to indicate interference from one signal path onto another one—transmitter to transmitter, transmitter to receiver or between two receivers. Channel isolation maps to system level MIMO performance characteristics.
- *Receiver sensitivity* needs to be enough for the system to be capable of decoding the weakest UE signals. Minimum requirements are set by 3GPP. The biggest challenge in adhering to the requirements is keeping the amount of noise in the system lowest possible and hence picking components with the best possible Noise Figure (NF).
- *Receiver dynamic range* is related to receiver sensitivity. The system needs to be capable of receiving both the weakest signals associated with a far away user in bad conditions (sensitivity and low extreme for dynamic range) as with strong signals from a nearby user in good conditions (high extreme for dynamic range). The receiver chain (LNA/LNAs, AGC, ADC) needs to accommodate the maximum spread between these two values.
- The *NF* of a component/system is a measure of degradation of the signal as it is processed, and an indication of the quality of the product. Its value (in decibel or dB) indicates the degradation of the signal as compared to an "ideal" receiver with the same gain/bandwidth characteristics. A lower number is better. The NF of a combined system can be calculated from the NF of each component through the Friis formula:

$$NF_{system} = NF_1 + \frac{NF_2 - 1}{G_1} + \frac{NF_3 - 1}{G_1 G_2} + \dots + \frac{NF_n - 1}{G_1 G_2 \dots G_{n-1}}$$

where G represents the gain from each component in the system. Friis shows that the dominating factor is the NF of the first component in the system (NF_1) associated with the first LNA in the receiver chain.

mmWave radio frequency module

The high frequencies of the mmWave RF bands allow for more bandwidth to be assigned to communication and promises to achieve the throughput and latency targets of 5G standards. mmWave RF solutions require in-depth RF knowledge to design and build. For this reason, mmWave RF components are not only sold standalone, but (even more so than in the <6 GHz band) integrated in complete RF subsystems by specialized companies targeting a specific band, output power and beamforming characteristics. The mmWave RF module integrates filters, modulator/de-modulator, PA, LNA, TDD switch, PLL, and other components into a single optimized Printed Circuit Board or even a single chip (Fig. 4–25).

Note that in the low- to mid-end market, beamforming for mmWave is implemented in the analog and not the digital domain. This means that rather than having a single analog interface for each antenna element, there are two interfaces (horizontal and vertical polarization) that are shared across all antenna elements.

Key specifications of the mmWave RF subsystem include:

- Effective Isotropic Radiated Power (EIRP) depicts the equivalent power if the RF subsystem would be radiating in all directions accounting for beamforming gains. High EIRP numbers are enabled by increasing the amount of PAs (and thus LNAs) in the system because an increased amount of PAs not only increases the absolute number of transmitted power, but also improves the beamforming gain—the Antenna Element (AE) count effectively gets counted double towards the EIRP. EIRP obviously impacts
- EVM. Target EVM numbers are dictated by 3GPP and need to be met by the RF subsystem.
- Scan range indicates the "reach" of the beamformer both in horizontal (azimuth) and vertical (elevation) directions. A 90-degree (−45. . .45) range implies that 4 RF subsystems need to be connected to support a full 360-degree range.
- Frequency band/range. Obviously, the target frequency band and range need to be wide enough to cover the application requirements. To our knowledge, there are no single RF solutions (yet) in the market that cover the complete set of options between 24 and 39 GHz as required by 5G/NR release 15.
- Power consumption/power efficiency. These are the key metrics for mmWave RF subsystems. mmWave PAs are inefficient due to the complexity of the technology. In addition, the analog nature of the beamformer (with some exemptions) means that a feedback based DPD algorithm can't be implemented, further reducing beamforming gains. mmWave RF subsystems require high EIRP to achieve coverage.
- Beamforming precision defines the minimum width and resolution of control of the radiated beam.

mmWave Link Budget Example

An example link budget to provide insight into mmWave RF Subsystem dimensioning is given here: We consider a gNB (base station) and UE (client) system with the following characteristics:

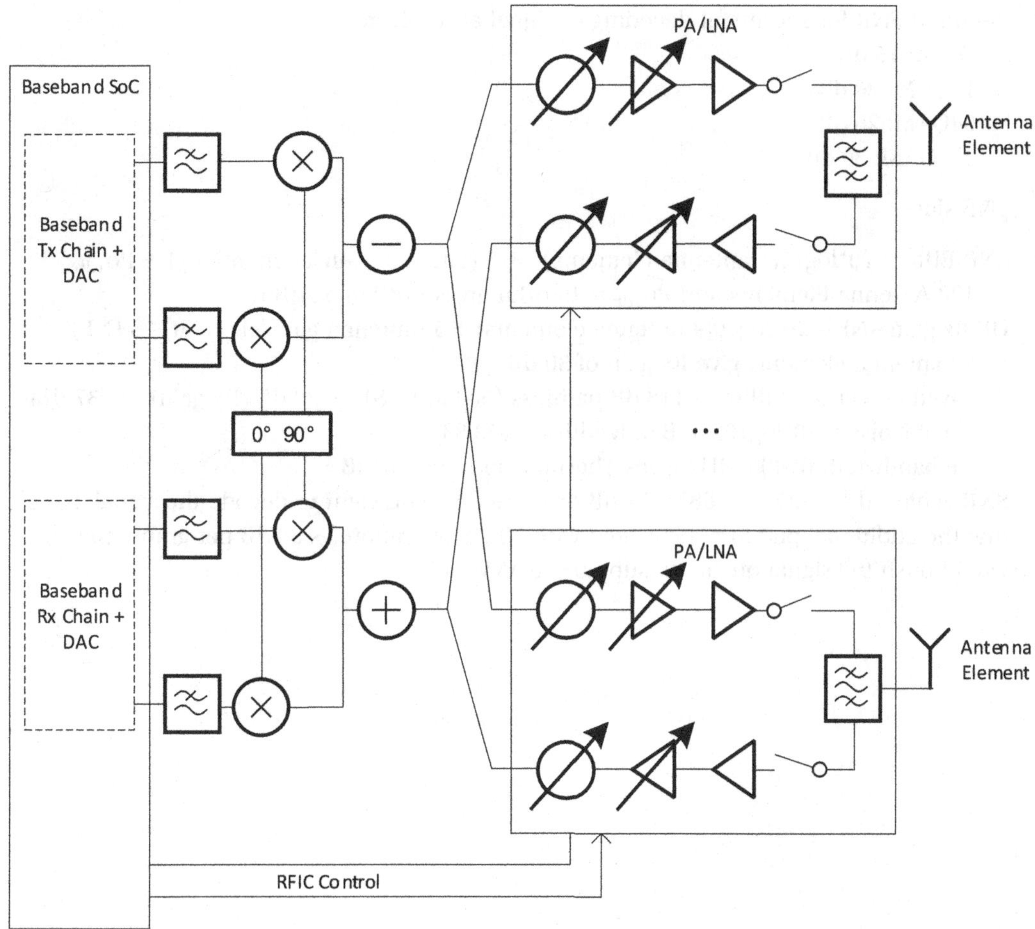

FIGURE 4–25 mmWave radio frequency (RF) system.

- RF Bandwidth = 400 MHz
- Receiver Noise Figure (gNB and UE) = 5 dB
- Number of gNB antennas is 256, antenna gain/element of 5 dB
- Number of gNB antennas is 32, antenna gain/element of 5 dB
- Transmit power/PA (Po_{1dB}) of 15 dBm (low end CMOS up to SiGe)
- Assume PAPR of 11 dB (moderate to no CFR applied)
- Assume no losses due to T/R switches, cable/connector, etc.—these can be added by the reader in a true system analysis
- 3GPP 38.803[11] defined path loss (see graph below in Fig. 4–26)
  - For UMi LOS, 500mtr: 118 dB
  - For UMi NLOS, 500mtr: 148 dB

- Required SNR for successful decoding of signal at receiver:
  - QPSK: 15 dB
  - 16QAM: 20 dB
  - 64QAM: 25 dB
  - 256QAM: 30 dB

  gNB side

- gNB EIRP = 20*log10(# antenna elements) + 5 (antenna gain/element) - 11 + Po_{1db}
  - 128 Antenna Elements and Po_{1db} = 15 dBm gives EIRP of 51 dBm
- UE Rx gain−NF = 20*log10(# antenna elements) + 5 (antenna gain/element)−5 (NF)
  - 32 antenna elements give Rx gain of 30 dB
- Received power is 51 dBm − 118 dB pathloss (500 m LOS) + 30 dB (Rx gain) = −37 dBm
- Thermal Noise = 10*log10(RF Bandwidth) − 173.83
  - RF bandwidth of 400 MHz gives Thermal Noise of −88dB
- SNR achieved = − 37 − (−88) = 51 dB or more than sufficient to decode the signal. Note that the additional pathloss associated with NLOS operation (~30 dB per graph above) would push the signal quality to support 16QAM only.

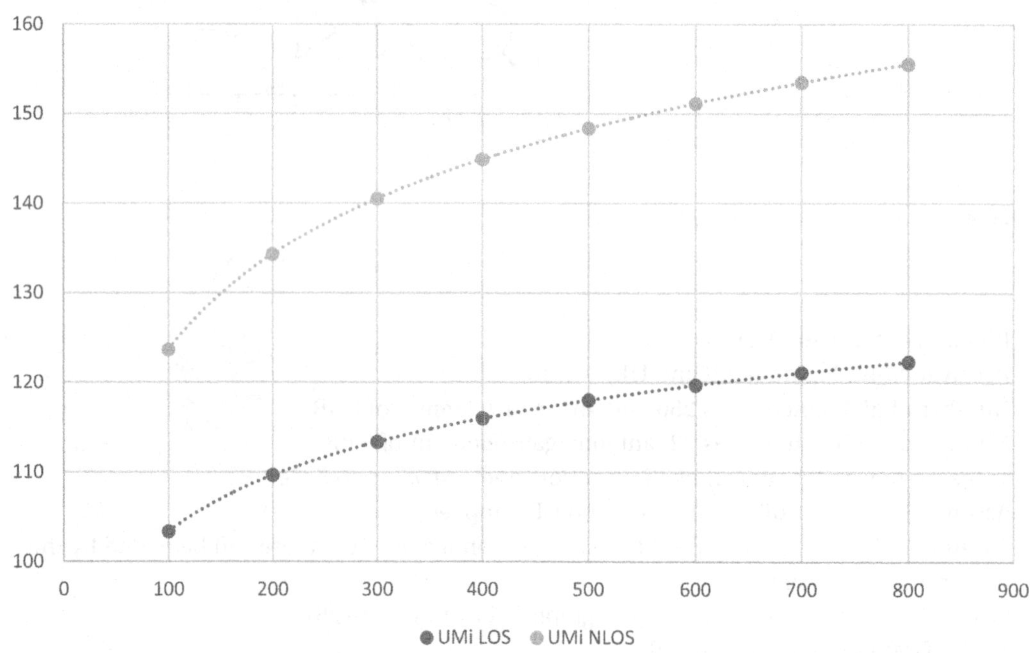

FIGURE 4–26 28 GHz pathloss per 3GPP.

Beam switch process steps

In case the system is being built from individual subsystems that are not provided by a single vendor, the topic of mmWave RFIC control and beam switching is important to cover. Beam switching involves multiple stages: Assigning the beam weight(s), beam selection, beam weight application:

Beam weight assignment involves writing a set of beam weights to RFIC internal memory. This operation is typically done through a set of SPI and/or I2C operations that to a sequence of write operations to the internal memory/register set. The internal memory will hold these beam weights to minimize the overhead associated with writing new beam weight for each beam selection iteration. Typically, this step involves writing a sequence of values (phase, amplitude, . . .) to the RFIC.

Beam selection is the step in which a specific (set of) beam weight(s) out of the RFIC internal memory is passed to the RFICs/PAs/LNAs in the system. This stage does not actually apply the beam weight but preconfigures the various components to hold a specific "new" beam. Typically, this step involves writing a single "index" value to the RFIC.

Weight application involves applying the selected beam weights as well as the settling time associated with the analog on-chip and off-chip components. We assume weight application to be done by means of a unique GPIO signal that allows for highest speed [targeting the 100–200 ns beam switching time or a fraction of the OFDM Cyclic Prefix (CP)].

Typical RF subsystems for mmWave include multiple RFICs, each supporting only 2–4 PAs. This allows the RFIC design to be simplified, whilst keeping the RF antenna close to the RFIC (minimize RF losses) and providing scalability by cascading multiple RFICs. But, each RFIC needs individual control signaling for Rx/Tx switching, as well as beam weight assignment/selection/application. Typical signaling is implemented by a combination of I2C (2 wires) or SPI (4–5 wires) for programming beam weights, combined with SPI or GPIO for Rx/Tx switching and weight application triggering. Assuming a 50 MHz SPI clock and 100 bits/weight update, each RFIC requires 2 μs programming time. This means SPI chaining cannot (or only to very limited extend) be used: the latency implication would directly impact beam weight assignment cost.

The obvious challenge with this approach is the impact on Baseband SoC I/O connectivity. The number of low-speed I/O between baseband SoC and RF subsystem can grow prohibitively large. For example, assuming a SPI controlled RFIC that implements 8x PA + LNA and a 512 AE subsystem would require ($512/8 \times 4 =$) 256 SPI I/O elements on the baseband SoC.

Typical systems solve this challenge by adding an FPGA to the system that provides the fanout by translating a single (or a few) SPI interfaces from the baseband SoC to a large amount of individual SPI connections to each RFIC. The FPGA can additionally implement the logic to calculate the appropriate individual weights to each RFIC from a single amplitude/angle output from the baseband SoC.

Integrated small cell

The Integrated Small Cell (ISC, Fig. 4–27) is typically defined as a 2–4 antenna, <6 GHz, <24 dBm/output port or relatively low output power mmWave (<35 dBm EIRP) indoor small cell. We assume the system to deploy an Option 2 or "Option 0" (i.e., full stack in Integrated Small Cell) where Physical Layer and Stack are executing in the ISC context. Optionally, other functional splits (Option 6) are possible.
Key components in the ISC include:

1. Layer 2 Stack
   a. This entails termination of the Ethernet/IPSec based fronthaul interface to the Core Network/Central Unit as well as support for the Layer 2 (MAC/RLC, PDCP/SDAP, RRC) stack
   b. Includes termination of network-based timing (IEEE1588/Synchronous Ethernet)
   c. Includes Operation and Maintenance (OAM) and related "slow path" functions.
2. Physical Layer
   a. Implements the Digital Signal Processing that converts the MAC/Transport Block information into time-domain analog I/Q that can be transmitted by the RF subsystem
   b. Typically includes Digital Front End (DFE) functions.
3. RF Subsystem
   a. Implements the analog conversion from baseband IQ to (IF and) RF and the antenna Receive (Rx) and Transmit (Tx) functions.

Notes: Even though Fig. 4–27 does not dictate a physical implementation, it closely resembles actual implementations, with key components being a Networking Processor host (hosting the L2 stack), a 5G modem (hosting the Physical Layer), and an RF subsystem. We will discuss these in the upcoming subsections. Some implementations combine the Networking Processor and Modem functions into a single chip. This integrated architecture was popular in 4G/LTE ISCs and some 5G options such as Octasic's[12]—specifically towards the lower end of the performance spectrum.

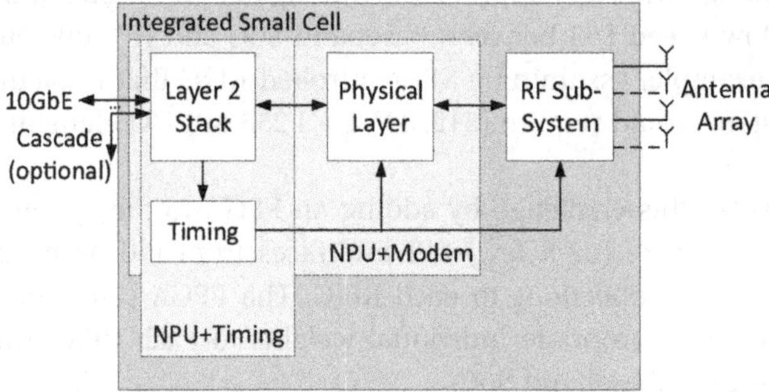

FIGURE 4–27 Key components in the integrated small cell.

Networking processor/host

The NPU is considered the central point of the architecture, providing Wide Area Network level connectivity to the CU or Core Network (e.g., 3GPP S1 interface) as well as hosting upper layer (control plane) stacks, 5G L2 U-Plane, modem drivers and application Virtual Machines. Typically defined as a custom networking centric SoC, this chip supports offloads for networking functions such as IPSec/GTP tunneling functionality, Ethernet switching/routing, and QoS. Depending on performance and cost tradeoffs, this is typically an SoC in the 5–25 W DC power range. It interfaces to the 5G modem (and/or other modems) through a PCIe interface (physically) and FAPI (logically, Application Protocol Interface or API).

If the ISC is not 5G-only but supports additional wireless air interfaces such as Wi-Fi (both 2.4 and 5/6 GHz), these are also connected over PCIe. Separation of the NPU from the 5G modem is new to the 5G era; in the 4G days, the host and modem were supported on a single device that would implement both the and Physical Layer stacks and provide embedded Ethernet connectivity as mentioned above. In a multimode 4G/5G scenario, the legacy 4G modem is connected over Ethernet (RGMII). Given Ethernet connectivity, routing of S1 frames to the 4G modem can be handled by the networking functions in the Networking Processor.

System timing is supported by Networking Processor integrated hardware timing logic that supports IEEE1588/PTP, GPS and Network Timing Protocol (NTP) based frame timing. This logic can also be used to control an external (30.72/122.88 MHz) VCTCXO that drives the 3GPP modems. Hardware support for Ethernet MAC layer or PHY timestamping is expected to be supported. Timing logic can optionally be handed in a separate FPGA or CPLD implementation, external to the Networking Processor.

Physical layer/5G modem

Given power and cost targets in the ISC, the Physical Layer is implemented as an ASIC/ASSP device that is highly optimized to the application. Most implementations follow a "Baseband on a Chip" architecture where the bulk of the Physical Layer processing is implemented using a combination of DSP cores and hardware accelerators—also known as a heterogeneous multicore integration. The hardware is proprietary and so is often the software; traditional implementations of ISC products assume the device (silicon) vendor to provide some form of a reference Physical Layer stack that is not required to be opened to the system integrator for product hardening or differentiation.

In the 4G product generation with its (never materialized) hype around Pico- and Femtocells, roughly a dozen small cell chipsets were developed by vendors such as Broadcom, Freescale (now NXP), Mindspeed (now Intel), Octasic, Qualcomm, and others. The 5G market has fewer vendors including NXP (LA1200), Octasic (OCT3032W), and Qualcomm (FSM100xx).

Radio frequency subsystem

The RF subsystem from the ISC is designed along the same criteria—if not re-using the exact same design—as that of the RU. Because of this, we refer the reader to the RU discussion for more information.

Multicore central processing unit selection criteria

All CU, DU, and RU systems we described here include a multicore processor, in one function or the other. The choice of multicore processor architecture is an important one, as it defines the software development environment that the implementation teams work with during the duration of the project, as well as any roadmap/follow-on projects that come after. Consider

- Software re-use. Software development constitutes the bulk of the development effort. Consider selecting an Instruction Set Architecture and/or processor family that is supported by 3rd party stack vendors and does not come with the penalties (cost, time) associated with porting efforts. Tooling support and developer familiarity with tools and libraries is key to quickly ramp up development. This is one of the most obvious reasons behind the success of the Linux operating system and for that reason, robust Linux support for the chosen hardware is key.
- Portfolio scalability. Well-written software scales from low-end to high-end deployments and thus allows for quick development of derivative projects. This is only relevant if scaled (up and down) products exist, or at least are presented as roadmap items by the device vendor.
- Price, power, and support are easily understood as decision points so no need to elaborate on these topics. Keep in mind that power consumption can be a make-or-break topic: active cooling (fans) can be a no-go for products deployed both inside (noise) and outside (ingress protection against moisture and dust as well as reliability concerns). Actively cooling devices consuming more than twenty Watts of power becomes costly.

In the remainder of the discussion here, we focus on the "engineering-centric" analysis of implementation options: Instruction Set Architecture and relevant benchmarks for CPU and Memory bandwidth. We assume this to be used to establish a "short list" of device candidates that can be further evaluated to the soft metrics above.

Instruction set options used to be multiple: Arm, MIPS, PowerPC/Power Architecture, x86, and other architectures have been used in embedded and server products alike. Nowadays, the bulk of new software development is done targeting Arm or x86 architectures, with RISC-V being the "hot newcomer" in the processor architecture market.

Single instruction multiple data

Especially relevant for the support of Digital Signal Processing algorithms is the support for a high-performance SIMD Instruction Set. Single Instruction Multiple Data (SIMD, Fig. 4–28) is

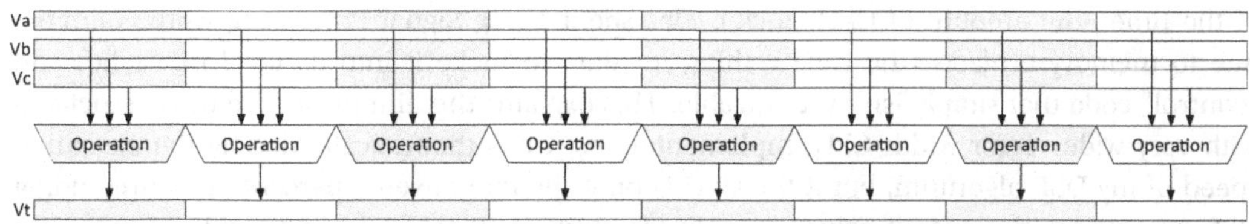

FIGURE 4–28 Single instruction multiple data processing concept.

a class of parallel computing that describes computers with multiple processing elements that perform the same operation on multiple data points simultaneously, exploiting data parallelism rather than concurrency. SIMD is particularly useful in DSP algorithms and included in most modern CPU architectures including ARM Cortex A-series, Power Architecture, and x86.

Operations in a SIMD pipeline are done on a fixed-length vector—typically 128, 256, or 512 bits wide, comprising of multiple data elements that are subject to the same operation. The element data types can vary but are typically 8-, 16- or 32-bit fixed point or 16- (half precision) or 32-bit IEEE single-precision Floating Point. Most implementations keep a separate vector register file.

Typical SIMD instruction sets are optimized for DSP with Intra and interelement arithmetic instructions, Intra and interelement conditional instructions, and Permute, Shift and Rotate, Splat, Pack/Unpack, and Merge instructions. Given that performance is key, most instructions are fully pipelined to support a single-cycle throughput.

Like for any other DSP, SIMD programming can become a specialized task. Support for SIMD can be achieved in multiple ways:

- Let Compiler do the job. Auto-vectorization is technically supported since many years. However, we will claim that there are no practical auto-vectorizers in existence that achieve the performance targets that are required. Even if auto-vectorization becomes a real thing, the software will still need to be written in a "compiler friendly way": they will still require certain expertise to work with.
- Using C intrinsics. There are standardized lists of intrinsics supported in all SIMD enabled compilers. Intrinsics can guarantee good level of control, up to the level of register assignment if need be, although generated code often only gets close to Assembly programming
- Using Assembly Language programming, which is the most effective, and the most laborious way. We can provide 100% performance extraction at the cost of software efforts.

In practice, C intrinsics is often the "way to go," striking a balance between effort spent and performance gained. Lucky enough, through manufacturer efforts, there are software libraries available not only for generic SIMD tasks but specifically for O-RAN and 4G/5G processing, as provided by Arm, Intel, and others.

It is important to keep in mind that when it comes to SIMD processing, "bigger is not always better." In these applications, the processor is not busy with "number crunching" all of the time—the amount of CPU clock cycles spent doing Signal Processing work is limited due to memory artifacts (the CPU waiting for data to make it into its local L1 cache) and "control" code that simply isn't vectorizable. This explains the diminishing returns associated with very wide vector width CPU implementations: I can theoretically double the execution speed of my DSP algorithm, but if the CPU is only spending twenty percent of its time doing DSP processing, the overall gains are only ten percent. Increased vector widths come at a cost of CPU frequency: Intel CPUs reduce the turbo boost frequency of their cores when the instruction pipeline is heavy in wide-vector SIMD processing.

x86

Looking at O-RAN ecosystems, we find that x86—or more specifically, Intel—is an obvious candidate, given the implied (but as we noted, not explicit) software-defined, COTS nature of O-RAN products. An x86 based reference design named FlexRAN[13] includes a reference implementation of a 4G/5G Physical layer, which is an obvious starting point for system design.

SIMD support is present in all x86 processors since the introduction of MMX (64b)/SSE (128b) by Intel, and 3DNow! by AMD in the late 90s. SSE evolved into SSE-2, SSE-3 and SSE-4 in the 2000s, increasing the supported instruction count and introducing more data types (e.g., double precision Floating Point).

Intel Sandy Bridge and AMD Bulldozer family of processors supported 256b Advanced Vector Extensions (AVX) providing additional vector with an even richer instruction set. AVX evolved into AVX2 in the mid-2010s and AVX512 in the late 2010s, first supported by the Knights Landing processor by Intel.

Arm

ARM A-series (just like Power architecture devices) provides high-performance and high-efficiency hardware support for Floating-Point (FP) and SIMD operations. The ARM A72 core is the "de facto" baseline benchmark core, from which scaling up and down to higher and lower core and frequency classes is done. It supports 128-bit SIMD using what is called the ASIMD (or often, NEON) unit, shown in the execution pipeline below Fig. 4–29:

Native Floating-Point (FP) and NEON/ASIMD instructions are executed in the same execution units running at core speed of about 2 GHz in the 16 nm or more advanced technology nodes.

Catching up to Intel efforts, Arm is extending the vector processing capabilities with implementation options for vector lengths that scale from 128 to 2048 bits in the Scalable Vector Extension (SVE) and SVE2 definitions. Rather than specifying a specific vector length, the SVE architecture scales the vector width between 128 and 2048 bits per register, with a Vector-Length Agnostic (VLA) programming model that allows a single piece of software to execute over the range of supported vector widths. The idea is that code no longer needs to be re-written as the processor implementation changes and this obviously reduces deployment cost and time-to-market.

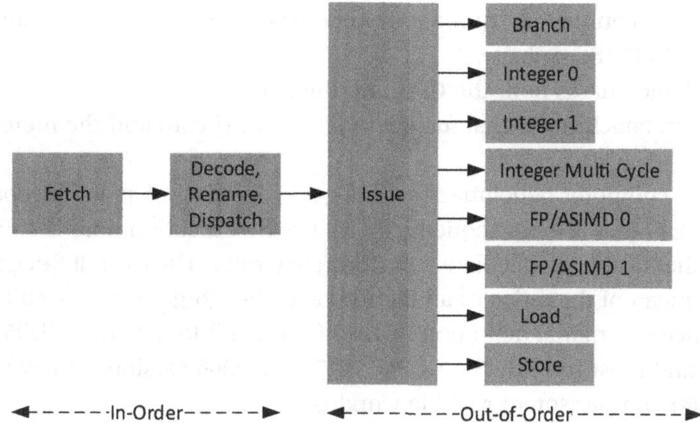

FIGURE 4–29 Arm A72 processing pipeline.

In order to enable the O-RAN ecosystem, Arm has released a set of reference libraries[14] for Physical layer processing that are optimized to ASIMD/NEON.

Performance benchmarking

Choosing among processors involves benchmarking to define which device provides the minimum processing power estimated to be used for the application, with some safety margin added. Performance benchmarking can be done using synthetic benchmarks to compare devices in generic applications, or application specific benchmarks that are unique to our O-RAN use-case:

- Synthetic core benchmarks
  - Dhrystone is a synthetic CPU benchmark developed in the mid-1990s to be a representation of CPU integer performance. Performance is calculated as the number of iterations/second of the main loop of the executed code. Its performance reports are not standardized but most often, performance is reported in "DMIPS" which is the amount of iterations/second multiplied by the core frequency (and the number of CPUs in the system). Dhrystone is considered as outdated (replaced by CoreMark) but is still widely used in the industry as a first indication of performance. Reasons for Dhrystone being considered outdated include its inclusion of library calls in the benchmark (making performance comparisons more difficult), its potential for compiler optimization (where clever compilers are tuned for delivery of Dhrystone performance) and the fact that (due to it's small footprint) it fits inside the processor L1 cache and hence does not benchmark any of the memory subsystem.

Dhrystone benchmarks both the CPU and the compiler.

  - CoreMark was developed in 2009 by EEMBC with the intention to replace the aging Dhrystone benchmark. It is fully written in C and contains several core centric low-level algorithms including list processing, CRC calculation, and so on. CoreMark

performance is counted as "number of iterations per second" of execution and published on EEMBC website.
CoreMark benchmarks both the CPU and the compiler.

- Synthetic system benchmarks that include both the CPU core and the memory subsystem.
  - SPECint is a computer benchmark for Integer processing of server payloads. It is a set of benchmarks that are run sequentially where the execution time is compared to a reference that is able to calculate a performance ratio. The overall SPECint score is a geometric mean of the ratios of all the reference benchmarks in the suite. SPECint has evolved since its original definition in 1992 ("CPU92") to include CPU95, CPU2000, CPU2006, and most recently SPECCPU_2017. The 2006 version is most used as a reference for comparison of real-life workloads.
  SPECint benchmarks both the CPU and memory subsystem as well as the compiler.
- Application specific benchmarking is the obvious best method to predict the performance of the system—it literally is a benchmark of the actual application. The challenge with application specific benchmark is its implementation time. The choice of (hardware) architecture is done before the software teams engage making an application specific benchmark an unrealistic luxury. Consider an intermediate solution of benchmarking a "proof point" application as we've explained in Chapter 10, Interoperability and Test, as a performance reference if it's possible to capture key performance elements as a part of this benchmark.

Memory and I/O dimensioning

Even before the application is implemented, we can estimate the required memory throughput by analyzing the packet flow through the system as we've shown in the performance chapter of the system. The theoretical performance number needs to be matched to the achievable memory throughput of the target devices used to make sure that the memory interface does not become a system bottleneck.

- STREAM is a synthetic low-level benchmark designed to measure sustainable memory bandwidth (in MB/s) for four kernels (Copy, Scale, Add, and Triad) with freely available source code.

STREAM benchmarks the memory subsystem of a system.

- LMbench is a series of synthetic low-level benchmarks designed to measure bandwidths for eight application use-cases (file read/write, socket IO, and memory IO) where the memory read and write benchmarks are the most reported ones. LMbench includes latency measurements.

LMBench benchmarks the memory subsystem (internal caches and external DDR) of a system.

Hardware offload

Some processor vendors offer support for hardware offload of compute intensive components of the processing stack. Networking-centric offloads often include:

- Acceleration of ciphering algorithms. This includes both IPSec acceleration as well as ZUC/SNOW/AES 3GPP f8/f9 encryption and authentication. Ciphering can be supported through optimized ISA support where modern Arm and x86 architectures both support dedicated AES instructions (but no ZUC/SNOW) or through a look-aside accelerator that performs these operations autonomously once its drivers have been integrated into the application. Some processors offload not only the cipher algorithms but in fact the complete (IPSec, networking) protocol stack through an embedded Network Processor that is customized per vendor and application. This offload provides best possible performance per Watt but comes at the cost of integration efforts and is unique to each vendor. This vendor lock-in may be undesirable.
- Acceleration of Networking Quality of Service. Software implementations of (wireline) Quality of Service algorithms such as Ethernet policing and shaping are notoriously CPU intensive and limited in their capabilities. Most Network Interface Cards or on-chip integrated Ethernet controllers include hardware capabilities for some QoS offload. Not all are easily mapped to 3GPP or application use-case requirements.
- Artificial Intelligence (AI), Machine Learning (ML) and other math-centric offload. The 5G application space is not only about implementing a wireless connection but about intelligent use of user data. Consider location-based services, low-latency edge compute and similar applications that fashionably use AI/ML. If the edge application is integrated with the networking processor (stack host), it requires efficient support for these applications, which implies access to (or integration of) AI/ML capabilities.

PCIe performance

Given that PCIe is a key interface used in O-RAN and so many other embedded applications, we will discuss this standard and related performance topics here.

Different versions of the PCIe standard define the baud rate and code rate of the SerDes interface that is used as follows: (Table 4–1)

Table 4–1 PCIe per-lane performance.

Generation	Baud rate (GTPS)/lane	Code rate	Net rate
1.0a, 1.1	2.5	8/10	2
2.0, 2.1	5	8/10	4
3.0, 3.1	8	128/130	7.88
4.0	16	128/130	15.75
5.0	32	128/130	31.51

All above numbers are for a single lane. By deploying multiple lanes ("PCIe gen 3 × 4" would indicate four-lane operation), rates can be scaled by an integer amount. As is shown in the table, the code rate reduces the actual number of bits transferred over the wire by about 20% (PCI gen 1 & 2) or approximately 2% (PCI gen 3, 4, & 5). Additional overhead is lost to packetization and protocol overhead.

Transaction layer packetization

Originator-to-target bits aren't just transferred over the SerDes link but wrapped into a protocol layer that allows the PCIe interface that provides the functionality required present itself as a reliable memory mapped entity to the user (SoC). Transaction Layer Packets (TLPs) are used for this encapsulation, for example shown for a PCIe gen 3 interface (Fig. 4–30):

The TLP header indicates the transaction type (e.g., Memory Write Request), payload length, target address, etc. Payload is optional—for example, not needed if the TLP header indicates a Memory Read Request. In the case of the Memory Read Request, the first TLP indicates from Requester to Completer what the requested address and size (length) are. The Completer responds back with a Completion TLP that contains the desired payload (memory content).

Note that this is a significant difference between write and read operations, write operations are "posted" or fire-and-forget; once the packet is formed, it can be ignored. A read operation needs the requestor to wait for completion. This is also called a nonposted operation.

TLP overhead is in the order of 20–30 bytes, depending on PCIe generation, optional Cyclic Redundancy Check (CRC), and so on. See the PCIe standards documents for more details.

The Maximum Payload Size (MPS) according to the standard is 4096 bytes as shown above. For (size/cost) efficiency reasons, SoCs rarely support this MPS. 256B would be a typical implementation choice. Assuming a 30B TLP overhead and a 256B payload, efficiency is reduced to 256/(256 + 30) = 90% of Net Rate in an ideal case. Larger transactions than the MPS are broken down into multiple smaller (MPS or less) sized ones. To provide system level efficiency when multiple devices are connected to a bus, Memory Read Requests can be configured to have a Maximum Read Request size. Note that systems that have a payload/packet transfer size smaller than the MPS will see degraded performance given that the various TLP, ACK/NAK, and overheads exist per transaction and hence need to be accounted for independent of the size of the transaction.

For every TLP that is sent, an acknowledgment (ACK, or negative acknowledgment, NAK) Data Link Layer Packet is expected back from the receiver. The TLP is kept in transmitter

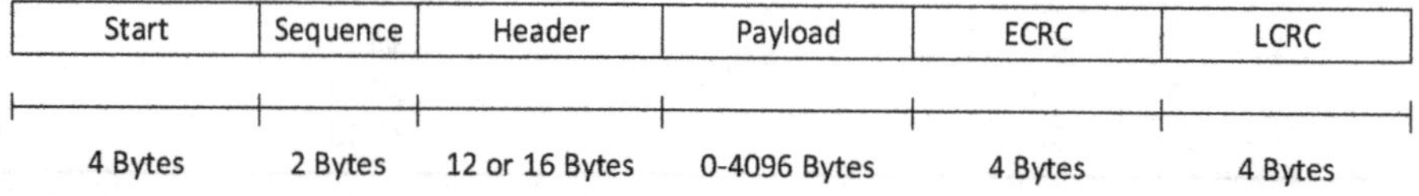

FIGURE 4–30 PCIe gen3 transaction layer packets (TLP).

memory until retransmitted and successfully acknowledged. This provides for the reliability of PCIe. Reduction of traffic amount on the link is done by collapsing acknowledgments—send a single acknowledgment for multiple TLPs.

To provide for better alignment to natural (for example, 64 bytes) boundaries, the Read Completion Boundary concept allows the data TLPs that return to a read request to divide the payload into chunks that are below the MPS.

Example calculation

The various overheads and packetization options in PCIe impact the achievable performance. For example, take a PCIe Gen3.0 × 1 (single lane) system. The transfer rate for Gen3.0 was set to be 8 Gb/s to double the effective bandwidth from Gen2.0 rate of 5 Gb/s. Also, 128B/130B encoding was used (for Gen3) over the 8B/10B (Gen1, Gen2) to reduce the loss characteristic from 20% (8B/10B) to 2% and a net rate of 7.88 Gbps (see above).
Adding TLP overheads: 32 Bytes

- TLP overhead including ECRC: 20 Bytes
- Link layer overheads: 12 Bytes
  - DL-HDR sequence number in memory read/write request: 2B
  - Link layer CRC added to memory read/write request: 4B
  - Link layer ACK from the other end: 6B

The overall efficiency can now be calculated as:

$$\text{Packet efficiency} = \frac{\text{Maximum Payload Size}}{\text{Maximum Payload Size} + \text{Overheads}} = \frac{256\text{B}}{256\text{B} + 32\text{B}} = 0.88$$

The theoretical maximum data throughput is the packet efficiency as a percentage of the Theoretical Bandwidth calculated above: Theoretical maximum data throughput = 0.88 × 7.88 Gbps = 6.95 Gbps.

In practice, we account for approximately 80% of theoretical PCIe performance to be achievable on a well-implemented practical system. Note that PCIe is a bi-directional interface, the throughput targets can be achieved in both directions simultaneously.

References

1. Broadcom. *Monterey/BCM56670 Series.* <https://www.broadcom.com/products/ethernet-connectivity/switching/strataxgs/bcm56670>.
2. Open Data Center Committee *OTII Server for 5G and Edge computing.* **2019**. <http://www.opendatacenter.cn/work-group/p-1143085492931629057.html>.
3. NXP. *B4860: QorIQ® B4860 Baseband Processor.* <https://www.nxp.com/products/processors-and-microcontrollers/power-architecture/qoriq-communication-processors/qonverge-platform/qoriq-b4860-baseband-processor:B4860>.

4. Frumusanu, A. *Marvell Announces OCTEON Fusion and OCTEON TX2 5G Infrastructure Processors.* **2020.** <https://www.anandtech.com/show/15572/marvell-announces-octeon-fusion-and-octeon-tx2-5g-infrastructure-processors>.

5. NXP. *Layerscape® Access LA1200 Programmable Baseband Processor.* <https://www.nxp.com/products/processors-and-microcontrollers/arm-processors/layerscape-multicore-processors/layerscape-access-la1200-programmable-baseband-processor:LA1200>.

6. 3GPP TS 36.104; Evolved Universal Terrestrial Radio Access (E-UTRA); Base Station (BS) radio transmission and reception.

7. 3GPP TS 38.104; 5G; NR; Base Station (BS) radio transmission and reception.

8. Andersen, O.; Björsell, N.; Keskitalo, N. *A Test-Bed Designed to Utilize Zhu's General Sampling Theorem to Characterize Power Amplifiers*, I2MTC 2009 - International Instrumentation and Measurement Technology Conference, Singapore, 5–7 May **2009**.

9. Zhu, Y.-M. Generalized sampling theorem. *IEEE Transactions on Circuits and Systems II: Analog and Digital Signal Processing* **1992,** *39*, 587–588.

10. Analog Devices. *AD9375 Digital Pre-Distortion.* <https://www.analog.com/en/applications/technology/sdr-radioverse-pavilion-home/wideband-transceivers/digital-pre-distortion.html>.

11. 3GPP TR 38.803; Study on new radio access technology: Radio Frequency (RF) and co-existence aspects.

12. Octasic. *OCT3032W.* <http://www.octasic.com/product/oct3032w/>.

13. <https://github.com/intel/FlexRAN>.

14. Arm Developer. *Arm RAN Acceleration Library.* <https://developer.arm.com/solutions/infrastructure/developer-resources/5g/ran>.

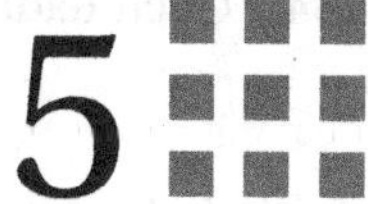

System software

We try to identify the minimum required Linux features that are needed to constitute a "Distributed Unit viable image," whilst keeping in mind that the actual features depend per instantiation—it is difficult to do a broad sweep requirements definition.

Operating system

Bare metal/RTOS versus embedded Linux

Before we delve into details on the Linux Operating System (OS), let's quickly evaluate what the options for device and (partially by implication) OS are. When building an embedded system, we have two options for choice of target device: design on a microcontroller—with a traditional Digital Signal Processor (DSP) as an example of a microcontroller—or design on a microprocessor. The difference between the two (Table 5–1) is that a microprocessor is defined to run a higher order OS (like Linux) that is characterized by a larger size image and near infinite (software) capacity. A microcontroller, on the other hand, typically executes OS and application image from an embedded (and size-bound) memory with a small footprint and a simple boot process. Microcontroller software is easier to comprehend with a smaller team (single person) of engineers but proprietary in nature and more complicated to expand in features and to maintain long-term. The advantage of a microprocessor and a Linux based system is that the OS is mature, long term maintainable and portable to future (roadmap) devices. This is a key reason to design on a microprocessor, and in fact one of the main justifications of moving to an O-RAN based system: making the 3GPP protocol stack into "just another application running on Linux" allows it to benefit from standard microprocessor roadmaps rather than custom chip designs from a few specialized (and expensive) vendors.

Over the last few decades, Linux has been ported to a plethora of processor architectures, including Power Architecture, Microprocessor without Interlocked Pipelined Stages (MIPS), Arm, RISC-V and obviously x86. These days, most support is centered on the latter three: Arm, RISC-V and x86.

Given that 3rd Generation Partnership Project (3GPP) stack application complexity and size are continuously grown (as we showed in the O-RAN overview chapter), only very few commercial base station implementations rely on a microcontroller/RTOS environment to run anything outside of the (low) Physical Layer (and even that is optional) stacks. The trend clearly is towards abstraction of software into portability, re-use and standardization, as supported by Linux.

Open Radio Access Network (O-RAN) Systems Architecture and Design. DOI: https://doi.org/10.1016/B978-0-323-91923-4.00010-0

Table 5–1 Microcontroller and microprocessor characteristics.

	Microcontroller	Microprocessor
Operating system	Bare metal or RTOS	Higher order OS such as Linux
Storage for Operating System and Application	On-chip	External (flash) memory
Complexity	Simple	Complex
Application space	Bounded	Almost any
Footprint	Small, limited external components for minimum operation	Large
Cost	Lower, both component cost and system cost including external components	Higher

Boot process and application load

A microcontroller boots using a relatively simple process where the boot image is stored in a known location on (internal) flash memory and during boot, the chip points the (primary) processor Program Counter (PC) to the location of this image. This is relatively simple and straightforward. In a microprocessor with an OS image that is stored in external (flash or drive) memory, the boot process becomes more complex. The initial boot from power-on is done from an on-chip Read Only Memory (ROM) that initializes key peripherals (say, I2C, PCIe, or other). The "real" OS is stored on a memory peripheral that is connected over one of these peripherals and is loaded from there. The second stage is loading the bootloader (Universal Bootloader aka U-Boot or UEFI being commonly used standards) which both creates the Ramdisk (that will hold the root file system) and loads the Linux kernel. From there, the Linux kernel boot take over.

In case of an x86 (Intel) central processing unit (CPU), the boot process starts from a Baseboard Management Controller (BMC—potentially part of the Platform Control Hub or PCH) which configures the main CPU for it to be able to come out of reset and load its image bootloader from an eSPI connected flash memory, in addition to monitoring hardware state (temperature, power supply voltages, remote management access, and so on). From there onwards, the boot process is identical to that of a microprocessor. BMCs can enable advanced features such as a dedicated management Ethernet interface that is intended not to be connected to the Internet.

As a side note: keep in mind that flash storage is limited in the number of times it can be written on—typically somewhere in the range of 10,000 s of times. Using flash storage for continued tracing/logging (as opposed to dumping hopefully rare crash reports) is not recommended for this reason with[1] being a well-known example in the automotive embedded world of "how not to do things".

Prevention of unauthorized (e.g., modified) firmware from running to eliminate threats from (1) loss of functionality (unauthorized re-purposing of the hardware for a different use), (2) loss of end-user data (e.g., providing third party access to unciphered end-user data), and (3) loss of uniqueness [protection of Intellectual Property Rights (IPR)] are supported by secure boot.

The concept of secure boot implies that during the boot process (which we learned has multiple boot stages for a modern OS such as Linux), each image verifies the validity of the next image that is loaded and executed. There are multiple varieties of secure boot, defined by chip or IP manufacturers such as Intel and Arm. The underlying approaches, though, are mostly common and work on the assumption that none of the images, that are being loaded (including the bootloader itself), can be trusted. To do so, the boot process relies on digital signatures (Fig. 5–1). A digital signature centers around having two keys. A private key that is only known to the originator (author) of the source and used to "sign" a message (or a software image). The signing process generates a unique value ("signature") associated with the message that is appended to it. The public key (that, as the name indicates can be made public) can be used to validate the correctness of the image by comparison to the signature value. It is important to keep in mind that the public key can't be used to *generate* the signature, but only to *validate* it. The combination of private and public key allows us to validate the origins of the image and to ensure that it was not modified—otherwise the signature would no longer be matching.

During the manufacturing process, the public key of the first image (say, the bootloader) is stored (directly or indirectly) in a one-time programmable (fuse) memory of the device. The initial boot code itself is also stored in memory that can't be modified (ROM). This ensures that even the initial boot process can't be modified.

Long Term Support

Commonality between a commercial (consumer PC) grade Linux and an embedded version is the advantage of being able to leverage a huge team of developers whose efforts are shared with many other applications. At the same time, a consumer Linux image is not intended for being used for many years in the field or supporting long term support (LTS) for security patches for a product deployed in the field a long time ago.

Whilst kernel.org maintains a strict regime on accepting functional and driver software components, there is a huge and fragmented ecosystem maintained by many software vendors supporting Linux distributions with features that are tuned to various markets. Consider

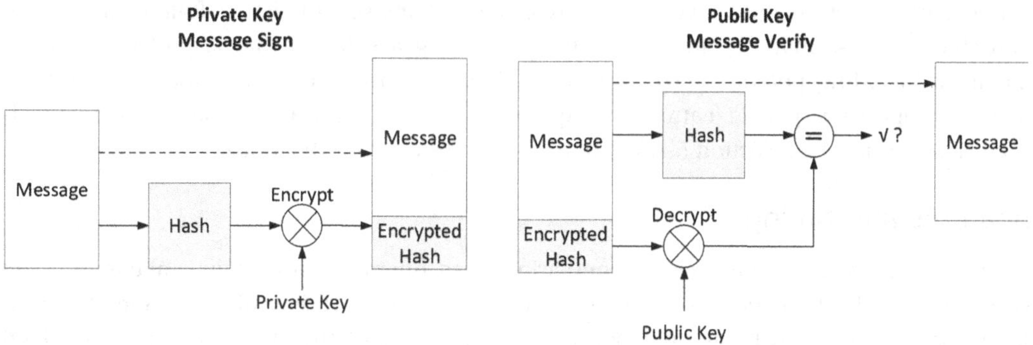

FIGURE 5–1 Private key signing and public key verification.

well-known vendors like Debian, Fedora, Red Hat, Ubuntu, and Wind River, as well as the Linux Board Support Packages (BSPs) that are delivered by most of the silicon/device vendors as a kick-start for development. Optimized low-level drivers are in fact often only delivered through these vendor BSPs or a small set of commercial Linux products.

As a result, many system integrators choose to build and maintain their own Linux distribution in which they can pick features as needed for their specific market. This choice of Linux kernel version is made in early development phases and rarely updated ("frozen branch").

The concept of LTS releases of Linux is brought in to remedy this challenge by tagging specific releases (versions) of LTS and support for a longer period than standard editions. Linux kernel.org versions v2.6.16 and v.2.6.27 have been unofficially supported as LTS before it was established in 2011 as a formal initiative. v4.4 is supported for 6 years (or longer, up to 10 years Super Long-Term Support). Commercial releases such as Ubuntu support 5 years LTS with a release cadence of 2 years which is typically enough to bring a product to market without having to switch OS releases halfway development or initial deployment.

Roll your own versus commercial grade

Roll Your Own (RYO) Linux means that you build your own Linux distribution from source/scratch including maintenance. This includes optional selection of which desktop [Graphical User Interface (GUI)] and software packages (applications and/or services) are to be included in the file system, allowing to tune the Linux image to the application. Projects such as Yocto help developers in the process of building (embedded) Linux systems for many hardware architectures and are supported by hardware vendors as an "out of the box" OS that is supplied with evaluation boards and commercial hardware.

A Linux distribution (or "distro") is made from a software collection that is based upon a Linux kernel and a package management system to define the software packages to be installed. Distros can be free-of-charge or commercially backed and are available tuned for a variety of systems—from embedded devices (OpenWrt is widely used in Wi-Fi routers) to personal computers (e.g., Ubuntu). A typical Linux distribution includes the kernel, tools and libraries, a desktop environment, and so on. It also comes with the piece of mind of having a support infrastructure and a level of security given that any security issues found will be fixed and shared across many customers, as well as having a solution that is protected against liabilities including potential IPR protection. Some commercial grade Linux distributions include support for specific features as required in telecommunication systems such as real-time (RT) support (see Section 5.1.5) and networking stack optimizations.

Realtime and timing

As we explained, Linux has gained popularity as an operating system in embedded (even consumer) applications because of its easy adaptability to a variety of processors and customization—enabled by being open-source in nature and potentially free of cost. Embedded versions of Linux are often stripped-down for unneeded features including the GUI and application support. This compactness and optimization of features to the application is what defines the embedded implementation of Linux. By implication of being small, compute requirements and implementation cost (say, disk size) are contained (but not necessarily small as they would be in a microcontroller environment). As Linux is gaining popularity not only in server and consumer PC application but also in an embedded environment, we observed that SIMD capabilities give modern General-Purpose cores reasonable "number crunching" capabilities as required to implement L1 processing. These two factors (Linux becoming an embedded capable OS and Moore's law supporting general-purpose processors becoming capable DSPs) define the popularity of Software Defined Networking in general and O-RAN in particular.

However, there is a challenge on how to support the latency and OS overhead challenges associated with using standard Linux as a development and deployment framework, replacing a typical DSP Implementation that heavily relies on custom RTOS environments designed to be low latency. In fact, there are several challenges to be solved:

Challenge 1: Operating System response time for real time applications

3GPP protocol implementation requires deterministic timing to slot and symbol boundaries, whereas Linux has been built to be a general-purpose operating system, not optimized for real time. There are a few different ways of tackling the OS latency topic in real time scheduling environments:

- Cokernel approaches (such as Xenomai) which are an alternative to PREEMPT_RT and allow the system a more realtime behavior. In different implementation-specific ways, there approaches combine a "pico kernel" [also known as cokernel, pico-kernel, nano-kernel or dual kernel (dk)] with the normal Linux kernel, operating between or in parallel with Linux and the hardware. Interrupts are intercepted by this pico kernel and directed to a real-time environment that is executing the latency critical applications. In the example of Xenomai, the pico kernel is called Cobalt, a small real-time infrastructure which schedules time-critical activities independently from the main kernel logic and communicates with remainder Linux processes through Portable Operating System Interface (POSIX) Application Programming Interface (APIs). See[2] for more information.
- PREEMPT_RT adds predictable latency to execution of a standard Linux environment by converting Linux into a fully preemptible kernel. But the latency bound is very high, in the order of magnitude 10 μs worst case latency between interrupt and corresponding user-space process execution time. Compared to an RTOS as deployed on a DSP (order of magnitude 1 μs response time to an event), this can be a prohibiting execution of a typical DSP like Layer 1 framework.
- A third alternative is to use Linux as an OS, but don't rely on any OS services directly by (for example) implementing all relevant drivers in a user-space environment as opposed to relying on the OS to provide these services for you. This concept relies on a few provisions that Linux makes
  - the ability to affine a core to a task.

- the ability to set the priority of a User Space task to a level that prohibits other tasks (Kernel or User Space) from interrupting it.
- the ability to steer interrupts (from timers, IO, . . .) to a (sub)set of available cores.
- the ability to provide access to the relevant portion of the system memory map to be able to support custom driver development.

Now, the application developer can design a proprietary and application-optimized "micro kernel" OS to schedule different tasks within this application. In a way, it can be looked at as running an RTOS as a (high priority, 100% active) User Space application. Developed either custom or as a general-market product, several of "micro kernel" environments exist today, either as standalone development environments (example[3]) or as part of a commercially available LTE/5G Layer-2 stack where the stack and the OS are effectively combined into a single image.

Combined, these three capabilities allow full "ownership" over CPU resources to a defined application, whilst continuing operation in an Symmetric Multi Processing (SMP) Linux OS and benefiting from its features, including development environment, trace/log capabilities and other OS services. Containerization (Docker) and Virtualization (KVM) can still be supported, including "nested" PREEMPT_RT operation, but typically with performance impact. It is important to understand which applications benefit from virtualization (typically, higher-level applications) and which ones are better suited for nonvirtualized deployment (typically, applications that have performance criteria that do not allow for scaling benefits in virtualization).

The second and last strategies (use Linux with preemption capabilities and build a custom application scheduler on top) are typically preferred by application stack vendors to promote software portability across platforms given that they rely only in limited fashion on the OS infrastructure features.

To support improved preemption performance (quantified as shown in Fig. 5–2), the Linux CONFIG_PREEMPT_RT patch set is applied. This patch allows nearly all the kernel to

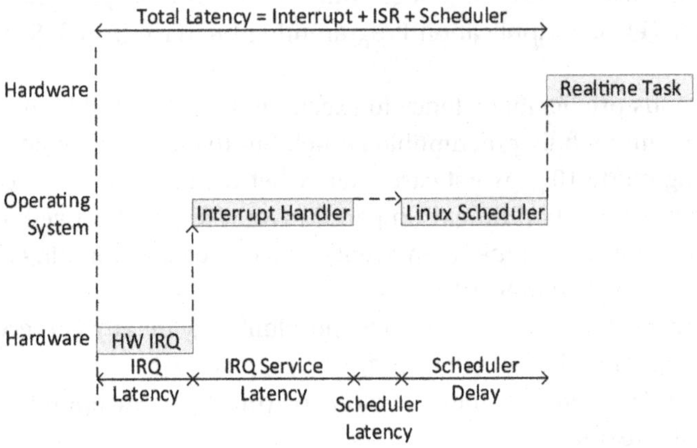

FIGURE 5–2 Operating system (OS) latency for interrupt handling.

be preempted, except for a few very small regions of code ("raw_spinlock critical regions"). This is done by replacing most kernel spinlocks with mutexes that support priority inheritance, as well as moving all interrupt and software interrupts to kernel threads.

Note that O-RAN WG6 defines the worst case benchmarked (with cyclic test,[4] the industry standard benchmark) PREEMPT_RT wakeup time to be better than 20 μs (with proposals to bring this down to 10 μs) for the DU application. This includes hypervisor overhead in the case of Virtual Machine deployment. In the CU application, PREEMPT_RT is optional, with proposals to define a maximum response time of 20 μs.

The wireless system often (such as in 3GPP systems) needs to be both frequency- and time-synchronous across physical units, which implies a requirement for a board level clock that is kept synchronous across the network. Time and frequency alignment are achieved through a combination of Synchronous Ethernet, GPS and/or IEEE1588 (Precision Time Protocol (PTP), Fig. 5–3) or similar protocols, or potentially from a (different) 3GPP network for which timing information is decoded over-the-air. Most 3GPP systems incorporate a board-level oscillator (e.g., running at 122.88 MHz) as well as a Pulse Per Second source to perform both time and frequency synchronization functions. Time and frequency synchronization details are covered in a separate chapter.

To allow the application software to function independent of any hardware timer functionality and to provide good (and bounded) application software performance, user plane modules have been designed to use a software timer which is based on a slot or frame timer, which is incremented once per frame or slot. Software timers can be implemented in many ways, for example, as an array of linked lists (Fig. 5–4), where each entry in the array represents a frame time interval, and each entry contains a linked list of nodes. Reference[5] provides an overview of timing wheel functionality (Fig. 5–4). The presence of a node in the linked list represents a timeout at that time.

As different timeout cases (e.g., packet retransmission timeout for ARQ implementations, initial packet transmission timeout for QoS, etc.) are supported through the single common timeout function described here, a common API is used to check for any timed-out items inside the linked list for a given frame number. This API is assumed to be called once per frame time, for example at the start of each frame. By keeping track of a node type field, the code can differentiate between different actions to be taken for different node timeouts.

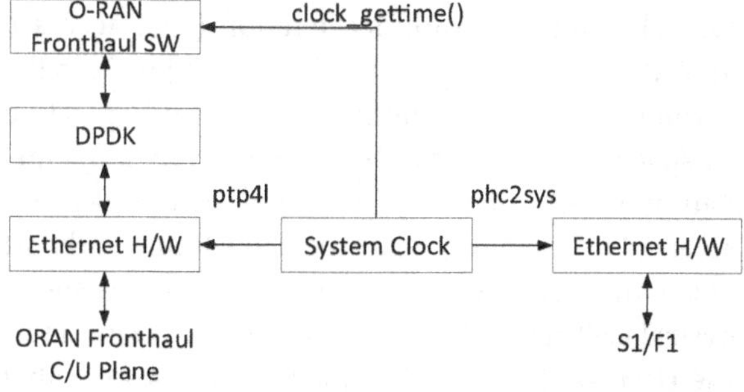

FIGURE 5–3 Time synchronization features in Linux.

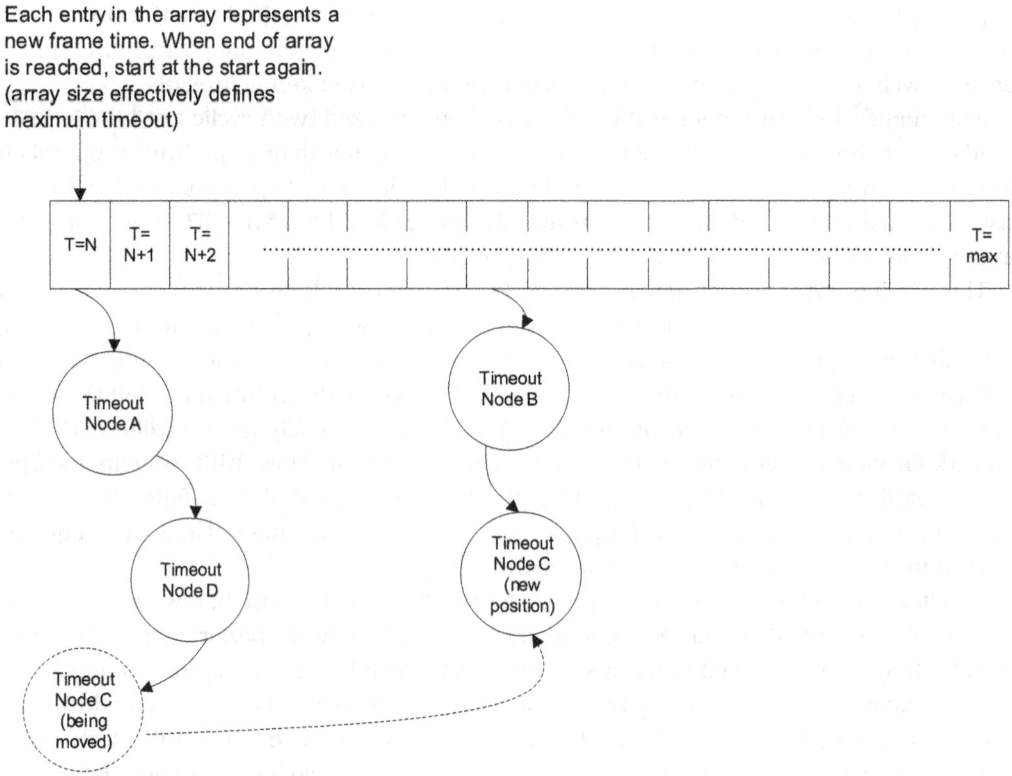

Each entry in the array represents a new frame time. When end of array is reached, start at the start again. (array size effectively defines maximum timeout)

T=N	T= N+1	T= N+2	....	T= max

Timeout Node A → Timeout Node D → Timeout Node C (being moved)

Timeout Node B → Timeout Node C (new position)

FIGURE 5–4 Timeout array.

The code assumes that all nodes that are not timed out (e.g., for ARQ: acknowledge received in-time for a transmitted payload) are removed before the timeout would take place. This means that if the linked list for the current array entry is nonempty, a timeout related action needs to take place. This saves processing power, as the code doesn't need to walk through any lists in order to check for any timeout to have happened. It is assumed here that the timeout case is less common than the nontimeout case.

Challenge 2: Low overhead peripheral and accelerator access from the OS

A layer-1 application relies on hardware acceleration (e.g., access to an on- or off-chip Forward Error Correction decoder) and peripheral (e.g., Radio Over Ethernet packet IO) for key real time and performance critical components. However, typical Linux drivers rely on a vendor provided kernel space driver that can be obscure in its implementation and has little understood determinism and performance. This challenge is solved by moving to the concept of user-space drivers. For example, take the UIO framework, which is intended for devices that are accessed through memory-mapped device registers, as is the norm for devices attached to Physical Cell ID (PCI) and similar buses. A simple in-kernel device driver can be written using the User Mode Input/Output (UIO) framework that allows a user-space program to map that register back into its own memory, and to respond to interrupts from the device. This does not provide generic access to any PCI device, but does make it easy to get user-space access to a device of interest, so that the bulk of the driver can be developed, debugged, and maintained outside of the kernel. OS frameworks like Data Plane Development Kit (DPDK) supported proprietary UIO (and in case of virtualized environments: VFIO) frameworks primarily focused on high-speed networking processing. The DPDK is an Open source software project managed by the Linux Foundation. It provides a set of data plane libraries and network interface controller polling-mode drivers for offloading packet and other processing from the operating system kernel to processes running in user space. DPDK has been embraced across the software ecosystem as the preferred abstraction layer for user-space software development, in combination with PREEMPT_RT as described above due to portability across platforms and Instruction Set Architectures.

Within DPDK, the Wireless Baseband library provides a common programming framework that abstracts HW accelerators based on FPGA and/or Fixed Function Accelerators that assist with 3GPP Physical Layer processing. Furthermore, it decouples the application from the compute-intensive wireless functions by abstracting their optimized libraries to appear as virtual BBDEV devices, where BBDEV is defined as the Wireless Baseband Device Library (https://doc.dpdk.org/guides/prog_guide/bbdev.html). The functional scope of the BBDEV library are those functions in relation to the 3GPP Layer 1 signal processing (channel coding, modulation, ...). The BBDEV framework currently only supports Forward Error Correction (FEC) functionality. More about this below.

Combined, the solutions to Challenge 1 and Challenge 2 are sufficient to implement a hardware-abstracted, high-performance framework that removes the overhead associated with Linux, even though it forces the programmer to be aware of the real time aspects of his/her application and as such disallows the programmer to use specific OS services. This is a bit of a paradox, but gives the application much needed platform portability, which is a key requirement.

Challenge 3: Application Programming to Real-Time requirements

A real-time environment poses several restrictions on the programmer. Typical examples include

- requirement for static memory management—dynamic data structures and associated memory management is (typically) unbound/unpredictable with regards to latency.
- requirement for time management. Remove unbounded loops and other algorithms that have a undefined maximum execution time. When needed, implement 'escape' options to terminate processing early.
- requirement for removal of recursive algorithms for similar reasons as in the previous point.

Buffer and memory management

Real time applications such as MAC (Medium Access Control)/RLC (Radio Link Control)/ PDCP (Packet Data Convergence Protocol) stacks typically implement a private buffer management scheme that instead of doing standard malloc() based memory allocations on an ad-hoc basis, rely on a single large and physically contiguous memory allocation (100 s of Mbytes in size, potentially). There are multiple benefits to this type of memory management, including:

- Determinism—potentially at the cost of flexibility, such memory management scheme is completely flexible in supporting a known cycle "cost" for buffer allocation and release, and no user- to kernel-space switches that consume physical time.
- Ease of memory address translation. Conversion from virtual to physical memory address can be done with single subtract of a base address.

Most stack implementations come with a (private) buffer management scheme that is optimized to the application. It relies on a single large memory allocation that is done at boot-time using the Linux Huge pages feature that is initialized at boot-time as defined in.[6] Within the application multiple processes can share this large allocation. In order to support zero-copy stack implementations, this memory is defined as sharable within the application domain.

The U-Plane component of most MAC/RLC/PDCP stacks are implemented using two logical memory domains:

- The first domain is defined by Transport, GTP, and portion of the PDCP stack that interfaces to the GTP stack (downlink: pre-PDCP security). This memory domain is characterized by:
  - Buffer size defined by transport packet size
  - Requirement for fast/optimized implementation of physical to virtual and vice versa address translation as (optionally) required for interfacing to accelerator hardware (security engine, Ethernet controller)

The transport block buffers can optionally be shared between user-space (UDP, GTP, PCDP) and kernel space in case the transport (IPSec) stack is implemented as a kernel-space component. Zero-copy operation and buffer sharing are enabled by use of user-space buffers for both domains.

Defined by typical maximum Ethernet packet size ($\sim$1500B), transport packets are smaller than standard Linux page size (4KB) and as such can be allocated through malloc() mechanism.

- The second memory domain is defined by MAC/RLC and the portion of the PDCP stack that interfaces to the RLC (downlink: postsecurity). This memory domain is characterized by
  - Buffer size defined by maximum transport block size (and/or other L2 defined maximum buffer size)
  - Multi-use buffers as required for optimized implementations of HARQ and ARQ
  - Requirement for fast/optimized implementation of physical to virtual and vice versa address translation as required for interfacing to accelerator hardware (security engine, DMA engine)

Defined by the maximum L2 packet size, typically L2 packet buffers are typically larger then 4KB and as such cannot be allocated as a physically contiguous buffer using standard Linux mechanisms.

Hardware acceleration model

In the discussion above ("challenge 2"), we discussed DPDK frameworks for IO and acceleration. We want to explore these concepts a bit more in detail.

There can be two model for HW acceleration usages—look-aside and inline.

Look-aside Acceleration Model (Fig. 5–5)

In this look-aside acceleration model where the host CPU invokes an accelerator for data processing and receives the result after processing is complete.

A look-aside architecture allows the application to offload work to a hardware accelerator and continue to perform other work in parallel—this could be to continue to execute other software tasks in parallel as shown in Fig. 5–6 or to sleep and wait for the accelerator hardware to complete. This model requires the API to support two operations, one for initiating the offload and another for retrieving the operation once complete.

This model requires careful software design to ensure that the accelerator overhead is appropriately parallelized.

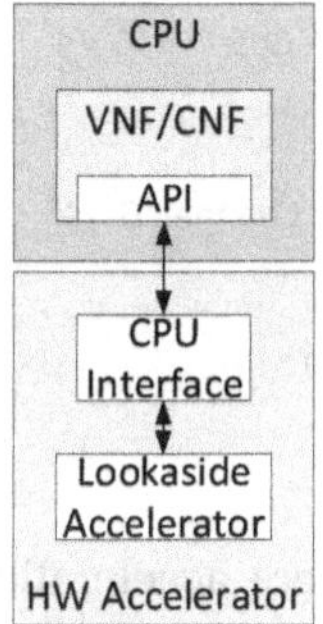

FIGURE 5–5 Look aside accelerator physical implementation.

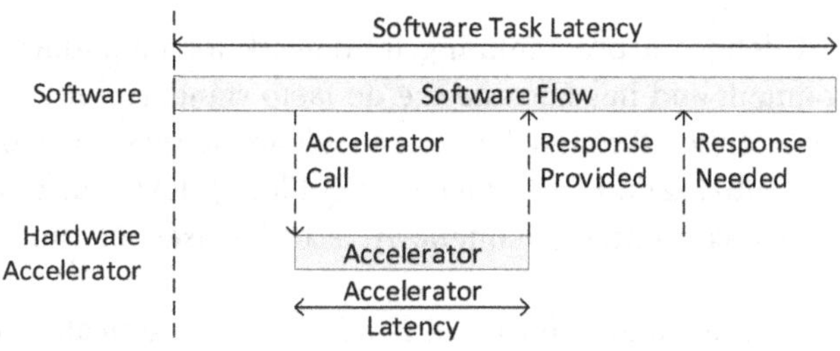

FIGURE 5–6 Parallelizing accelerator calls.

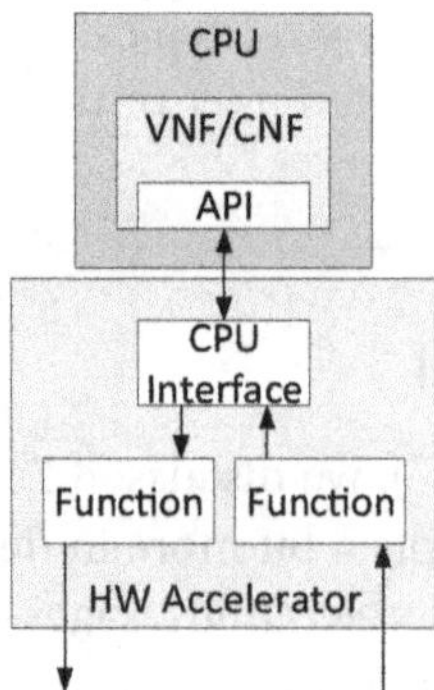

FIGURE 5–7 Inline accelerator.

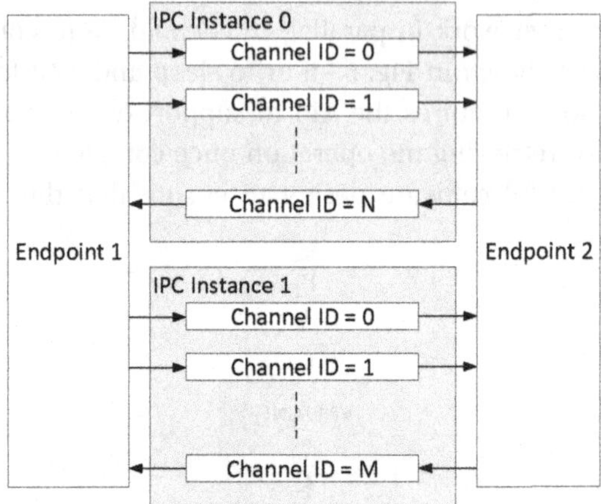

FIGURE 5–8 Inter processor communication model.

Inline Acceleration Model (Fig. 5–7)

In the inline acceleration model where acceleration by function and I/O-based acceleration are performed on the physical interface as the packet ingresses/egresses the platform.

Bbdev and inter process communication

The BBDEV standard defines a programming framework for offloading wireless workloads from a DPDK environment and has become the de-facto standard for offload of Linux-based stack implementations, either through hardware offload or (assumed optimized) software drivers. It includes an initialization stage including (offload) device discovery and configuration as well as offload task enqueue/dequeue APIs to be used in runtime by the wireless application itself.

BBDEV and similar acceleration and IO APIs rely on communication queues that implement Inter Processor Communication (IPC, Fig. 5–8) between two end points, typically physically connected via a PCIe link. IPC is a software mechanism for communicating between two entities but important to implement efficiently to achieve high performance. It is typically vendor provided through a set of libraries by the BSP.

It consists of multiple channels (or bbdev queues) that define a 1:1 (unlikely 1: many given performance targets dictate no use of software locks for enqueue or dequeue operations) communication path between the two end points. The channel/queue is identified by a unique ID. It is up to the application, how it is using the queues, whether they are sharing the same queue for different message type or using different queues for different message types.

A model of an IPC channel implementation itself is shown below in Fig. 5–9:

The characteristics of this channel are:

- For each channel, one end point is the producer and the other end point is the consumer. The producer sends the message and the consumer receives it.
- A channel has a single direction: from the producer to the consumer.
- One end point can be producer for one channel and consumer for another channel. So, for bidirectional communication at least two channels are needed.
- The channel is modeled as a circular queue (FIFO) of buffer descriptors (BDs), called a BD ring. A BD contains a pointer to the message data, and the message length. So, an IPC channel does not carry data buffers, but only pointers to them.
- A channel has a configurable depth, representing the BD ring size. Different channels can have different depths.
- A channel has a producer index, telling the index in the BD ring for the next message to be sent by the producer.
- A channel has a consumer index, telling the index in the BD ring for the next message to be received by the consumer.
- A channel is empty if the producer and consumer indexes are equal.
- A channel is full if the linearized difference between the producer and consumer indexes is equal to the channel depth.

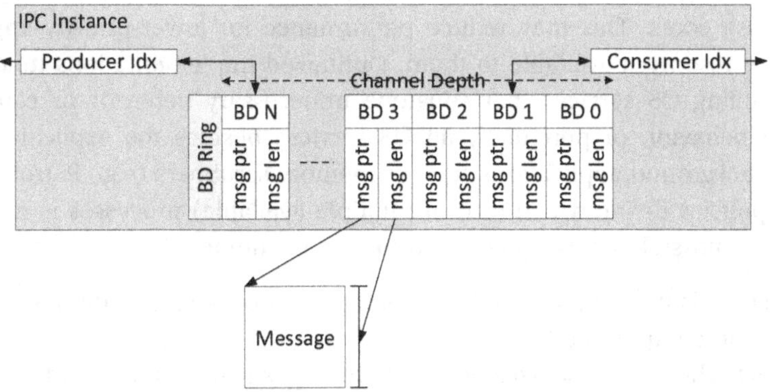

FIGURE 5–9 Inter processor communication channel model.

The producer performs the following process for sending a message over an IPC channel:

- writes the data to be sent in a buffer (this step can be done by the application either independently or through an IPC API function).
- writes the buffer pointer and length in the BD located at the current producer index in the channel BD ring.
- increments the producer index.

The consumer performs the following process for receiving a message over an IPC channel:

- reads the buffer pointer and length from the BD located at the current consumer index in the channel BD ring.
- increments the consumer index.
- reads the received data from the buffer (this step can be done by the application either independently or through an IPC API function).

The consumer can poll the channels using an API function or can receive an interrupt to know on which IPC channel it received a message.

Linux and processor performance tuning

Modern processors all contain multiple cores where, by default, the Linux OS scheduler allocates core (CPU) resources to processes that need them. For CU, DU, and RU use-cases that rely on aggressive U-Plane performance targets, binding application threads to CPU cores can help achieve deterministic latency (reduce task switching) and increase performance by ensuring application code and data remains local to the core cache.

Processor binding can be done using the Linux <*taskset*> command which is used for setting or retrieving a process' CPU affinity. Input parameters are the Linux process ID (pid) and a mask that defines the allowed processor cores this process can execute on. More advanced features are supported through the *cpuset* concept.

Note that binding applications implies that other applications are no longer capable of running on these cores. This may reduce performance for lower-priority applications that have fewer core resources available to them. Optimized implementations tune performance further by disabling OS services that consume unnecessary behavior or can create unexpected system behavior, or potentially all OS services besides the explicitly needed ones. This includes background applications such as memory scrubbers (e.g., Patrol Scrub).

Scheduler policies define which (out of multiple available) processes is run on the core. There are several possible Linux scheduler priorities including:

- SCHED_OTHER is the default scheduling policy for processes that do not have any priority/real-time requirements.
- SCHED_FIFO schedules processes until they relinquish control back to the scheduler and can only be interrupted by other, higher-priority processes.

- SCHED_RR is like SCHED_FIFO but each process is allocated for a certain portion of physical time (its quantum) before it is placed at the end of the scheduling list for a given priority.

Process priority can be set in software through the <*sched_setscheduler()*> API or through the Linux command line with <*chrt*>.

Some processors support a Turbo mode of operation given thermal and DC power headroom—potentially enabled by Dynamic Voltage and Frequency Scaling. For telecommunications-grade systems with high reliability and performance predictability requirements, this mode is often disabled. On the opposite side of the power spectrum, processors implement various power-down (sleep) modes that are also typically disabled, most often through boot time configuration.

On multithreaded processors, incremental gains from enabling additional processor threads can vary depending on the application. Careful benchmarking can establish any potential gains or losses due to cache pollution.

Networking stacks

CU, DU, and RU implementations require networking stack support for termination and forwarding of U-Plane and C-Plane traffic but also for Management and OAM operations. These stacks can be supported by the Linux (kernel) implementation, specialized open-source or third party solutions, such as[7,8] or a combination thereof where performance-critical pieces are optimized independently from remainder components.

Features

Target networking features that are expected to be supported by cell-site type of products include

- L2 protocol support: 802.1q (VLAN), 802.1ad (QinQ), QoS/traffic prioritization, Link Aggregation, MPLS, Jumbo Frame support (9KByte).
- IP protocol support: Access Control Lists (ACL), IPv4/IPv6, IPSec, UDP, GTP, ingress policing based on traffic type (e.g., Control Plane policing), DiffServ.
- Routing and Management: Command Line Interface (CLI), Simple Network Management Protocol (SNMP) v2 and v3, Segment Routing, Border Gateway Protocol (BGP), NETCONF/YANG, Secure Copy Protocol, Syslog, Network Timing Protocol, Dynamic Host Configuration Protocol (DHCP) and DHCP relay, Network Layer Reachability Information, Ethernet Operation and Maintenance (OAM).

Security

TS 33.210 directs that the base station platform shall support network layer security protocols defined by IETF IPsec security protocols specified in RFC-4301 and in RFC-2401.[7] Details of platform specific IPsec implementation are unique per implementation, some aspects covered in later chapters. The minimum requirements of IPsec are listed below.

- ESP shall be supported according to RFC-4303.
  - Extended sequence numbers may be supported.
  - For compatibility with earlier 3GPP it shall be possible to communicate with nodes supporting RFC-2406.
    - See annex E of TS33.210 for main differences.
- ESP encryption transforms and authentication algorithms marked with "MUST" in RFC 8221 shall be followed.
  - AES-GMAC with AES-128 shall be supported.
  - NULL authentication algorithm is explicitly not allowed for use unless an authenticated encryption algorithm is used.
- Where random values are needed, RFC 4086 shall be used as guidelines for hardware and software pseudorandom number generators.
- Internet Key Exchange protocol IKEv2 shall be supported for negotiation of IPsec SAs as defined by RFC 7296 and RFC 6311.
- For IKE_SA_INIT exchange
  - Following algorithms shall be supported:
    - Confidentiality: AES-GCM with a 16 octet ICV with 128-bit key length
    - Pseudo-random function: PRF_HMAC_SHA2_256
    - Integrity: AUTH_HMAC_SHA256_128
    - Diffie-Hellman group 19 (256-bit random ECP group)
  - Following algorithms should be supported:
    - Confidentiality: AES-GCM with a 16 octet ICV with 256-bit key length
    - Pseudo-random function: PRF_HMAC_SHA2_384
    - Diffie-Hellman group 20 (384-bit random ECP group)
- For IKE_AUTH exchange:
  - Authentication method 2—Shared Key Message Integrity Code shall be supported.
  - IP addresses and Fully Qualified Domain Names shall be supported for identification.
  - Re-keying of IPsec SAs and IKE SAs shall be supported as specified in RFC.
  - In addition to the requirements defined in RFC 7296, rekeying shall not lead to a noticeable degradation of service.
- For the CREATE_CHILD_SA exchange
  - A DH key exchange should be used, and the session keys should be changed frequently.
- For reauthentication
  - Reauthentication of IKE SAs as specified in RFC 7296 Section 2.8.3 shall be supported.
  - An Network Element (NE) shall proactively initiate reauthentication of IKE SAs, and creation of its Child SAs, that is, the new SAs shall be established before the old ones expire.

- An NE shall destroy an IKE SA and its Child SAs when the authentication lifetime of the IKE SA expires.
- Reauthentication shall not lead to a noticeable degradation of service.

Performance

Off-the-shelf, Linux provides a feature-rich networking environment including functional support for most components required for deployment of a CU/DU or RU product. This networking solution is part of the Linux kernel. Whilst performant enough for initial product development, there are a few challenges to using the Linux kernel stacks for commercial product development, all of them related to performance:

- Memory management. All networking related components in the kernel use a common data structure for packet processing: *sk_buff*. This data structure maintains all required information for packet processing such as pointers to header and payload and packet queueing capabilities implemented as double linked lists. The sk_buff datastructure is not natively visible to Linux user given that Linux, by design, strictly separates kernel and user space components. This means that user space programs that need access to the packets end up using a memory copy operation to convert the sk_buff structure to some other structure that the user-space accesses. This is exactly what is being done when you open a socket from a user space application and do packet receive/transmit operations.

The challenge here is that this memory copy operation is costly, both because the CPU is wasting time reading and writing (copying) packet payload that it has no visibility into, but also from a memory bandwidth perspective, because each read/write operation loads the memory bus more heavily.

- Accelerator access. Moving towards the embedded processor space, processors often include specialized accelerators to speed up processor-intensive functionality such as ciphering (say, IPSec) or egress QoS/Traffic Management. Access to these accelerators is proprietary in nature and supported in a limited fashion by the Linux kernel—either feature-limited or delayed till much after the processor is released to the market. As a result, the accelerators are often found underutilized.
- Latency. Especially for fronthaul traffic, the undeterministic nature of the Linux networking stack is enough to eliminate it from being used.
- As we mentioned, the Linux kernel is feature-rich in its networking support. This is both an advantage and a pain, the rich feature support is sometimes not needed and can cause performance limitations. A light-weight implementation of networking components that implement the full processing chain from Ethernet receive/transmit up to detection of GTP-U is, for example, all that is needed to implement the transport components associated with U-plane traffic.

We thus make the case for an optimized implementation for (components of) the transport stack, that is feature-lean (support only those components that are required for the product), integrates with provided hardware acceleration and eliminates data copying where not needed.

Such stacks are available commercially from vendors such as 6Wind,[7] or in an open-source implementation such as Vector Packet processing (VPP)[8] that aims to provide a production quality packet processing stack on commodity CPUs. Alternatively, the System Integrator can build a fully customized version of the pieces of the networking stack that are performance critical.

The chart below in Fig. 5–10 shows performance for an IPSec termination application between (on the right) a default Linux stack, in the middle a feature-rich commercial grade stack and on the left a low-level optimized implementation that is application-tuned. We see how the latter two options are ~5–8× more performance optimized compared to the Linux default.

Keep in mind that choosing a performance-optimized implementation for a networking stack does not mean that Linux is eliminated from the networking product altogether. Modern Network Interface Cards and on-chip networking IO blocks include hardware classification capabilities that can be used to separate incoming traffic based on header fields such as IP protocol (IPSec) or UDP port (2152 for GTP). After classification, packets can be directed by hardware to use a specific buffer memory (Linux kernel: sk_buff or a proprietary buffer for optimized stacks) and a separate queue that interfaces to either the optimized implementation or to the Linux kernel for "all other traffic." This gives the system integrator the best of both worlds: Features and ease-of-use associated with Linux with hardware optimized performance (Fig. 5–11).

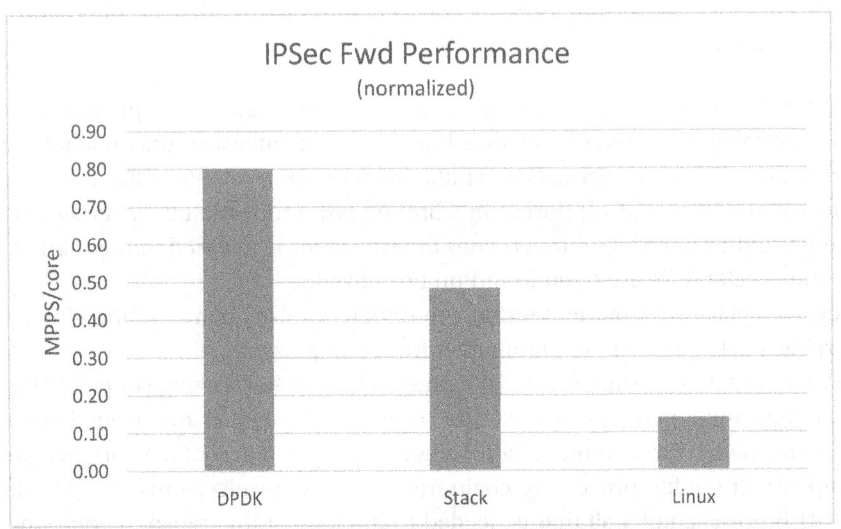

FIGURE 5–10 IPSec performance comparing Linux to more optimized implementations.

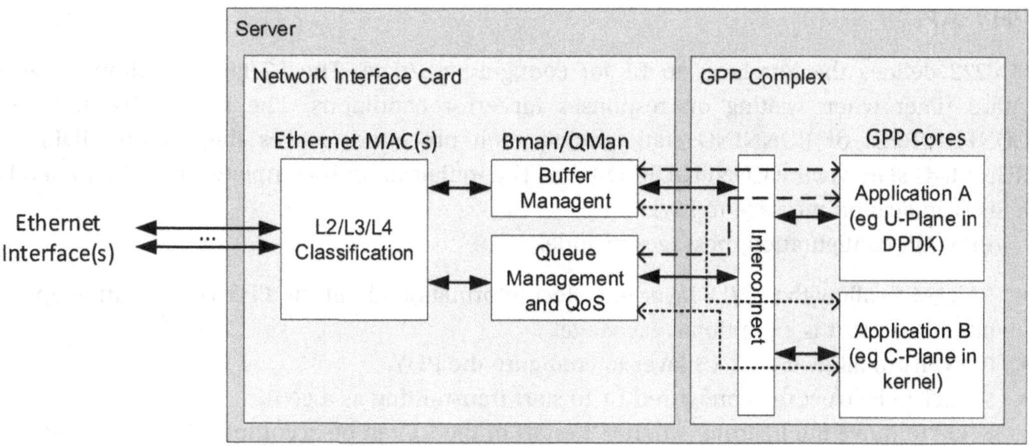

FIGURE 5–11 Hardware traffic bifurcation.

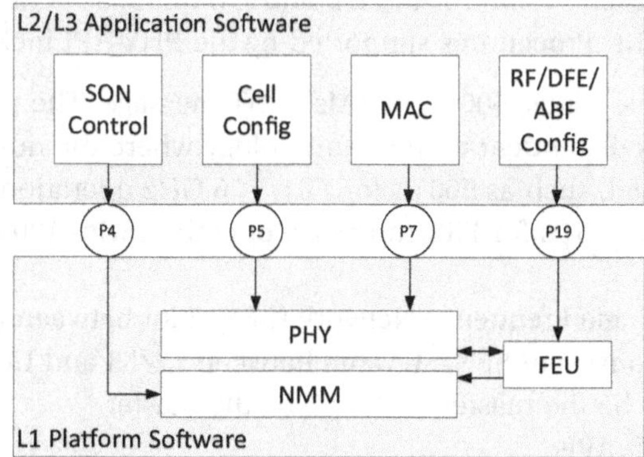

FIGURE 5–12 Functional application platform interface internal APIs. *Copied from small cell forum.*

Functional application platform interface

The Functional Application Platform Interface (FAPI) specification suite is an initiative within the Small Cell Forum to create interoperability and innovation for providers of hardware/ software platforms and application software by providing a common API. With a common API, those two parts can be "interchangeable."

The FAPI suite includes three specifications:

- SCF222, '5 G FAPI: PHY API' - main data path (P7) and PHY mode control (P5) interface
- SCF223, '5 G FAPI: RF and Digital Front End Control API' - (P19) for Frontend Unit control
- SCF224, 'Network Monitor Mode API' - (P4) for 2 G/3 G/4 G/5 G

The FAPI interface (Fig. 5–12) defines internal interfaces to (in O-RAN language) the DU and is agnostic to O-RAN or other RAN architectures.

PHY API

SCF222 defines the interfaces to L1 for configuration/data. The L2/L3 layer should use a guard timer when waiting on responses for error conditions. The L1 can be in IDLE, CONFIGURED, or RUNNING state. Initialization procedure moves the L1 from IDLE to RUNNING state via the CONFIGURED state. The initialization is complete when the L1 sends a SLOT.indication message to L2/L3.

PHY API configuration messages include:

- PARAM to allow the L2/L3 layer to collect information about the PHY configuration and current state. It is an optional exchange.
- CONFIG to allow the L2/L3 layer to configure the PHY.
- START to instruct the configured L1 to start transmitting as a gNB.
- STOP to move the L1 from running. Restart of the L1 can be accomplished by a STOP procedure followed by a START procedure using the same configuration.

Slot procedures are used to control the DL and UL frame structures and to transfer data between the L2/L3 and L1. Procedures supported by the PHY API include:

- Transmission of 125us, 250us, 500us, or 1Ms SLOT message. The periodicity of the SLOT. indication message is dependent on the numerology where the number of messages per subframe is highlighted, such as 500 μs for FR1/ < 6 GHz operation with 30 KHz Subcarrier Spacing or 125 μs for FR2/mmWave operation with 120 Khz subcarrier spacing.
- Synchronization of Single Frequency Network (SFN)/Slot between the L2/L3 and PHY. Used to maintain a consistent SFN/SL value between L2/L3 and L1. Configuration can be set to have the L2/L3 be the master or the L1 be the master.
- Per-Channel Transmit APIs
  - Transmission of the Broadcast Channel (BCH) transport channel used to transmit the Master Information Block on the Physical Broadcast Channel (PBCH).
  - Transmission of the PCH transport channel. PCH transport channel is used to transmit paging messages to the User Equipment (UE). The UE has specific paging occasions where it listens for paging information. The L2/L3 is responsible for calculating the correct paging occasion for a UE while the L1 is only responsible for transmitting PCH PDUs when instructed by the DL_TTI.request message.
  - Transmission of the Downlink Shared Channel (DL-SCH) transport channel. The DL-SCH transport channel is used to send data from the NB to a single UE. HARQ is applied on the DL-SCH transport channel and it's up to the L2/L3 stack to schedule enough uplink bandwidth for HARQ (Physical Uplink Shared Channel (PUSCH) or Physical Uplink Control Channel (PUCCH) based) for acknowledgment (ACK)/ negative acknowledgment (NACK) feedback to be returned. Transmission of the DL-SCH transport channel includes the TX_Data.request messages that defines the MAC PDU data—this message is special in that it contains large (bulk) payload whereas all other (DL) messages or message components are small. For that reason, TX_Data.

  - request is often implemented in a hardware-optimized manner (using DMA engine rather than a memory copy) that reduces CPU loading.
  - Transmission of Downlink Control Information (DCI).
  - Transmission of the Channel State Information (CSI) reference signal.
- Per-Channel Reception APIs
  - Reception of the Random Access Channel (RACH) transport channel. The RACH procedure begins when the L1 receives an UL_TTI.request message indicating the presence of a RACH. If no RACH preamble is detected by the PHY then no RACH. indication message is sent.
  - Reception of the UL-SCH transport channel. The UL-SCH transport channel is used to send data from the UE to the gNB. HARQ is always applied on the UL-SCH transport channel and the ACK/NACK value is indicated when the next transmission for the HARQ process is scheduled via DCI including in the next iteration of this procedure.
  - Reception of Uplink Control Information.
  - Reception of the sounding reference signal.

Defined to be flexible and support varying levels of sophistication, beamforming APIs are defined in tables loaded into the L1 at configuration time. In slot messages, a precoding index and/or beamforming index is included in each PDU.

nFAPI

Designed to support operation for an Option-6 split with MAC and Physical layers connected over Ethernet, nFAPI is a packetized version of the FAPI interface, defined as (but not limited to) an environment where the L2/L3 stacks are implemented as a Virtual Network Function (VNF) on a server environment and the Physical Layer is implemented as a Physical Network Function (PNF) on dedicated hardware.

The nFAPI implementation allows the L2/L3 and Physical Layer components to "speak" FAPI to each other but through two translation layers (FAPI→nFAPI and nFAPI→FAPI).

Zooming in on the key FAPI interface (P7), we note some differences:

- FAPI uses a SLOT.indication message sent from the PHY at the start of every slot (125/250/500/1000μs). This message is not used in nFAPI as the jitter in the fronthaul connection between VNF and PNF prevents this message from being a suitable mechanism for signaling a slot interval. Independent timing for the two components through GPS/IEEE1588 or other mechanisms are used instead for establishing time, and the PHY sync procedure combined with API message timing is used for slot level communication.
- The SFN/SL FAPI procedure does not apply in nFAPI. PHY synchronization and delay management mechanisms are used for synchronization and timely messages between VNF and PHY instances.
  - API message timing is achieved as follows: the nFAPI messages from VNF to PNF must arrive at the Physical Layer instance a minimum Timing Offset before the slot

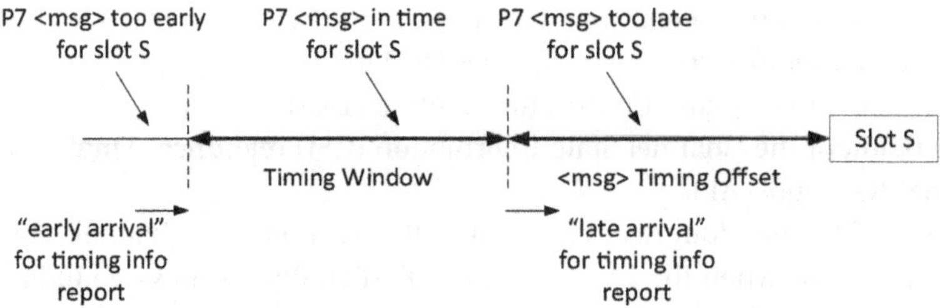

FIGURE 5–13 Timing window management.

they are configured starts transmission on the air interface as shown in Fig. 5–13. If the downlink transmit request misses this window, the Physical Layer would not have sufficient time to execute it's functions and transmit over the air interface. This is obviously not desired. As such, if the nFAPI message arrievs to the Physical Layer late, it returns a message indicating the lateness in μs and the transmit offset can be adjusted for next transmission.

- The nFAPI delay management mechanism can be used to establish the timing reference differences between the VNF and PNF. It can also, optionally, be used by the VNF to instruct the PHY instance to update its slot number based on the offset defined by the VNF.

Network monitor mode API

Well-described in,[9] Self-Organizing Networks (SON) are introduced to reduce operator Capital Expene (CAPEX) and Operational Expense (OPEX) in deployed 3GPP networks like LTE and 5G. This is driven by the increasing complexity in configuration parameters associated with the wireless network—think of transmit power, operational band, TDD pattern and many more. Especially when having to operate multiple network generations (2G, 3G, 4G and 5G) across many operational bands, all interfering with each other and becoming more and more dense to support increased demand for performance, network configuration becomes incredibly complex. There are many papers written on SON.

SONs can be organized centralized where organizing algorithms are executed in a central network location (the operator cloud), distributed where each node (Base Station/Access Point) configures itself, or in a hybrid mode where the configuration is done partially from the cloud and partially by each node itself. The associated SON algorithms target to optimize the performance of the network with minimum human interference and are typically considered proprietary/secret as they impact system performance greatly. Performance can be measured in aggregate throughput across all users, minimum downtime, best coverage and other parameters.

SON targets

- Configuration of PCI as defined in 3GPP Release 8 ("eNB self-configuration")—the Cell ID needs to be unique to each base station so that UE can distinguish between cells for the purpose of handovers.

- Management of Automatic Neighbor Relations (ANR)—each cell needs to own a database of neighbor cells to be able to manage handovers to these neighbors. Ideally, this database is continuously updated so to be able to detect neighbors being switched off and on. Neighbors can operate either in-band (i.e., in the same band as the SON cell) or out-of-band (i.e., in a different band as the SON cell).
- Inter Cell Interference Coordination (ICIC)—targets to coordinate transmissions between cells in such a way that interference is mitigated as much as possible. For example by allocating time, frequency and power resources orthogonally between cells for UEs that are subject to this interference. Advanced ICIC schemes evolve to a SFN where frequency planning is not preconfigured but implemented dynamically on a slot-by-slot (in 5G: on a submillisecond) basis.
- Mobility Robustness Optimization—involves automated optimization of handover related parameters, to remove negative impact on user experience, for example, from handover ping-pong.
- Cell outage detection and self-healing networks—the capability to detect that cells are not functioning properly and to mitigate these effects by steering user traffic and coordinating cell repair (even if this is as simple as triggering a device reset procedure).

For SON to work, it relies on

- Communication between cells—typically over a wireline (backhaul) connection that allows them to coordinate.
- Sensing capabilities to detect other transmitting cells, both in-band (obvious in the ICIC example) and out-of-band (obvious in the ANR example). This sensing is typically referred to as Network Listening Mode or Network Monitoring Mode (NMM).

NMM has been standardized by the Small Cell Forum in the "LTE Network Monitor Mode Specification" document. For the LTE standard, the scope is defined as: to allow a femto cell to support cell RF self-configuration and interference management, which amongst others include (1) carrier and cell id selection, (2) DL TX power setting, (3) UE TX power setting, and (4) frequency reference. This is a relatively simple set of targets but defines a baseline for NMM development. This standard defines the following measurements for a given carrier frequency:

- Received Signal Strength
- Search for any cells (within range) that are transmitting
- Read (decode) the PBCH
- Read (decode) the System Information Block 1
- Read (decode) remainder Broadcast Channel information blocks.

Operation can be supported for in-band (only on the channel that this cell is operating at) or out-of-band (also on other channels) operation. Operation can also happen either only at boot time or also during run-time, and either whilst base station operation is ongoing or only in when idling from normal user operation. As long as requirements are simple, NMM can be implemented in hardware that is also used for communication operation or with dedicated hardware: the NMM unit. Dedicated hardware allows for more advanced modes such as out-of-band operation, nonboot-time operation, and so on.

Security aspects

For commercial and private networks both, security aspects are important but often overlooked. Telecommunications equipment is seen as critical infrastructure and thus needs to be protected against intruders. Given the software centric nature of modern telecommunication equipment, we can treat it as any web-connected application. O-RAN and other networking equipment is often installed (like any other networking equipment) on private or semipublic networks and subject to similar attack vectors, both from the Core Network side where the networking equipment is connected to the Internet as well as from the client side where the network is accessed over the wireless air interface which makes all local applications subject to attack from any malicious user within the wireless network range. In this chapter, we discuss some common attack scenarios and appropriate protection mechanisms against them, in the form of a checklist of best practices. We discuss them in the sequence in which they are commonly executed, focusing on network accessed attack as opposed to attack vectors that are possible with physical access to a target system. Security should not be treated as a separate process but be part of the development flow.

Assessment of product system security is typically done through third party "Ethical Hacker" engagements (such as[10]) who have in-house up-to-date knowledge on security topics.

Profiling

Profiling (passive information gathering) is the first phase in all cyber-attack scenarios. Attackers can use a variety of sources to learn as much as possible about the target business and how it operates, such as internet searches, social engineering, research of public sources such as Domain Name Systems (DNS) and IP-registrations, dumpster diving (searching the trash) and nonintrusive network scanning. It includes a wide range of techniques to cast a broad information gathering net, including

- Network Blocks. Network blocks owned by the organization can leak a lot of information about the current and previous infrastructure used.
- DNS host name enumeration, by brute force guessing of common subdomains.
- Port scanning is also another useful tool of identifying platforms. Part of the effort of port scanning is to develop a fingerprint to identify your targets. With small cells/femto cells/RUs sitting on the open internet they become targets through these fingerprinting activities. Once a malicious party has a database of all devices attached to the internet, they can search that database for characteristics to identify particular target devices and move from there.
- Social media sourcing for identification of persons.

- Validation of certificate transparency. Certificate Transparency is a system that keeps track of issued certificates in publicly available logs. When the browser opens a connection to a web application, it will check these logs to validate whether the Secure Sockets Layer (SSL) certificate of the server is valid. If the used certificate is not in this list, it could mean that the certificate is issued incorrectly or fraudulently, which can allow the browser to take preventive actions.
- Identification of IP-ranges and domain names used by the organization. By scanning the ranges and DNS-systems, running services can be discovered. Outdated services or misconfigurations can be abused to obtain access to the infrastructure of the organization in a later attack stage.
- Zone walking is a technique which allows attackers to read the full content of DNSSEC-signed (Domain Name System Security Extensions) DNS zones. As a result, sensitive data like customer lists (subdomains) and IP addresses of servers can leak.
- Shodan is a search engine that lets attackers find specific types of hosts, such as embedded webservers, connected to the internet using a variety of filters.
- Knowledge of the IP address(es) of the internet router is essential for a targeted attack on a company network. This information can be retrieved, for example, by analyzing emails or via access to the wireless guest network.
- In addition to data about servers/applications and people, extra information can be found on the internet that can be useful for a cyber-attack, such as technical documentation, financial data and previously leaked account (and password) information.
- File enumeration is a technique for retrieving publicly accessible files in order to gather information about the target. It is not required that there are references to the files on, for example, a website. This attack is also known as Predictable Resource Location, Directory Enumeration, and Resource Enumeration.
- Web archives search are used by attackers to look back in time at the various stages of the website or application. This can provide information about the changes made on a website (e.g., vulnerabilities that have been fixed, an old CMS moved to a different path), but it also makes clear what growth a company has had. Additionally, archives also can be a source to identify email addresses, names, phone numbers, addresses, and so on.
- Some sources contain information about performed attacks on systems. This information can often be publicly accessed and can contain information about hosts and infrastructure.
- Names, e-mail addresses, and other personal data of employees can be useful for phishing and brute-force attacks. This information is often available in publicly accessible source.
- A data breach is a confirmed incident where sensitive, confidential, or otherwise protected information has been accessed and/or disclosed in an unauthorized manner. Common data breach exposures include personal information such as credit card numbers, social security numbers, passwords as well as company information such as customer lists, production processes, and software source code. This data is often used for future attacks or blackmail.

- The more information an attacker can discover, the easier it is to target a machine or service. Therefore it is important to leak as little information as possible about the hardware, operating system, and other software in use. For instance, it should not be possible for an attacker to use the load balancer to obtain sensitive information about the network.
- Internet searches, social engineering, research of public.

Network reconnaissance

In this phase ("active information gathering"), attackers map the structure and configuration of the wired and wireless side of the wireless network equipment and remainder components on the network. The wireless network side is visible for all devices that are located within the range of the network. An incorrect configuration could allow users to get unauthorized access to the internal network.

- Many firewalls do not make it mandatory to change the default login details of the administrator account and are therefore vulnerable. Default admin credentials are always checked by intruders and should be disabled.
- Virtual private network (VPN) (IPSec) tunneling is used on O-RAN and other networking equipment such as Firewalls. It is required to use secure encryption and authentication but not always enabled, or only enabled for a portion of the traffic, driven by compute resource limitations or the false assumption that the equipment is connected to a secure network. VPN support also applies to associated networking equipment such as front haul and cell site routing equipment which should include secure configuration and services.
- Firewalls protect network services by only allowing specific connections and blocking malicious traffic. It is vital that firewalls are configured carefully, as a misconfiguration could allow bypassing the access control mechanisms. ACLs are a network filter utilized by routers and some switches to permit and restrict data flows into and out of network interfaces. When an ACL is configured on an interface, the network device analyzes data passing through the interface, compares it to the criteria described in the ACL, and either permits the data to flow or prohibits it.

Attackers are sometimes able to circumvent firewalls, for example, by splitting Transmission Control Protocol (TCP) packets in such a way their header does not fit into one IP packet, by using specific network source ports or by using specific TCP flags. These configuration errors might lead to more services being accessible than desired.

- The network layout is important when it comes to security. All hosts must be assigned to the correct zones. For example, if an externally accessible web server is in the same zone as the internal servers, this is a major risk. If the web server has a vulnerability, this can have consequences for the entire internal infrastructure.

- It is important that proper security decisions are made during the network planning phase. Consider the use of virtual local area networks (VLANs) and subnets to isolate traffic domains.
- Access to the network may be done by means outside of the 4G/5G network itself. Consider a parallel operated Wi-Fi network with common equipment for Edge Computing. Companies often set up a separate wireless network for guests. For optimal security, this network should not be accessible to all guests without any limits. Wi-Fi networks have known vulnerabilities that needs to be protected against independently.
- Environments based on containers (such as Docker), often use tooling for central management (container orchestration). These tools must be set up carefully to prevent attackers from taking over the entire infrastructure by breaking into one application.
- Ensure that there are no unnecessary applications running on the systems, and that when any applications/services are started, they are properly secured.
- Remote management aspects
  - SNMP is based on UDP, a simple, stateless protocol, and is therefore susceptible to IP spoofing, and replay attacks. In addition, the commonly used SNMP protocols 1, 2, and 2c offer no traffic encryption, meaning SNMP information and credentials can be easily intercepted over a local network. Traditional SNMP protocols also have weak authentication schemes and are commonly left configured with default public and private community strings.
  - The remote management of devices in a network are managed with different protocols such as, RDP (Remote Desktop Protocol), VNC (Virtual Network Computing), X11, SSH (Secure Shell), FTP (File Transfer Protocol), etc. There may be known vulnerabilities due to not updating this software. Also, misconfiguration for these protocols can cause serious problems.

Security updates

A secure computer requires a secure base. It is therefore crucial that the used operation systems are fully patched, and no security updates are missing.

- In server software, as in all other software, security vulnerabilities are discovered continuously. Therefore it is essential to patch software as soon as possible. Patching is needed even more in case of programs or guides which make it easy for less skilled attackers to carry out an attack that are available on the Internet. To resolve security vulnerabilities, vendors, and communities actively maintain software. This is only done until a point of time at which the software becomes unsupported, usually referred to as end-of-life. Particularly in a quasiembedded solution like an O-RAN RU, this is an aspect that is often overlooked. We discussed Linux LTS and RYO versus commercial grade solutions before (Sections 1.3 and 1.4 respectively).
- SSH is the standard service used to manage a server remotely. Therefore it must be configured securely.

- Software often provides standard accounts at installation time. Also, account names like 'admin' or 'guest' are widely used. These accounts are an easy target for attackers, and are therefore best deleted or renamed, or at least provided with a strong password.
- Server software often is installed with demonstration or test scripts are used to test if the software is functioning as expected. When the software is deployed, such components should be disabled because they often provide unneeded information which may be used by an attacker. Such components are usually poorly secured.
- Detailed error information (logs sent over the network to debug any issues) may aid an attacker getting a full view on a target host's configuration and application code. Error information therefore should be transmitted through a secure link and not contain unnecessary technical information.
- Beyond keeping the OS updates, it is important that the platform keeps all components updated. Having a Bill Of Material (BOM) of all components included in the system and knowing what their Common Vulnerabilities and Exposures (CVE) status is with MITRE (the MITRE Corporation is a non-profit organization managing R&D for cyber security and similar fields) is crucial. Having a development/test/release process flow that keeps security central will reduce the friction needed to ensure security.

Application architecture

The security of custom software is largely dependent on the application architecture. The correct use of a modern framework and software techniques can prevent a lot of application weaknesses from becoming exploitable vulnerabilities. Standard software development guidelines are defined (for example) by MISRA for C-programming for specifically embedded systems. Tools such as Coverity can analyze software and flag typical errors such as buffer overflow but are often ignored under software delivery pressures. Topics include

- Applications use memory to store transient data on the stack and heap. Care must be taken to prevent programs from reading from or writing to invalid memory addresses.
- Memory that was allocated to programs must be released to the OS properly to prevent vulnerabilities.
- Memory that was de-allocated must not be accessed by the program anymore.
- A NULL pointer dereference occurs when the application dereferences a pointer that it expects to be valid, but is NULL, typically causing a crash or exit.
- Denial-of-service can occur when a program allocates memory for a certain action but never releases it to the OS.
- If there are random values that must be generated, it is important that these values are not predictable. So, make sure that all random values are generated with a sufficiently secure random number generator, either in software or a hardware accelerator/external security processor.
- Buffer overflow and underflow conditions should be managed. A buffer overflow condition exists when a program attempts to put more data in a buffer than it can hold or

use pointers to memory that has not been allocated to the program (anymore). Writing outside the bounds of the allocated memory can corrupt data, crash the program, or cause the execution of malicious code. Attackers might exploit such conditions by supplying tainted input to the program.
  - The program overwriting data after the end of the memory buffer (which can include the return address of the calling function) or before the start of the memory buffer.
  - The program reading data after the end of the memory buffer or before the start of the memory buffer.
- If the application does not properly control the allocation and maintenance of a limited resource, such as memory or disk space, an adversary might be able to exhaust the available resources, leading to a denial-of-service condition.
- Application variables are limited in range. A byte, for instance, can only represent 256 different values. If "7" is added to "250," then this will result in a byte value of "1." Attackers might abuse these unexpected results.
- If an application accepts serialized objects (e.g., ASN.1), great care must be taken when the objects are deserialized to prevent attackers from executing code unauthorized.
- Sometimes developers build backdoors into an application in order to facilitate debugging. Attackers can also make use of them, which is why this kind of functionality should not be present in a production environment.
- The security of an application that loads code or uses functionality from a source that is not under the full control of the developer, relies on the good intentions and the security of the party who provides the code. The invoked scripts often run with the same level of access as the code of the application itself.
- When applications evolve, functionality is sometimes re-implemented. The obsolete implementation should be removed from the application to prevent abuse.
- Applications and apps can interface with external systems. If the communication is protected with SSL/TLS (Transport Layer Security) then it is required that both the server and client check whether the used certificates are valid. They should not be expired or revoked and signed by a reputable certificate authority (this also include a company's own CA).
- Advanced application components can require careful configuration to be deployed securely.
- In application components, as in all other software, security vulnerabilities are discovered continuously. As soon as such vulnerabilities are publicly released, or get otherwise publicly known, they may be exploited by attackers. Therefore it is essential to patch software as soon as possible with a secure software update mechanism (refer to the discussion on secure boot). To resolve security vulnerabilities, all vendors including third parties and communities actively maintain software. This is only done until a point of time at which the software becomes unsupported, usually referred to as 'end-of-life.'
- Ciphering and authentication algorithms are only effective if the applied algorithms are classified as safe and are used properly. Certifications (such as FIPS) can be required for deployment in secure environments.

- Some applications temporarily require elevated permissions. Care must be taken that the lowering of the privilege level always is executed correctly.
- Applications should run at the minimum access level required for correct operation. A webserver should for instance never run as "root." This limits the level of access gained by an attacker in case of a compromised service or component.
- File permissions on operating system level should be configured as restrictively as possible. Especially configuration files and data store deserve careful configuration.
- An application (or extension/plugin that integrates into the application) could use (bundled) binary files for its working. If so, the integrity of the code that is executed within these binary files cannot easily be reviewed, forming a risk for the user that executes the plugin.
- Monitor and manage all third-party components (Linux Kernel, Open Source Software (OSS) projects, etc.).
  - Know what components are included in the release package (BOM).
  - Know what CVE numbers may be assigned by MITRE and monitor MITRE for any new CVEs.
  - Know the security risks associated with the third-party components.

3GPP and International Electrotechnical Commission security requirements

Security aspects of 3GPP are defined by the SA3 Working Group and specified in the 3GPP 33.401 for 4G/LTE (Table 5–2) and 33.501 standard for 5G/NR (Table 5–3). It covers mid/backhaul, DU internal, and fronthaul security aspects.

- *Mid/Backhaul.* The support of security associations is required between the Evolved Packet Core (4G/LTE) or Next Generation Core (5G) and the base station as well as between base stations. These security association establishments shall be mutually authenticated and used for user and control plane communication between the entities.
- *DU Internal.* 3GPP defines requirements that the base station shall execute authorized data/software, and that this shall execute with the help from a secure environment. This environment is defined in the secure boot section earlier in this document. Keys stored

Table 5–2 Security algorithms for 4G/LTE.

Algorithm	Type	Description	Recommendation
EEA0		No encryption	
128-EEA1/128-EIA1 (Snow 3G)	Privacy (EEA1) and Integrity (EIA1)	TS 35.215/TS33.401	Mandatory
128-EEA2/128-EIA2 (AES)	Privacy (EEA2) and Integrity (EIA2)	NIST 800–38A/TS33.401	Mandatory
128-EEA3/128-EIA3 (ZUC)	Privacy (EEA3) and Integrity (EIA3)	At the request of Chinese operators	Optional

Table 5–3 Security algorithms for 5G/NR.

Algorithm	Type	Description	Recommendation
EEA0		No encryption	
128-NEA1/128-NIA1 (Snow 3G)	Privacy (NEA1) and Integrity (NIA1)	TS 35.215/TS33.401	Mandatory
128-NEA2/128-NIA2 (AES)	Privacy (NEA2) and Integrity (NIA2)	NIST 800–38A/TS33.401	Mandatory
128-NEA3/128-NIA3 (ZUC)	Privacy (NEA3) and Integrity (NIA3)	At the request of Chinese operators	Optional

inside base stations are not allowed to leave a secure environment. This implies that, technically speaking, the cipher keys are not allowed to be stored in insecure memory. Whether (unprotected) external volatile memory is defined as secure is left open for interpretation. The CU converts packets from a secure (IPSec) link from the fronthaul network to (secure) packets. As part of this, the IPSec frame is decrypted to plaintext and re-encrypted using the PDCP cipher protocol (discussed below). As part of this operation, the plaintext packet is made visible and likely stored in volatile memory, external to the processor executing the cipher operations. 3GPP defines this user plane data cipher and integrity processing to take place within a secure environment—the same one as where keys are stored.

- *Fronthaul.* Algorithms include AES, SNOW, and ZUC with 128-bit key support for encryption (ciphering) and authentication (integrity protection). Ciphering is mandatory to be supported for both U-Plane and C-Plane traffic that goes over-the-air. Integrity protection has been introduced as a mandatory feature to be supported for the U-Plane traffic by both gNB and UE in 5G/NR, which is a change from 4G/LTE that only mandated ciphering for the U-Plane. Whilst support is mandatory, feature enablement is optional on a per-operator basis and can be limited to traffic throughputs (rates) to enable implementations on older platforms with limited hardware support.
- *Setup and Configuration.* Setting up and configuring a base station by Operation and Maintenance (O&M) systems shall be authenticated and authorized by the small cell so that attackers shall not be able to modify the settings and software configurations via local or remote access.
  - Certificate enrollment mechanism specified in TS 33.310 for base station should be supported.
  - Communication between the O&M systems and the base station shall be confidentiality, integrity, and replay protected from unauthorized parties. The security associations between the small cell and an entity in the core network or in an O&M domain trusted by the operator shall be supported. These security association establishments shall be mutually authenticated. The security associations shall be realized according to TS 33.210 and TS 33.310. In practice, this defines the requirements for a VPN/IPSec tunnel between the O-RAN components as well as

between the CU and the Core Network. We discussed some of the high-level performance aspects in the CU System Component.

- The base station shall be able to ensure that software/data change attempts are authorized.
- The base station shall use authorized data/software.
- Sensitive parts of the boot-up process shall be executed with the help of the secure environment.
- Confidentiality of software transfer towards the small cell shall be ensured.
- Integrity protection of software transfer towards the small cell shall be ensured.
- The base station software update shall be verified before its installation (subclause 4.2.3.3.5 of TS 33.117).

For the application use-case of industrial communication networks (private 5G), International Electrotechnical Commission (IEC) standards 62443 covers security aspects and should be used as a guideline at the system level, both for technical and process related aspects. In addition, IEC 63074 guides use of IEC 62443 with regards to functional safety.

References

1. https://insideevs.com/news/376037/tesla-mcu-emmc-memory-issue/.
2. https://xenomai.org/.
3. https://www.enea.com/products/operating-systems/real-time-accelerated-linux/.
4. https://github.com/LITMUS-RT/cyclictest.
5. Varghese, G.; Lauck, A. Hashed and hierarchical timing wheels: efficient data structures for implementing a timer facility. *IEEE/ACM Trans. Netw.* **1997,** *5* (6).
6. http://lwn.net/Articles/374424/ (Huge pages).
7. https://www.6wind.com/6windgate/.
8. https://en.wikipedia.org/wiki/Vector_Packet_Processing.
9. Østerbø, O.; Grøndalen, O. Benefits of self-organizing networks (SON) for mobile operators. *J. Computer Netw. Commun.* **2012,** *2012.*
10. http://www.whitehats.nl.

Further reading

SCF. 5G FAPI: PHY API Specification, https://www.smallcellforum.org/5g-phy-api-release/.

CPRI. eCPRI Specification v2.0. 'Common Public Radio Interface: eCPRI Interface Specification' May 2 2019.

ETSI. NR; Base Station (BS) radio transmission and reception, 3GPP TS 38.104.

ETSI. NR; Multiplexing and channel coding, 3GPP TS 38.212.

ETSI. NR; Physical channels and modulation, 3GPP TS 38.211.

ETSI. NR; Physical layer measurements, 3GPP TS 38.215.

ETSI. NR; Physical layer procedures for control, 3GPP TS 38.213.

ETSI. NR; Physical layer procedures for data, 3GPP TS 38.214.

ETSI. NR; Requirements for support of radio resource management, 3GPP TS 38.133.

6

User-plane application components

This chapter outlines the key application components in the CU, DU and radio unit (RU) systems, focusing specifically on the U-Plane components that are relevant from a central processing unit (CPU)/accelerator/memory footprint and performance perspective and therefore particularly of interest to the systems architect.

GTP Protocol

Two versions of the GPRS Tunneling Protocol (GTP) exist, GTPv1 and GTPv2. GTPv1 was used in pre-LTE communication interfacing between base station and General Packet Radio Service (GPRS) Support Nodes within a Public Land Mobile Network. GTPv2-C (defined in 3GPP TS 29.274) is used on Evolved Packet Core (EPC) interfaces and used User Datagram Protocol (UDP) port number 2123. GTP version 2 is different to version 1 only in GTP-C, because of required enhancements for bearer handling.

GTP-U (defined in 3GPP TS 29.281) encapsulates user plane for Long Term Evolution (LTE)and 5G using UDP port number 2152. A GTP-U tunnel is identified by a Tunnel Endpoint Identifier (TEID), an IP address, and a UDP port number as shown in Fig. 6–1.

A short summary of the header fields is:

- Version—defines the version of the GTP-U protocol and set to 1 for GTP.
- Protocol Type is set to 1 for GTP. Note that the 0 version is reserved for GTP' (GTP prime) that is used for carrying charging data in GSM/UMTS protocols
- Extension Header Flag indicates the presence of a meaningful value of the Next Extension Header field. Next Extension Header Type (last byte in the regular GTP header) indicates the type of extension header that follows this field in the Payload Data Unit (PDU). In 5G networks, the 6-bit QoS Flow Identifier (QFI) field is carried inside the GTP-U extension header that we copied (from 3GPPP 38.415) below in Fig. 6–2. We explained the use of TEID and QFI mapping to Quality of Service (QoS) flows earlier in this book.
- Sequence Number Flag indicates the presence of a meaningful value of the Sequence Number field. When present (as indicated by this flag), the Sequence Number field is incremented for each transmitted packet for those flows whose transmission order needs to be preserved.
- N-PDU Number flag (PN) indicates the presence of a meaningful value of the N-PDU Number field. When present, this field is used in legacy (2G/3G) networks for support of RLC Acknowledged Mode (AM) mode handovers.
- Message Type—this field indicates the type of GTP-U message. 3GPP 29.060 defines the complete list of message types. Message types are included for path management (echo

Open Radio Access Network (O-RAN) Systems Architecture and Design. DOI: https://doi.org/10.1016/B978-0-323-91923-4.00009-4

8 bits

Version	PT	*	E	S	PN
Message Type					
Length					
Length					
TEID					
TEID					
TEID					
TEID					
Sequence Number (					
Sequence Number					
N-PDU Number					
Next Extension Header Type					

FIGURE 6–1 GTP-U PDU Format.

8 bits

PDU Type (DL=0, UL=1)		Reserved
Spare	RQI (DL only)	QoS Flow Identifier
Padding to make PDU (n*4– 2) octets (0..3 Bytes)		
...		
Padding to make PDU (n*4– 2) octets (0..3 Bytes)		

FIGURE 6–2 PDU session information format.

request/response), tunnel creation, modification and deletion, and various error and other management uses.

- Length indicates the length of all payload (including extension headers) beyond the 8 bytes of the 8 mandatory header bytes.
- TEID identifies a tunnel endpoint in the receiving GTP-U protocol entity. The receiving end side of a GTP tunnel locally assigns the TEID value the transmitting side shall use, in a randomized manner. The "all zeroes" value shall not be used.

The GTP protocol defines payload a T-PDU, typically an IP packet (datagram) to be transmitted between the client (UE) and the Internet (external packet network). A G-PDU is the definition of the combined T-PDU + GTP header (the complete datagram including GTP encapsulation).

Note how GTP (and IPSec) headers add to the size of the IP packet being transported over the backhaul interface, which may (when Maximum Transfer Unit (MTU) sizes are badly configured) lead to high percentages of fragmentation of IP traffic over the mid/backhaul interface. Fragmentation is both costly to implement (CPU cycles) and prone to packet loss (the whole IP packet is discarded when a single fragment is lost) and as such should be avoided. The GTP protocol implementation should as such support Path MTU Discovery (MTUD) in its implementation.

GTPv1-U tunnel endpoints do not need to change the hop count/or Time To Live (TTL) values in the IP header or to perform any IP routing.

PDCP protocol

As we explained before, the main role of the Packet Data Convergence Protocol (PDCP) layer is conversion of end-to-end (typically, IP, but could be Ethernet or other protocols such as in URLLC) to a packet that can be transmitted over-the-air, including header compression, authentication/ciphering and re-ordering and duplicate detection through sequence number addition. A PDCP header is added that contains the following fields:

- Sequence Number (SN)—this allows the receiver to support in-sequence delivery of the packet to the upper layer including duplicate detection. PDCP sequence numbers 12 bits wide [UM DRBs, AM Data Radio Bearers (DRBs), and Signaling Radio Bearers (SRBs)] or 18 bits wide (UM DRBs and AM DRBs).
- Data/Control (D/C) bit to differentiate Data and Control PDUs. Control means PDCP control packets and not SRBs.
- The MAC-I field which carries a message authentication code calculated for integrity protection using the AES, SNOW, and ZUC algorithms as specified in 3GPP 33.501. For SRBs, the MAC-I field is always present and set to 0 if integrity protection is not configured. For DRBs, the MAC-I field is only present when the bearer is configured for integrity protection.
- Additional bits associated with PDCP CONTROL PDUs that are used to carry either PDCP status reports of interspersed RoHC feedback
  - PDU Type—used to separate PDCP status report from interspersed ROHC feedback
  - Frame Missing Count (FMC) indicates the COUNT value of the first missing PDCP SDU within the reordering window
  - Bitmap to define missing SDUs

8 bits

D/C	R	R	R	PDCP SN
PDCP SN (cont.)				
Data				
...				
Data				
MAC-I (optional)				
MAC-I (cont.) (optional)				
MAC-I (cont.) (optional)				
MAC-I (cont.) (optional)				

FIGURE 6–3 PDCP PDU example (data radio bearer, 12b SN).

An example of a PDCP header for a DRB with a 12-bit Sequence Number (SN) is shown below in Fig. 6–3. Remainder PDU formats including SRB and different SN lengths are shown in the 3GPP standard.

The PDCP SN number range is configurable between 12 and 18 bits. The reasoning is explained in the RLC section below (Section 6.3).

Receive side PDCP operation includes a state machine/process associated with each radio bearer that contains amongst others:

- a set of COUNT values that indicate next SDU to be transmitted, next SDU expected to be received and first SDU not delivered to upper layers. These values are used to support the in-sequence deliver and associated reordering process. The COUNT value is a concatenation of the hyper frame number and the PDCP SN.
- a set of timers for transmit (Discard Timer) and receive (Reordering Timer) operation supporting timeout operation when lower layers stop performing expected functions (i.e., packet transmission and reception) and associated buffers need to be emptied and reset.

The PDCP protocol does not include any fragmentation and reassembly and thus operates at the GTP/IP packet rate and not at the Transport Block cadence dictated by the Physical Layer and driving the RLC and MAC layers. Implementation of PDCP protocols in software is therefore typically done on IP packet buffers and associated buffer management algorithms as shown in Fig. 6–4. A typical memory management scheme is shown below.

We show two memory domains in different shades here, one presecurity PDCP (IP/IPSec, GTP, PDCP) and one postsecurity PDCP (PDCP, RLC, MAC).

- The presecurity PDCP memory domain This memory domain is characterized by:
- Buffer size defined by transport packet size (IP centric buffer size optimization)
- Single-use buffers—depending on Dual Connectivity implementation at PDCP level

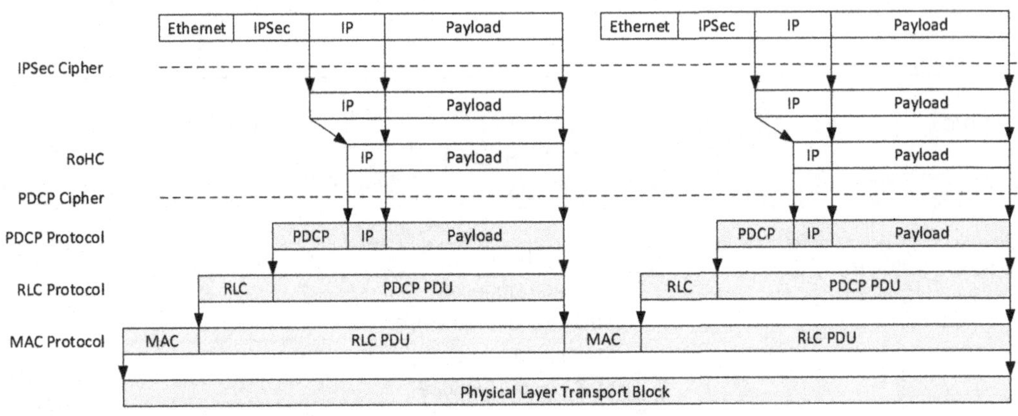

FIGURE 6–4 L2 Stack memory domains.

- Requirement for fast/optimized implementation of physical to virtual and vice versa address translation as required for interfacing to accelerator hardware (hardware security accelerator if used, Ethernet controller)
- The postsecurity PDCP memory domain. This memory domain is characterized by:
- Buffer size defined by maximum transport block size (and/or other L2 defined maximum buffer size)
- Multiuse buffers as required for optimized implementations of HARQ (Hybrid Automatic Repeat ReQuest) and ARQ (Automatic Repeat ReQuest)
- Requirement for fast/optimized implementation of physical to virtual and vice versa address translation as required for interfacing to accelerator hardware (hardware security engine, DMA engine)

Data copy between the two memory domains is handled by the security component, either implemented in software (effectively becoming a memory copy operation) or by means of a lookaside hardware accelerator which has an integrated Direct Memory Access (DMA) controller.

Robust header compression

The "Data" field in the PDCP PDU example shown below (Fig. 6–5) contains the (Ethernet or) IP packet that is delivered over the transport interface to the core network. These packets are optionally compressed with Robust Header Compression (RoHC) which is an Internet Engineering Task Force (IETF) standard (technically, multiple RFCs, starting with RFC3095) that defines a set of header compression schemes for Internet Protocol (IP), Real Time Protocol (RTP), User Datagram Protocol (UDP), IP Security (IPSec) and other headers, allowing for more efficient transmission of small IP packets (e.g., VoIP) over bandwidth sensitive links (e.g., wireless). Although the RoHC specification defines multiple compression schemes, VoIP is the most targeted example (and as such, RTPoUDPoIP). RoHC is a complex protocol (two state machines, various field encoding methods, Cyclic Redundancy Check (CRC), etc.) and as such gives a high loading on cores when implemented in pure software.

The initial version is called ROHC v1 and based on RFC 3095 and RFC 4815 (with clarifications in RFC 3759). It defines UDP/IP, RTP/UDP/IP, ESP/IP and Uncompressed Profiles but is a highly complex standard with many challenges on interpretation and by implication on interoperability.

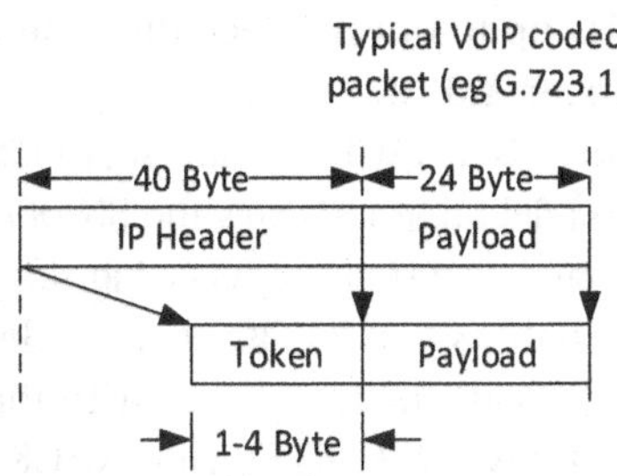

FIGURE 6–5 Robust header compression (RoHC) for PDCP PDU payload size reduction.

Because of this, a simplified version was defined in 2007 in RFC 4995. This version clarifies many of the concepts used but doesn't change the RoHC framework originally defined in RFC 3095. Building on RFC 4995, RFC 4996 defines a TCP profile.

RoHC has been incorporated in 3GPP standards since release 8 (first LTE release) as part of the PDCP standard.

RLC Protocol

As we explained before, the main role of the Radio Link Control (RLC) layer is transferring of upper lay SDUs to the MAC layer for transmission over the Physical Layer. This includes three modes of operation: Transparent Mode (TM), Unacknowledged Mode (UM) and Acknowledged Mode (AM). UM and AM modes both add an RLC header to the packet that contains the following fields:

- SN—this allows the receiver to support in-sequence delivery of the packet to the upper layer. In AM mode, the SN is also used to convey acknowledgment information back to the transmitter for the purpose of triggering retransmission. RLC sequence numbers can be 6 or 12 bits wide for UM mode, or 12 or 18 bits for AM mode operation.
- Segmentation Offset (SO) field which allows the receiver side to know which piece of a segmented RLC SDU packet buffer the received PDU belongs to. This field is combined with the SI field.
- Segmentation Indicator (SI) bits are used to indicate whether this PDU is a segment of an RLC SDU (or not), and if it's a segment, whether it's the start, middle or end of a packet.
- D/C indicator bits that are used to indicate whether this PDU contains RLC SDU data or is used (instead) to transfer RLC layer control signaling (such as acknowledgment information)
- The Polling (P) field is used to trigger transmission of a STATUS report from the RLC peer.
- Additional bits associated with RLC STATUS reports that are used to indicate correct (or incorrect) reception of RLC PDUs by the receiving peer entity, triggering retransmission if required. These fields are relevant for AM mode operation only.
  - Control PDU Type (CPT) that indicates whether this is a STATUS PDU (or, alternatively, reserved type)
  - Acknowledgment SN (ACK_SN) field indicates the SN of the next not received RLC SDU which is not reported as missing in the STATUS PDU. The original transmitter interprets that all RLC SDUs up to but not including the RLC SDU with SN = ACK_SN have been received by its peer AM RLC entity
  - Extension 1 bit to indicate there is a NACK_SN and set of E2/E3 bits following, and Negative Acknowledgment (NACK_SN) field that indicates the SN of the RLC SDU (or RLC SDU segment) that has been detected as lost at the receiving side of the AM RLC entity.
  - Extension 2 bit to indicate there is a n SO Offset set of fields following SO start and end (SOstart, SOend). The SOstart field (together with the SOend field) indicates the portion of the RLC SDU with SN = NACK_SN (the NACK_SN for which the SOstart is related to) that has been detected as lost at the receiving side of the AM RLC entity.

8 bits

D/C	P	SI	R	R	SN
SN					
SN					
SO					
SO					
Data					
...					
Data					

FIGURE 6–6 RLC header example (acknowledged mode, 18b SN).

- Extension 3 bit to indicate there is a NACK range field following and the NACK range field itself. This NACK range field is the number of consecutively lost RLC SDUs starting from and including NACK_SN
- An example of an AM mode PDU with 18 bits SN is shown below in Fig. 6–6. Remainder PDU formats are shown in the 3GPP standard.

The RLC SN number range is configurable as we said above. The appropriate width to be used depends on the target Round Trip Time of the RLC protocol and the target bit rate of the air interface, which together define the maximum amount of outstanding/open RLC sequence numbers for either acknowledgment (AM mode) or receive side re-ordering (UM mode). For example, assuming a maximum air interface latency associated with HARQ of 10Ms (multiple retransmissions in a worst-case scenario) and a 1 Gbps connection with 390B packet size, we can estimate the re-ordering window to be 1 Gbps/[390B x (8 bit/Byte)] x 20Ms ~ = 3200 packets. A 12b SN would be appropriate for this use-case.

Note: The AM mode 18b sequence number option can complicate "straightforward" implementations where the retransmission algorithm is implemented using software array implementations that rely on SN indexing to find a PDU context. 18b corresponds to 256 K entries into this array which (in a DU) can lead to unrealistic memory dimensioning requirements.

Receive side RLC operation includes a state machine/process associated with each logical channel that contains amongst others:

- a window of SN that are not or partially received. On arrival of a PDU that is out-of-sequence, this will either be ignored (if this SN has already been received and is thus interpreted as a duplicate) or the window is advanced to include the latest SN (if the received PDU is ahead of the window).
- a reassembly timer for each RLC (channel) context. Reception of a PDU that contains a partial SDU, this reassembly timer is started.

The associated state machines support fragmentation/reassembly and reliability through retransmission of lost SDU (fragments). The segmentation process involves unpacking an RLC PDU into RLC SDUs, or portions of SDUs. The RLC PDU size is based on transport block size and thus dictated by the MAC layer (scheduler). The MAC PDU (Transport Block) size is defined according to the characteristics of the Orthognal Frequency Division Multiplexing (OFDM) frame and the allocated time/frequency resources by the scheduler and the selected Modulation Coding Scheme (MCS). Large SDUs may be beyond the allocated MAC PDU size and need to be segmented to multiple RLC PDUs. If the RLC SDU is small, or the allocated resources are large, several RLC SDUs may be transmitted in a single MAC PDU. In 4G/LTE systems, this packing (concatenation) is performed by the RLC layer. In 5G/NR, this is done by the MAC layer.

MAC protocol

The Medium Access Control (MAC) scheduler if often interpreted independently from the MAC (and RLC) PDU generation protocol aspects which are different in nature (packet processing centric as opposed to algorithm centric) from the scheduler itself. References like[1] include abstractions of the interface between the scheduler and the remainder components of the MAC layer to allow for implementations by separate teams or companies.

Input to the scheduler are:

- Cell configuration information including bandwidths, antenna count, MIMO configuration
- Client configuration information such as User Equipment (UE) identity (RNTI), allocated logical channels, supported MIMO configurations, Discontinuous Reception (DRX) and MCS
- HARQ, RLC, MAC Control Element, Paging, RACH, Scheduling Request and similar buffer information that can be used to identify requirements for user time/frequency allocation
- Physical layer information such as Signal to Interference and Noise Ratio (SINR), HARQ retransmission rates and Channel Quality Indicator (CQI)
- Triggers to start operation

The output of the MAC scheduler is a downlink and uplink slot definition:

- A trigger to the remainder MAC to start building of Downlink (DL) MAC PDUs and the associated slot configuration
- A set of uplink scheduling decisions in the format of DCIs or an equivalent metaformat

With this information set, the MAC scheduler can be implemented using algorithms that we describe elsewhere in this document, whilst operating under the practical constraints imposed by HARQ turnaround times, bandwidth constraints, and so on.

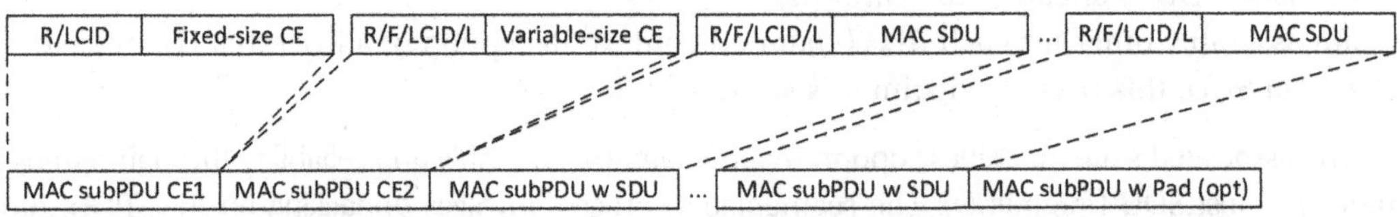

FIGURE 6–7 MAC PDU example (downlink).

The MAC PDU processing component is a traditional packet processing component. MAC PDUs are equivalent to Physical Layer Transport Blocks (TBs) and consist of a sequence of "subPDUs" which can be:

- MAC subheaders with padding or including padding.
- MAC subheaders with a Control Element (CE). The MAC layer defines dozens of Control Elements, used for buffer status reporting, DRX configuration and similar control components that can be communicated through small messages without going to the Radio Resource Control (RRC) layer control stack
- MAC subheaders with a MAC Service Data Unit (SDU). These MAC subPDUs carry RLC PDUs and therefore end-user payload.

The MAC PDU is a concatenation of multiple MAC subPDUs, shown below in Fig. 6–7: Each subPDU has a header which includes the following fields:

- Logical Channel ID (LCID) field identifies the logical channel instance of the corresponding MAC SDU to enable transfer to the appropriate RLC entity. This can optionally be extended to extended Logical Channel ID (eLCID)
- Length (L) field that indicates the length of the corresponding MAC SDU or variable-sized MAC CE in bytes
- A Format (F) field that indicates the length of the Length field (8 bits of 16 bits)
- A Reserved (R) bit, set to 0

eCPRI protocol overview

In LTE macro cell deployments, Common Public Radio Interface (CPRI) is the dominant form of sample transport between BBU and RRH. The CPRI standard defines a GSM/WCDMA/LTE optimized synchronous framing mechanism that carries user data, management and control. There is extensively flexibility for supporting multiple carriers and different

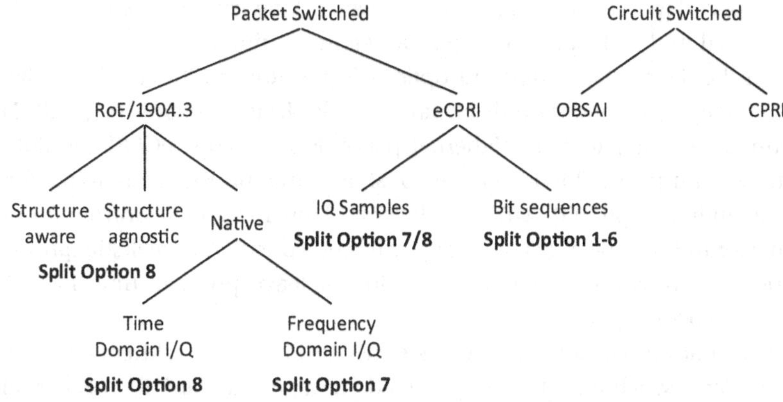

FIGURE 6–8 Fronthaul transport options.

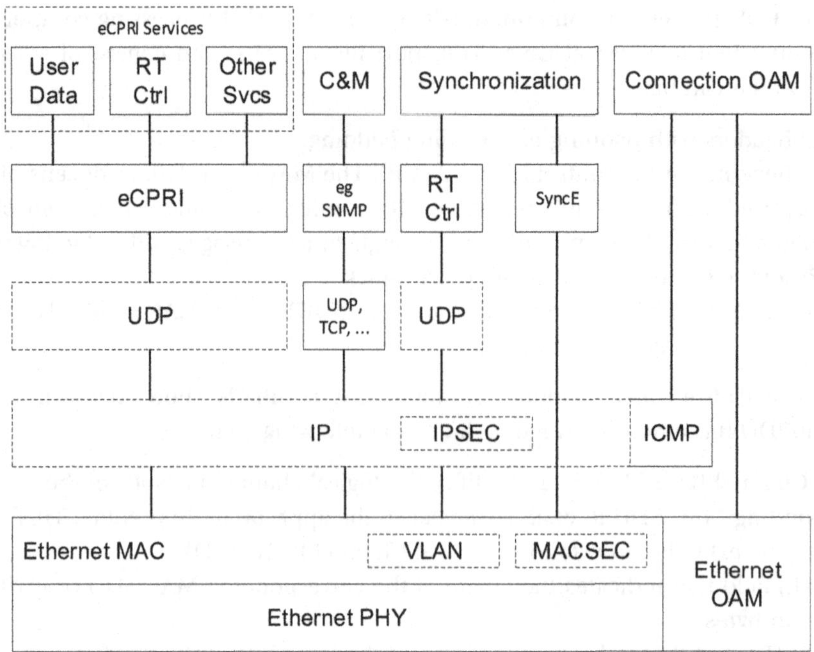

FIGURE 6–9 Fronthaul transport protocol encapsulation options.

sample widths over a single link. However, CPRI is defined for (mainly) 3GPP I/Q transport use cases and as such does not benefit from the flexibility and economy of scale provided by Ethernet.

Given 5G trends, flexibility in partitioning and economy of scale, the interface between the Digital Front-End (DFE) and the Baseband processor is Ethernet/IP based. Connectivity options include custom frame format, eCPRI or RoE as shown in Fig. 6–8 below. Ethernet based I/Q transport depends on Time Sensitive Networking capabilities in the Ethernet standard (not discussed in this document) to guarantee frame timing.

As shown in the figure above, different options in the RoE/eCPRI standards allow for mapping of split option 7 to standardized formats. Standardization efforts (e.g., xRAN Fronthaul Working Group) are ongoing to define Ethernet/packet format across eCPRI and RoE.

For illustrative purposes, below Fig. 6–9 shows the protocol mapping for eCPRI or IEEE1914.3, including optional UDP/IP layers. Transmission Control Protocol/Secure Transmission Control Protocol (TCP/SCTP) are not considered realistic protocol options given the strict requirements for latency that do not leave physical time for ack/nack and retransmission protocol support.

The eCPRI standard protocol layering relies on UDP/IP/Ethernet addressing for source/destination routing. Time synchronization is provided through standard IP mechanisms such as

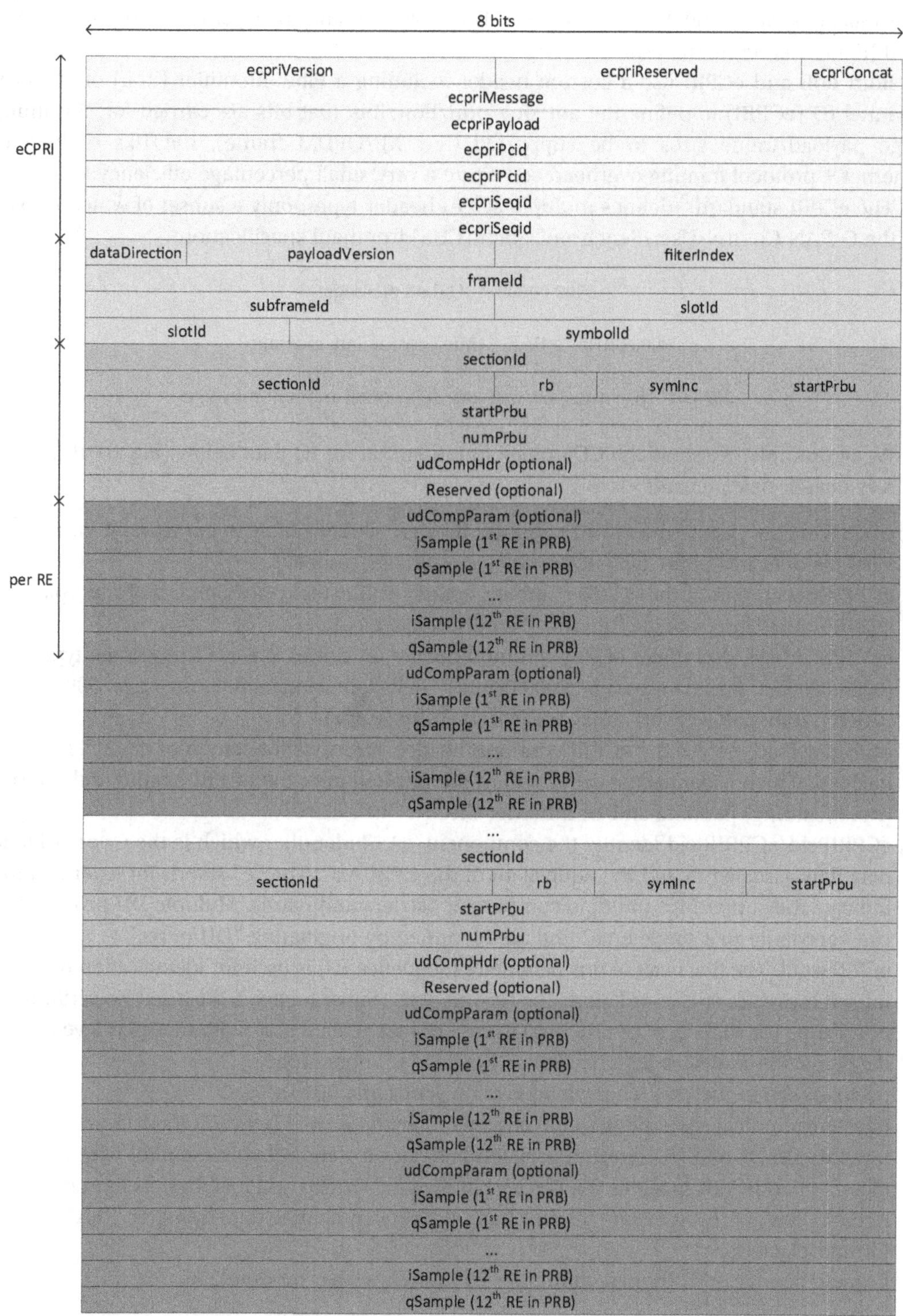

FIGURE 6–10 eCPRI U-plane message format.

IEEE1588. Operation and Maintenance (OAM) and control and management also rely on standard IEEE/Ethernet mechanisms.

Both RoE and eCPRI use a custom header including a Flow Identifier (RoE) or Physical Channel ID (eCPRI) to define the antenna/port/flow/link that bits are carried for. Assuming large payload/frame sizes to be supported (few KB/OFDM frame), the 10 s of Byte of Ethernet + protocol framing overhead constitute a very small percentage efficiency loss.

The eCPRI standard [2] defines multiple eCPRI header types, only a subset of which is used by the O-RAN Control/User/Synchronization (CUS) fronthaul specification:

$$0000\ 0000b = \text{IQ data message}$$

$$0000\ 0010b = \text{Real} - \text{time control data message}$$

$$0000\ 0101b = \text{transport network delay measurement message}$$

As an example, we show the CUS plane frame format for IQ data below (Fig. 6–10):

Key header fields include:

- eCPRIVersion, part of the eCPRI common header. 0001b is interpreted as valid for all versions of eCPRI up to 2.0. Other values are initially reserved.
- eCPRIConcat, part of the eCPRI common header. Indicates concatenation of multiple eCPRI frames into a single Ethernet frame when set to 1.
- eCPRIMessage, part of the eCPRI common header, indicates the eCPRI message type (0000 0000b = IQ data message; 0000 0010b = Real-time control data message; 0000 0101b = transport network delay measurement message)
- eCPRIPayload, part of the eCPRI common header, indicates the length of the eCPRI payload, which is defined as the length of the payload after the eCPRI header and up to (not including) padding and/or Ethernet CRC
- eCPRIPcid/eCPRIRtcid identify the component_eAxC identifier which is the C- or U-Plane data flow association and the equivalent of the CPRI AxC. The "e" stands for extended to indicate the support for multiple component carriers and bands. Multiple DU processes can contribute to a single eAxC and are identified by originating "DU ports."
- eCPRISeqid. The first octet of this parameter (Sequence ID) is used for identification of missed messages and re-ordering. The second octet (Subsequence ID) is used for verification of ordering (or support of re-ordering) in case of fragmentation at radio transport level.
- dataDirection indicates Rx (UL, "0") or Tx (DL, "1") operation.
- payloadVersion, defines the payload version (currently 001b).
- filterIndex indicates an index to the channel filter to be used between IQ data and air interface, in DL and UL. filterIndex 0000b indicates a standard channel filter where the channel bandwidth matches the carrier bandwidth. When used for Physical Random Access Channel (PRACH) in UL, the channel filter can change depending on PRACH format selected.
- frameId counts 10Ms frames, frameId = 20 frame number modulo 256.

- subframeId counts 1Ms subframes within the 10Ms frame.
- slotId counts the slot number within the subframe.
- symbolId counts the symbol number within the slot.
- sectionId identifies individual data sections described within a C-Plane message. The purpose of the section ID is to map U-Plane data sections to the corresponding C-Plane message associated with the data. Two or more C-Plane data section descriptions with same Section ID

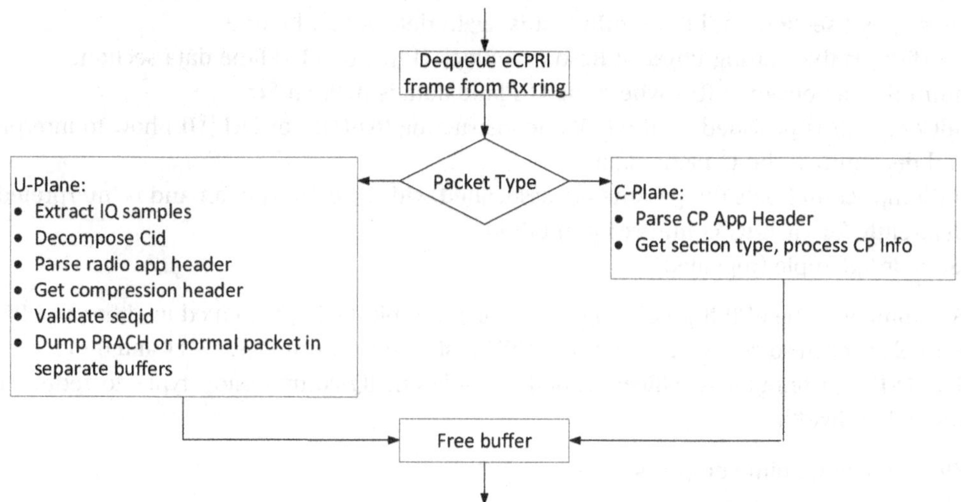

FIGURE 6–11 eCPRI processing flow, receive from ethernet.

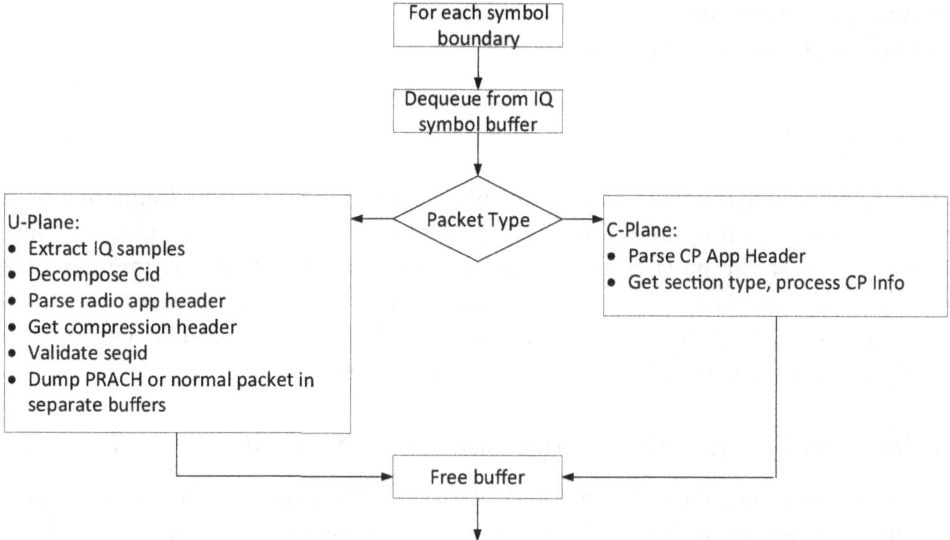

FIGURE 6–12 eCPRI processing flow, transmit to ethernet.

may be cited corresponding to a single U-Plane data section containing a combined payload for both sections (for example when supporting mixed CSI RS and PDSCH).

- Resource Block (RB) used to indicate if every RB is used or every other RB is used.
- symInc used to indicate which symbol number is relevant to the given sectionId. Each C-Plane message maintains a symbol number is and starts with the value of startSymbolid. The same value is used for each section in the message as long as symInc is zero. When symInc is one, the maintained symbol number should be incremented to the next symbol, and that new symbol number should be used for that section and each subsequent section until the symInc bit is again detected to be one.
- startPrbu is the starting Physical Resource Block (PRB) of a U-Plane data section.
- numPrbu defines the PRBs where the U-Plane data is defined for.
- udCompHdr is provided on the U-Plane instructing RU (DL) or DU (UL) how to interpret and decompress the U-Plane data.
- udCompParam holds the parameters associated with IQ compression and is interpreted differently for each IQ compression method.
- iSample/qSample (repeated).

A summary of the eCPRI processing flow, which is typically implemented in either an FPGA/ASIC IP block or in software on a multicore CPU is shown below in Figs. 6–11 and 6–12:

The O-RAN fronthaul specification defines different IQ compression types to reduce the fronthaul bandwidth:

- Block Floating Point compression
- Block Scaling compression
- μ-Law compression
- Beamspace compression and decompression
- Modulation compression
- Selective RE sending compression

Low physical layer

The Low Physical Layer (PHY) is responsible for the conversion of frequency to time domain and antenna preprocessing to prepare the samples for transmission over the air. These steps are the most compute-intensive and therefor typically implemented in Field Programmable Gate Array (FPGA)/Application Specific Integrated Circuit (ASIC) or vector Digital Signal Processor (DSP) centric systems that excel in computational efficiency rather than simplicity of code or other aspects.

Beamforming and time/frequency domain conversion

These components are part of the Physical Layer processing chain and discussed in broad terms elsewhere in this book. One important aspect on beamforming is the centralized nature of the beamforming operation: given that this involves a matrix multiplication between all layers (which are carried over a single eCPRI link to/from the DU) and all antennas (which are egress from the RU), beamforming is an operation that is difficult to split over multiple components. On the other hand, the remainder operations (Fast Fourtier Transform (FFT), Digital Front-End (DFE)) are unique per antenna and hence scale in a manner that allows them to be implemented in a scalable fashion using multiple devices, each of which is allocated to a subset of the RU antenna count. Especially in Massive MIMO systems, this topic becomes an architectural challenge given the contrary directional "pull" between centralized beamforming and distributed remainder processing.

PRACH extraction and filtering

As we said before, the reason for partial offload of PRACH processing to the RU is both to reduce compute resources in the DU as well as to reduce bandwidth between the RU and the DU. A typical PRACH processing chain is shown below in Fig. 6–13:

The PRACH time domain (baseband) signal contains a sequence of PRACH signatures, each prepended by a Cyclic Prefix. The PRACH signal is first filtered to select the target Resource Elements that are dedicated to the PRACH physical channel. This filtering can be done in time domain (Finite Impulse Response (FIR) filter) or in frequency domain by using a large FFT and frequency domain subcarrier selection. After converting the relevant PRACH samples to the frequency domain, they are transmitted from RU to DU for correlation with potential root sequences and detection of the transmitted PRACH sequence.

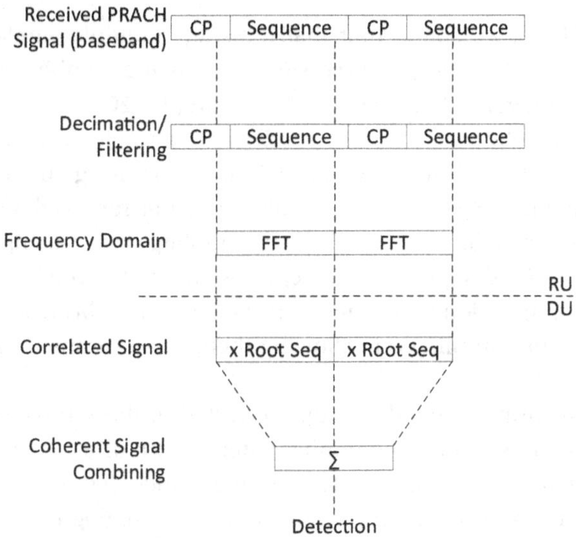

FIGURE 6–13 PRACH processing chain.

Digital front-end

DFE components include (digital) signal conditioning to convert the 3GPP-defined baseband signal to a conditioned signal that compensates for inaccuracies in the analog transmit chain. Functions include:

- Digital Upconversion (DUC) (also known as channelization), is a group of signal processing blocks which create and condition the composite signal built by multiple individual carriers at defined offsets in the spectrum. DUC operates per antenna per carrier. Its operation depends on the targeted output sample rate and carrier bandwidth.
- Crest Factor Reduction (CFR) aims to reduce the Peak to Average Power Ratio (PAPR) of the aggregated carrier signal. The NXP implementation operates per antenna on the composite signal of all carriers for this antenna.
- Digital Predistortion (DPD) is the block which implements a complex nonlinear function that approximates the inverse transfer function H^{-1} of the power amplifier (PA) baseband transfer function H.
  - It is assumed that the DPD feedback loop is implemented as part of the receiver chain in what is a TDD system.
- IQ/Compensation, Gain/DC control to compensate for inefficiencies in external baseband↔RF converters (in case of a baseband I/Q interface and external up/down conversion)
- Interpolation before the DAC in the Tx path and decimation after the ADC in the Rx paths is provided to simplify antialiasing filter requirements.

Signal aggregation and digital up conversion

Assuming that the RU is a wideband unit that transmits multiple wireless communications channels at once, the first step in the processing chain is the combining of these individual signals that can have different characteristics (for example, 2G/GSM, 3G/WCDMA, 4G/LTE and 5G/NR; different amplitudes; different PAPR) into a single combined baseband signal that is processed as one through the next stages in the processing chain (Fig. 6–14). In case of single-carrier operation, this step can be greatly limited or removed altogether.

DUC functionalities include channel filter, upsampling, and mixing. The channel filter aims at limiting Out of Band signals and meeting Adjacent Channel Leakage Ratio (ACLR) requirements. Upsampling is done for multi carrier operation to bring all carriers to the composite (sum of all carriers) sampling rate allowing them to be summed in the digital domain: mixing.

The mixed signal is further filtered and up converted to the sample rate of the data converter. This up conversion stage can be implemented before or after CFR. Filtering supports removal of aliasing signals. Given that analog filtering is included to perform the same function, up sampling filters are designed in conjunction of the analog filter so that both provide a good aliasing rejection. System modeling with Matlab or similar tools is required.

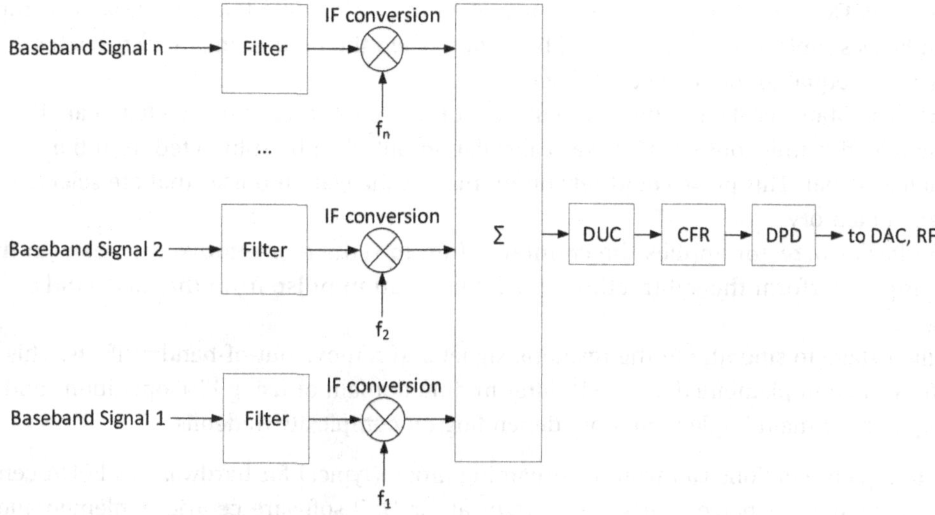

FIGURE 6–14 Signal aggregator.

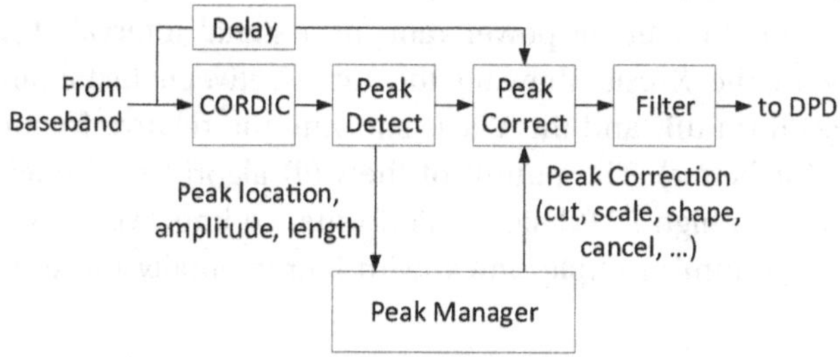

FIGURE 6–15 Example crest factor reduction (CFR) implementation.

Typical filter implementations are Half-Band because of their lower implementation complexity. Note that the receiver operation includes a similar filtering stage to bring the oversampled signal bandwidth down to baseband rate.

Crest factor reduction

CFR algorithms are a relatively simple mechanism to increase the PA efficiency (and hence achievable output power) of the system, even if (in low RF output power environments), DPD is not used to save system complexity. Many implementation options exist, we discuss a very basic one here (Fig. 6–15).

- Power measurement. Typically implemented as a Coordinate Rotation Digital Computer unit dedicated that converts the input signal (Z = A + jB) to polar.

- The Peak Detector detects the occurrence of a peak PAPR situation. The Peak Detector can be as simple as detecting a hard boundary when the input signal X(n) becomes greater of equal to the defined peak level A.
- The Peak Manager defines the corrective action to be taken on the detected peak. For example, this unit could generate a "cancellation pulse" to be subtracted from the original signal. This pulse could rely on premade cancellation pulses that are selected from a memory.
- The Peak Corrector applies the counter action to reduce or remove the peak (in this example, perform the subtraction of the cancellation pulse from the baseband signal).
- A filter stage to smooth out the resulting signal and remove out-of-band artifacts. This filter can be implemented as an FIR filter in time domain or using FFT operations and a frequency domain implementation, depending on complexity tradeoffs.

CFR implementations can work in streaming mode (typical for hardware or FPGA centric implementations) or batches of samples (typical for DSP software centric implementations) where batch processing uses overlapping batches of samples to ensure that peaks that occur in batch boundary areas are also detected.

The Complementary Cumulative Distribution Function is a representation of the distribution of the instantaneous to average power ratio in a signal interval. Typically, it is represented graphically with the X-axis showing the ratio between instantaneous and average transmit power or PAPR (in dB) and the Y-axis showing the relative frequency of occurrence of that specific ratio (or better). The output of the CFR algorithm should yield a curve that has fewer occurrences of a high PAPR signal. Performance improvements include the option to execute the CFR algorithm multiple times, with incrementally more aggressive detection thresholds.

Digital Predistortion

DPD is the block which implements a complex nonlinear function that approximates the inverse transfer function $G = H^{-1}$ of the PA baseband transfer function H. The DPD subsystem is implemented through two components that we discussed before and are shown below in Fig. 6–16:

1. The Actuator or Feed-Forward path—this component applies the DPD adaptation weights in the runtime system path of the system and hence performance critical

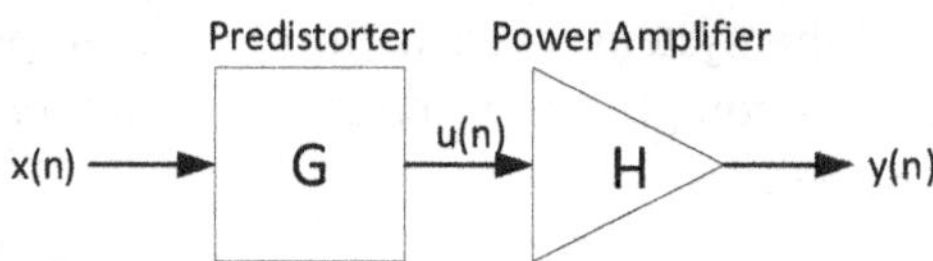

FIGURE 6–16 Feed-forward digital predistorter.

2. The DPD adaptation algorithm—this component calculates the DPD adaptation coefficients or weights and typically runs in a nontime-critical subsystem (for example, on host ARM cores).

The job of the feed-forward path is to apply the predistortion. We differentiate between algorithms for feed-forward in a few dimensions. First, consider the difference between the application of a memoryless or a memory model. A memoryless model assumes (as the name says), the PA not to have any memory—the current output signal is by no means impacted by previous signals transmitted by the PA. The output at each time as such depends only on the current input, adjusted for nonlinearity. The nonlinearity itself is characterized by AM/AM (amplitude of output as a function of the amplitude of the input) and AM/PM (phase of the output as a function of amplitude of the input) curves.

A memory modeled PA becomes relevant for higher output powers and higher bandwidth PAs such as used in enhanced Mobile Broadband (eMBB) 5G/NR applications. Memory effects mean that the output of the PA is not only impacted by the current input signal amplitude but also by the amplitude of previously transmitted signals, represented by previous (digital) samples in the digital domain. Causes for memory effects include thermals and other parasitic effects. A PA that exhibits memory behavior can only be limitedly compensated by a memoryless algorithm. By implication, DPD can become much exponentially more complex to implement for high-end PAs that support higher bandwidth—both the sample rate increases, and memory effects need to be compensated for.

Both memoryless and memory models are commonly pre-distorted by one out of two techniques: polynomial algorithm or Look-Up Table (LUT) based algorithm models. Within the *polynomial models*, Volterra series are used, including derivatives that reduce computational complexity. Derivatives include Wiener, Hammerstein, Wiener–Hammerstein and parallel Wiener structures, or as a Generalized Memory Polynomial[3]:

$$y(n) = \sum_{k=0}^{K_a-1}\sum_{l=0}^{L_a-1} a_{kl}x(n-l)|x(n-l)|^k + \sum_{k=0}^{K_b}\sum_{l=0}^{L_b-1}\sum_{m=1}^{M_b} b_{klm}x(n-l)|x(n-l-m)|^k + \sum_{k=1}^{K_c}\sum_{l=0}^{L_c-1}\sum_{m=1}^{M_c} c_{klm}x(n-l)|x(n-l+m)|^k$$

Where K_aL_a are the coefficient counts for the aligned signal and envelope including memory, $K_bL_bM_b$ are the coefficients for the lagging signal and $K_cL_cM_c$ are the coefficients for the leading signal. Not all coefficients are required, for example, even coefficients are often eliminated altogether. The number of nonzero coefficients defines the compute complexity (and associated power consumption). The implementation of the DPD forward path is like that of a filter and can be optimized with similar techniques including polyphase implementations that allow more efficient parallelization of the math.

Alternatively, an *LUT* can be used where the amplitude of the input sample is used as an index into a memory stored table that contains associated amplitude and phase coefficients. A separate table is used for every lagging or leading memory stage. Interpolation can be used to further increase performance between entries in the table.

The advantage of the LUT approach is that it consumes less multiply-accumulate compute resources as compared to the polynomial methods but is more memory intensive. Both implementation options are used in deployed systems using either FPGAs or (vector) DSP architectures.

DPD coefficient calculation is implemented to support dynamic calculation of the parameters going into the DPD actuator. DPD coefficient calculation is done through least-squares-type algorithm. Assuming (or given) that the PA model changes with parameters like temperature and aging, this piece can be implemented in a nonreal-time environment, including computation on General-Purpose CPU cores.

Parameters like power envelope, aging, and temperature are monitored by sensors embedded in the RF Front End Module. Transmit power envelopes can also be calculated in the digital baseband domain, like Received Signal Strength Indicator (RSSI) measurements as explained for Receive Side Gain Control as discussed below.

Performance of the resulting DPD implementation is measured as algorithmic (ACLR performance meeting or beating 3GPP and regulatory requirements) and in terms of PA DC power efficiency. State-of-the-art implementations near a 50% PA efficiency when meeting 3GPP −45dB ACLR targets.

Receive side gain control

The goal of Gain Control algorithms is to manage the ADC dynamic range, so it operates at highest possible efficiency - avoid both saturation and underflow and thus utilize the available effective number of bits precision in the data converter.

Gain setting can either be manual, automatic or combination of both. Manual implies the receive chain being fully under user control. This requires user to have perfect knowledge of the receiver and wireless signal in real-time. This is typically possible only in case of static environments such as Fixed Wireless Access deployment. This can be looked at as a quasiopen loop approach.

On the other hand, automatic is a closed loop approach where Rx power is measured, and the receiver chain is tuned accordingly. Gain control needs to be balanced between quick conversion to the target value whilst avoiding overshoot. In OFDM system, gain changes should be applied on symbol boundaries (within the OFDM Cyclic Prefix) and thus requires timing control. Signal processing chains needed for gain control include digital Receive Signal Strength Indication (calculate the sum of $I^2 + Q^2$).

Bursty traffic, dynamic user scheduling can drive very fluctuant power between slots. If user has some knowledge of this profile, it can use it to help ADC gain setting, which can lead to combined manual and automated gain control.

References

1. LTE MAC Scheduler Interface Specification v1.11.
2. CPRI eCPRI Specification v2.0, "*Common Public Radio Interface: eCPRI Interface Specification*," May 2 2019.
3. Morgan, D., Ma, Z., Kim, J., Zierdt, M., Pastalan, J. A generalized memory polynomial model for digital predistortion of rf power amplifiers. *IEEE Trans. Signal. Process.* **2006,** *54* (10).

Further reading

ETSI. NR; Physical channels and modulation, 3GPP TS 38.211.

ETSI. NR; Multiplexing and channel coding, 3GPP TS 38.212.

ETSI. NR; Physical layer procedures for control, 3GPP TS 38.213.

ETSI. NR; Physical layer procedures for data, 3GPP TS 38.214.

ETSI. NR; Physical layer measurements, 3GPP TS 38.215.

5G FAPI: PHY API Specification, https://www.smallcellforum.org/5g-phy-api-release/.

ETSI. NR; Requirements for support of radio resource management, 3GPP TS 38.133.

ETSI. NR; Base Station (BS) radio transmission and reception, 3GPP TS 38.104.

https://github.com/LITMUS-RT/cyclictest.

https://www.enea.com/products/operating-systems/real-time-accelerated-linux/.

https://www.6wind.com/6windgate/.

https://wiki.fd.io/view/VPP/What_is_VPP%3F.

7

Wireless scheduling and Quality of Service optimization techniques

Based on the work of Adam,[1] we start with a short summary outlining the unique aspects of the orthogonal frequency division multiple access [OFDM(A)] waveform and its impact on wireless scheduling algorithms from a theoretical point of view. Beyond theoretical analysis of typically employed algorithms, we include two things. First is an architectural framework for a practical scheduler implementation that balances compute complexity, performance, and the various aspects of scheduling algorithms [time domain, frequency domain, multiple input multiple output (MIMO)] that are typically are discussed independently in literature.

Second, we include several software optimization techniques that can be employed to improve scheduler performance in typical general-purpose processor (GPP) software-defined radio implementations, thus enabling the reader to achieve real time performance for the algorithms discussed.

This text includes a long list of literature references. This is done on purpose to allow the reader to investigate algorithmic enhancements and system performance optimizations that are outside the scope of this text.

Orthogonal frequency division multiple access

Orthogonal frequency division multiplexing (OFDM) was developed specifically for wireless applications because it offers significant advantages over legacy spread-spectrum solutions such as direct-sequence spread spectrum used for example in 802.11b. These advantages include excellent resistance to selective fading and robustness to temporal dispersion.[2] OFDM is also used in nonwireless applications, such as DSL (DMT or Discrete Multi-Tone is very similar to OFDM) and powerline networking (e.g., HomePlug) for similar reasons.

OFDM helps resolve wireless applications issues by breaking one high-speed data stream into lower speed data streams, which are then transmitted simultaneously in parallel. Each lower speed stream is used to modulate a subcarrier. This creates a multicarrier transmission by dividing a wide frequency band (or channel) into narrower frequency bands (or subchannels) (Fig. 7–1).

Frequency is shown on the vertical axis and time on the horizontal axis. Both single-carrier and OFDM modes occupy the same channel bandwidth and, under ideal conditions, give the same data rates. It is the robustness of OFDM under less than ideal conditions that gives it a practical advantage over the wideband single-carrier mode.

Open Radio Access Network (O-RAN) Systems Architecture and Design. DOI: https://doi.org/10.1016/B978-0-323-91923-4.00004-5

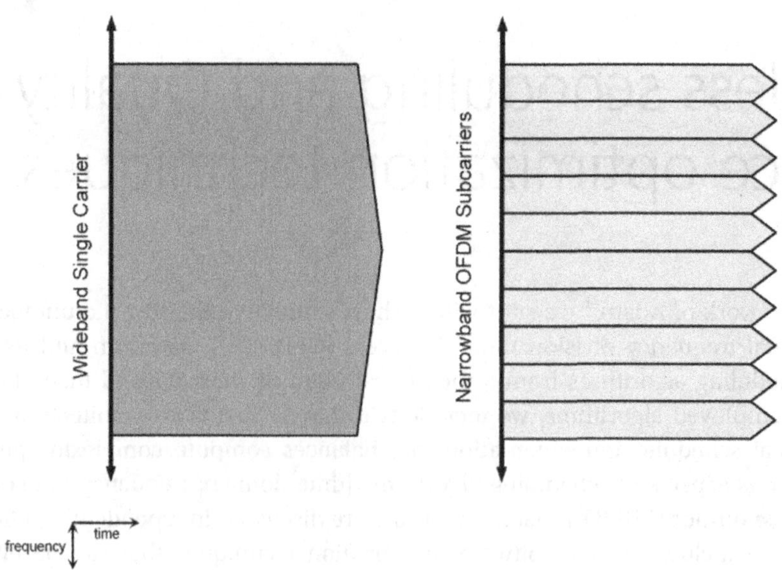

FIGURE 7–1 OFDM conceptualized. *OFDM*, orthogonal frequency division multiplexing.

In OFDM, the subcarrier pulse used for transmission is rectangular. This shape is chosen so that the pulse-forming and modulation can be performed by a simple inverse discrete Fourier transform (IDFT). The IDFT can be efficiently implemented by an inverse fast Fourier transform (IFFT). At the receiver, a fast Fourier transform (FFT) is performed to reverse the operation.

The IFFT leads to a sin(x)/x type of spectrum of the subcarriers, shown in Fig. 7–2.

Although the spectra of the subcarriers overlap, the information transmitted over the subcarriers can still be separated due to the orthogonality relation. Orthogonality is the ability to distinguish every subcarrier at the receiver without any interference from other subcarriers. The mathematical property of orthogonality states that the integral of the product of any two subcarriers is zero. By using an IFFT for modulation, the spacing of the subcarriers is chosen such that at the frequency where the received signal is evaluated, all other signals are zero. However, to preserve this orthogonality, the following conditions must be met:

- The receiver and transmitter must be perfectly synchronized. This means they must both assume exactly the same modulation frequency and the same timescale for transmission. This requires a sophisticated receiver.
- The analog components of the transmitter and receiver must be of very high quality.
- There should be no multipath channel.

The last point is particularly disappointing because this approach was chosen specifically to combat the multipath channel (increasingly problematic below 11 GHz). Fortunately, there is a solution to this problem in the artificial prolongation of the OFDM symbol. This is done by periodically repeating the "tail" of the symbol and preceding the symbol with this

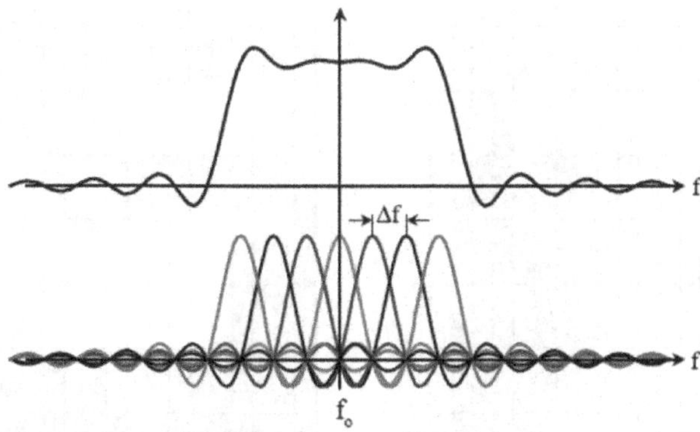

FIGURE 7–2 IFFT to implement OFDM efficiently. *IFFT*, inverse fast Fourier transform; *OFDM*, orthogonal frequency division multiplexing.

repetition. At the receiver, this so-called cyclic prefix (also known as guard interval) is removed. If the length of this interval Δ is longer than the maximum channel delay, all reflections of previous symbols are caught within the cyclic prefix and orthogonality is preserved. The cost of this solution is the extra time spent repeating the symbol cyclic prefix, resulting in less time available for transmitting information and therefore a lower data rate.

The channel response can be measured by inserting known "pilot" subcarriers at regular intervals in time or frequency at the transmitter. The subcarriers are monitored at the receiving station and compared to their known expected value. By dividing the received value by the expected one, the channel response for that pilot time-slot and pilot subcarrier can be obtained. Each individual time-slot and subcarrier channel responses can be found by interpolation. The original signal can be determined by dividing through the interpolated channel influence.

The principal advantage of dividing up the frequency band into many narrowband subcarriers in OFDM is that the effect of wireless channel impairments can be reduced. The most critical of these problems, multipath fading, is the result of receiving a single transmitted signal multiple times at the receiver antenna. This is due to reflections caused by the multiple paths that a radio signal can travel as it meets physical obstructions while it propagates through the air. Variations in amplitude, phase, and arrival times are all consequences of reflections. Multipath can also result in frequency-selective fading, which is a degradation of specific frequencies or frequency ranges across a channel, as particularly out-of-phase signals cancel each other out. While this can have a dramatic effect on a wideband single carrier signal, OFDM counters this problem relatively easily.

OFDMA is an enhanced version of OFDM which divides a signal into subchannels (i.e., groups of subcarriers), allowing different subscribers to be allocated different subchannels as well as different time intervals as shown in Fig. 7–3. Subchannels are combined from subcarriers chosen either pseudo-randomly or sequentially. Each subscriber is treated separately according to location, distance from the base station, interference, and power requirements. Different modulations can be used for each of the carriers in the system to improve coverage and throughput.

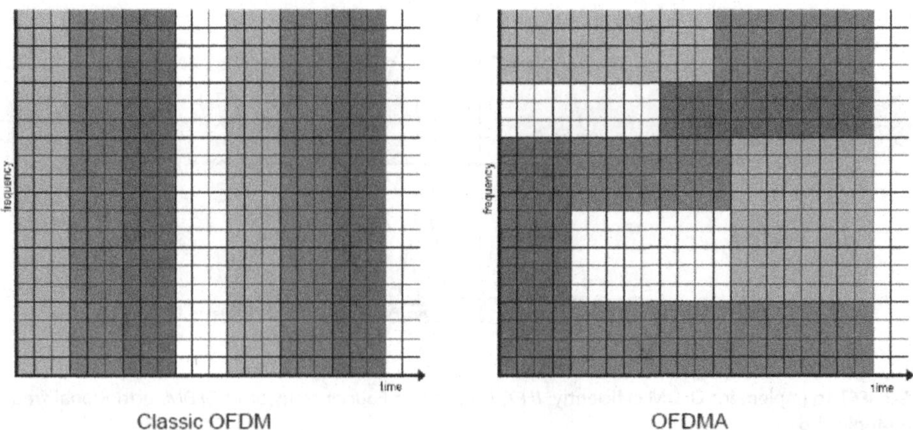

FIGURE 7–3 OFDM versus OFDMA. *OFDM*, orthogonal frequency division multiplexing; *OFDMA*, orthogonal frequency division multiple access.

The figure shows allocations to different users with different shades of gray for each user. Subchannels are represented as rows, OFDM symbols are represented as columns. OFDMA allows the transfer of data by variably allocating both subchannels and symbols, whereas OFDM is classically limited to only variably allocating symbols. The ability to variably allocate data on both the frequency and time domains is what allows OFDMA to be more flexible and efficient.

Orthogonal frequency division multiple access subcarrier allocation

As previously stated, an OFDMA symbol is made up of multiple subcarriers. The size of the FFT used to convert frequency-domain to time-domain signals determines how many are present. Typically, FFT sizes of 512, 1024, 2048, and 4096 points are used, yielding corresponding numbers of subcarriers. However, not all subcarriers are used for data transmission. A subcarrier can be assigned one of three functions: transmission of data (data subcarriers), channel estimation (pilot subcarriers), or guard band (null subcarriers). Data and pilot subcarriers are organized into subchannels (4G: resource block; 5G: resource element), where one subchannel may be assigned dozens of subcarriers. The subcarriers forming one subchannel may or may not be adjacent, depending on the air interface standard. 4G and 5G implement adjacent subcarrier allocation. The Medium Access Control (MAC) scheduler must be aware of the proportion of data subcarriers to null and pilot subcarriers as this will affect how much data can be mapped onto the OFDMA frame.

Adjacent subcarrier allocation results in different subchannel responses to different users, due to frequency-selective fading as shown in Fig. 7–4 the figure. As frequency-selective fading rarely affects all users identically, adjacent subcarrier allocation allows users to be allocated the subchannels which give them best channel quality. This topic has been extensively discussed in literature, and performance gains of up to 200%–300% have been achieved

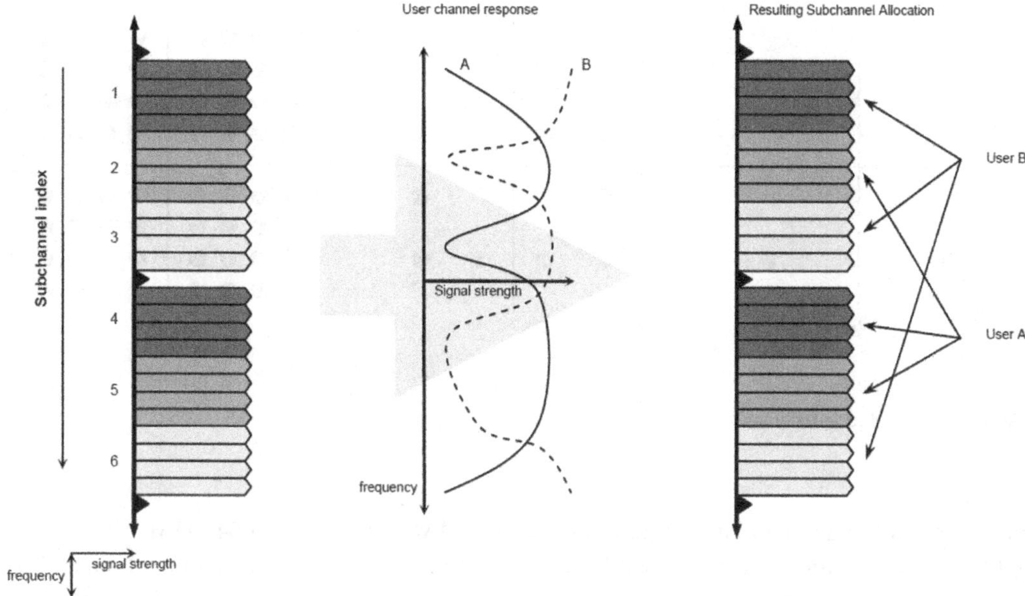

FIGURE 7–4 Simplified illustration of frequency-selective fading with adjacent subcarrier permutation.

using this type of subcarrier allocation.[3,4] The downside of the adjacent subcarrier allocation is a higher level of complexity: In order to allocate better subchannels to a user, an up-to-date knowledge of the exact subchannel response is necessary.

Note that Carrier Aggregation (CA), where a user is communicating to the base station over multiple allocated frequency channels concurrently, can be seen as a variety of OFDMA where each frequency channel can have a different channel characteristic/response and thus can benefit from scheduling/allocation schemes as shown above.

Modulation

The OFDMA air interface allows different modulation schemes to be applied to each subcarrier (Fig. 7–5). Modulation schemes include quadrature phase shift keying (QPSK), 16 quadrature amplitude modulation (QAM), and 64-QAM.

The phase of a dot is indicated by the angle a line from it to the origin makes with the positive x-axis. The amplitude is the length of this line. Thus, for QPSK, four different phase–amplitude combinations are possible (2-bit). While 16 are possible for 16-QAM (4-bit), 64 are possible for 64-QAM (6-bit). This figure also illustrates the greater susceptibility to noise for 64-QAM versus 16-QAM and QPSK modulations: a smaller amount of noise will cause a misinterpretation of one symbol for another.

A wireless network can dynamically adjust the modulation method depending on varying channel conditions. If the channel is detected as having a lower signal-to-interference ratio

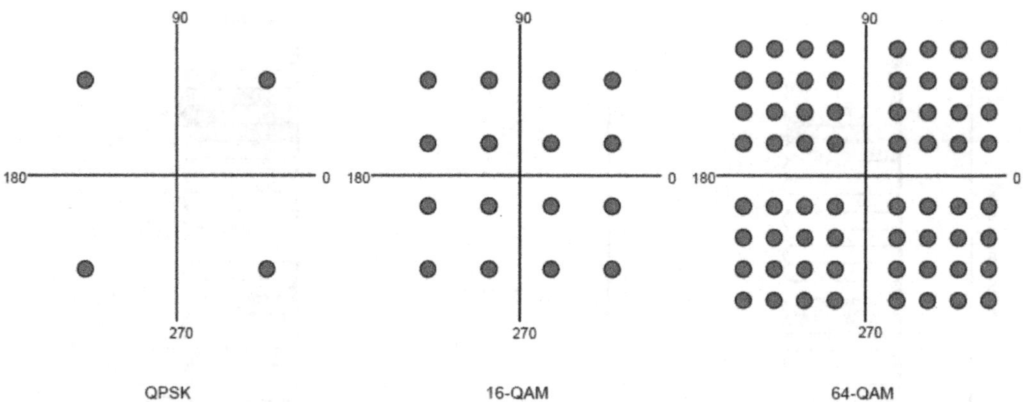

FIGURE 7–5 Modulation schemes.

(SIR), the modulation can be dropped down to 16-QAM or QPSK from 64-QAM. The bit rate will be lower, but the connection will have fewer errors. As such, modulation is adjusted by the physical layer to maintain an error ratio below a given level (e.g., 10e-6 bit error rate). Modulation can be individually adjusted per client, ensuring greater flexibility and efficiency. The MAC scheduler controls the modulation used, which impacts the amount of area the data takes up on the OFDMA frame.

Base station scheduler algorithms

Broadband Wireless Access (BWA) technologies have been studied and implemented since a long time, with 3G, 802.16/WiMAX, LTE, 5G, and 802.11/Wi-Fi standards as examples. Current-generation BWA systems rely on OFDMA and CA as methods to enable multiple user access. Note that a lot of the early research work has been done targeting 802.16/WiMAX networks as this was an early commercial system that implemented OFDM/OFDMA multiuser access.

The air interface typically being the bandwidth bottleneck, bandwidth allocation optimization is an important topic. Scheduling is the job responsible for the allocation of (time, frequency) resources across users. Its job is to decide which user can transmit/receive data at which point in time and where in the frequency domain, using a defined modulation and coding scheme (MCS).

There are two main approaches to the scheduling problem in wireless communications systems: the Quality of Service (QoS)-centric approach that operates regardless of the user channel state information (CSI), and the opportunistic approach that use CSI to optimize for spectral usage. Both have merits depending on the objectives. The following subsections give an overview of both approaches. Further details can be found by referring to the research papers directly.

Opportunistic scheduler

Opportunistic schedulers are schedulers that take advantage of physical layer (CSI) information to attempt to optimize to (typically) aggregate (system-wide) wireless capacity, optionally combined with adherence to minimum QoS. The motivation behind opportunistic scheduling is that since bandwidth is a limited commodity, efficient use of the available resources should be maximized. Research on previous attempts at channel capacity maximization reveals that there are three main ways of achieving this aim. The first is to schedule transmission to a user only when channel conditions are "good" for that user, taking advantage of the time-varying channel. The second option is to schedule transmission to a user on the frequency band (subchannels or subcarriers in the OFDMA signal) that yields a good transmission level. This takes advantage of frequency-varying channel conditions. The third alternative is to increase frequency reuse by allocating transmission power to subchannels dynamically, such that the resulting signal-to-noise ratio (SNR) is only just tuned to maintain the transmission in an optimized fashion, minimizing total radiated power, reducing cell size, and increasing frequency reuse. The following paragraphs provide additional details on the various implementations of these types of approaches.

The traditional execution of the opportunistic scheduler is to schedule data transmissions to users for which the channel conditions are good. This is known as multiuser diversity. Various research papers investigate the performance benefits of this approach. A landmark paper written by Knopp and Humblet explains that, as the number of users grows, the spectral efficiency can increase due to the multiuser diversity.[5]

Temporal multiuser diversity can be taken advantage of by selectively allocating bandwidth in the time domain. This is done by scheduling the user with highest channel quality the entire channel bandwidth at any point in time. This type of multiuser diversity gain was demonstrated by Bhagwat et al., who achieves a 15% performance improvement by applying this approach to the physical layer.[6] This scenario is different from the one a 4G/5G OFDMA scheduler is faced with, as the MAC layer in the presented paper is not aware of any preferential transmissions. Smith et al.'s approach is a popular one, with other authors such as Fragouli et al. using it as the basis for their own slightly improved algorithm.[7]

Andrews et al. also describe that the quality of a wireless channel is typically different for each user and changes randomly in time, both on slow and fast timescales.[8] The algorithm in this paper takes advantage of channel variations by giving some form of priority to users with temporarily better channels. Another paper where temporal multiuser diversity is taken advantage of is Ref.[9] where an algorithm for opportunistic transmission scheduling yields 20%–150% performance improvements over those which do not consider this. Viswanath et al. even suggest using multiple antennas to create artificial variations in channel quality over time, in order to better exploit them.[10] The exploitation of temporal multiuser diversity is investigated by Knopp and Humblet whose results suggest that in order to attain highest capacity, only one user should transmit at any given time over the entire bandwidth.[11]

Eryilmaz et al. present a particularly interesting approach to taking advantage of temporal multiuser diversity.[12] Their scheme attempts to do so without any explicit channel quality information, relying instead on traffic flow statistics. Considering that channel quality information is fed back and thus always delayed and inaccurate, this approach seems a reasonable one.

Zhimei Jiang and Shankaranarayana demonstrate why in most cases, it is preferable to favor users with good channel qualities.[13] The argument is that attempting to assign more resources to users with lower channel quality may degrade good users' performance in a disproportionate manner. This line of reasoning is highly relevant in the case of 802.16 scheduling, where a user with "bad" channel quality (e.g., QPSK modulation) will consume much more bandwidth compared to a user with "good" channel quality (64-QAM modulation) for the same amount of transmitted data.

Another type of multiuser diversity can be found in Ref.[14], where the authors outline a strategy to maximize the total packet throughput on an OFDMA network by making use of judicious subcarrier allocation (and implies users to be allocated adjacent OFDMA carriers as is done in 4G/5G systems). This is taking advantage of the frequency-selective nature of multiuser diversity. The system the authors consider assumes perfect knowledge of users' CSI, which is perhaps not always the case. In this paper, Guoqing Li and Hui Liu also refer to Pottie, who has demonstrated that this judicious subcarrier allocation can outperform interference-averaging techniques by a factor of 2–3 in spectrum efficiency.[3] The frequency-selective aspect of multiuser diversity is widely discussed in literature and taken advantage of in papers such as Ref.[15] (where the authors varying subcarrier and power allocations simultaneously), Ref.[16] (where an algorithm allowing a reduction in transmission power by 6–10 dB is presented), and Ref.[17] (where a real-time algorithm implementing a subcarrier allocation scheme is presented).

Kittipiyakul and Javidi disagree with this purely instantaneous throughput maximization technique because it causes rate-instability in an otherwise stabilizable system (i.e., queues face unbounded buildups).[18] Their alternative proposal is more optimal in the long term, which is a trade-off between two competing goals: the desire to get maximum throughput now and the desire to get maximum throughput in the future. This approach is one step toward combining the QoS and opportunistic scheduled approaches.

Another parameter to consider when maximizing total throughput is radiated power. As broadcast power is a limited resource, a sensible approach is to investigate the effects of varying allocated broadcast power depending on various factors. Some schemes rely on the minimizing of overall transmit power to reduce interference to other users, allowing greater reuse of available spectrum, for example. This is often termed the "water-filling" method. Such approach is outside the scope of this project but remains an interesting one, nonetheless. Further information can be found in Ref.[19] (a computationally inexpensive water-filling algorithm) and Ref.[20] (an uplink transmission algorithm minimizing total power, following a first-come first-served basis).

Grossglauser and Tse present an interesting and novel approach also taking advantage of multiuser diversity, by which they show that mobility (counter to intuition) increases the capacity of wireless networks.[21] Their algorithm assumes a mesh network topology, where users relay data to one another instead of relying on a central base station. The authors demonstrate how user mobility allows taking advantage of what can be termed spatial multi-user diversity. However, the authors do concede that their approach is probably not well suited to real-time applications such as voice communications.

It is now clear that significant performance gains can be achieved using an opportunistic approach. However, despite it revealing many interesting possibilities, such an approach alone is not suited for a 4G/5G scheduler, due to its inability to meet QoS guarantees.

The Quality of Service scheduler

Data scheduling from a QoS-centric approach concentrates on maximizing fairness and meeting QoS requirements. This method is generally used in communications systems that are relatively reliable (i.e., have a low bit error ratio (BER)), and where all users have a similar, unchanging connection. However, this approach is more difficult if channel conditions are worse or vary with time, frequency, and/or location. The reason for this is that scheduling transmissions in the future cannot be guaranteed: the channel may have changed, or retransmissions may be necessary. This section begins with a description of traditional wireline scheduling algorithms, such as for Asynchronous Transfer Mode (ATM) networks, followed by a discussion of various adaptations of these algorithms for wireless networks.

All traffic is not created equal: some traffic types are more susceptible to certain network degradations than others. For example, some traffic may be very sensitive to delay, while other traffic may be more vulnerable to overall network throughput. A connection-oriented network such as ATM attempts to allow for these different traffic types by implementing multiple service categories. Depending on their requirements, different scheduling algorithms may be chosen for different service categories. These algorithms can generally be separated into two categories: priority-based and fair-share schedulers. Priority-based schedulers manage traffic from different queues according to some defined priority criteria. Lower priority traffic is transmitted after higher priority traffic. In contrast, fair-share schedulers attempt to distribute bandwidth among the different traffic types such that some notion of fairness is preserved. Various scheduling algorithms falling into either of these two scheduler categories, such as the earliest deadline first (EDF) algorithm, or the Round-Robin (RR) algorithm, are summarized by Giroux and Ganti.[22] It is assumed that the reader is familiar with these basic types of scheduling algorithms.

Most research published on the topic of QoS scheduling is based on fair scheduling, and how to adapt this type of scheduling to wireless networks. A landmark paper by Lu et al. in this domain describes the adaptation of the wireline fluid fair queuing algorithm to handle location-dependent error bursts.[23] This work is also cited in another significant paper by Ramanathan and Agrawal,[24] where the difficulty of adapting fair queuing algorithms to wireless networks is considered. Their proposal aims to provide a long-term fairness guarantee while making efficient use of the air interface. This is achieved mainly by providing supplemental bandwidth to sessions that, due to poor quality of the wireless channel, have not received satisfactory service in the short term. The fairness is ensured by keeping track of the supplemental bandwidth granted in a "fairness server." This research is cited and built upon in numerous papers, including work by Nandagopal et al., which evaluates the performance of seven wireless fair queuing algorithms.[25] The same authors later write an excellent summary of the issues and approaches to fair queuing in wireless networks[26] and present a further improved algorithm taking Automatic Repeat Request (ARQ) into account for error-prone wireless channels in Ref.[27] Ng et al. discuss other algorithms aiming to counter location-dependent errors in wireless networks in Ref.[28] Work in this field actively continues: Namgi Kim and Hyunsoo Yoon have recently presented another fair queuing algorithm with link level retransmission (ARQ) in Ref.[29]

Unfortunately, fair queuing is expensive to implement in processor resources. This issue has been tackled by Shreedhar and Varghese, who achieve almost perfect fairness using a modified RR algorithm.[30] Other variations on the topic of fair scheduling include adapting fair scheduling to multicarrier systems such as OFDM,[31] and to systems with multiple transmit and receive antennas (MIMO).[32]

QoS priority-based scheduling has received comparatively little attention. Priority scheduling is particularly adapted to real-time traffic such as streaming audio or video. Shakkottai and Srikant have attempted to modify the earliest deadline due (EDD, like EDF algorithm) to function over a wireless channel.[33] Their algorithm uses a finite-state Markov channel model[34] to represent “good” and “bad” channels. In contrast, Veciana and Yang have discussed a way of improving user perceived performance for best effort traffic.[35] Yaxin Cao Li and Zhigang Cao have combined real-time and best effort traffic scheduling.[36]

In the scheme of joining multiple algorithms to serve different traffic types, we also find research by Wongthavarawat and Ganz, which deals specifically with traffic scheduling for an 802.16/WiMAX network, from a QoS perspective.[37] The paper is novel in merging together multiple wireline algorithms for 802.16 traffic scheduling, and in considering the uplink scheduling, which has thus far been largely ignored. However, no attempt is made to enhance any of these algorithms specifically for wireless systems, or to take advantage of multiuser diversity. A similar paper that compiles three different algorithms to serve different traffic types for an 802.16/WiMAX network is presented by Chu et al. in Ref.[38] Their proposal combines a wireless packet scheduling algorithm for real-time services with a weighted-RR for nonreal-time services and a First-In-First-Out for best effort services.

A paper that focuses on real-time VoIP traffic is presented by Lee et al.[39] The proposal uses a special bit in the MAC header to implement a proprietary scheduling scheme for VoIP traffic in an 802.16 network. The authors claim that this method achieves better performance than if VoIP traffic were mapped on real-time connections with implied grants.

Lee et al.'s algorithm could be combined with a scheme presented by Kim et al., which focuses on a nonreal-time scheduler with a minimum bit rate.[40] This proposal could be used to manage nonreal-time services flows. The authors claim that their algorithm significantly increases the number of nonreal-time services without compromising much cell throughput.

Despite many authors having presented solutions implementing QoS scheduling to transmit data over a wireless channel, few have combined algorithms guaranteeing QoS while making efficient use of channel resources. The following section discusses the efforts made in this direction.

Combined Quality of Service and opportunistic schedulers

Certain authors attempt to combine a QoS-approach to scheduling, while attempting to achieve efficient use of spectrum. A combination of adjacent subcarrier allocation and fair scheduling is presented by Ergen et al.[41] The authors admit that although an optimal solution to the problem is unlikely to be implementable in real time, their suboptimal solution allocates data fairly and converges close to meeting the optimal QoS criteria per symbol. It is, however, limited by the characteristic problem of not knowing CSI between the transmitter and receiver.

Wu and Negi also attempt to combine multiuser diversity and efficient QoS support over fading channels.[42] The authors claim a substantial increase in channel capacity, but only when delay requirements are not very tight.

Recently, Khattab and Elsayed have presented a scheduling scheme based on the EDD algorithm that exploits multiuser diversity and can provide statistical guarantees on delays.[43] The scheme achieves high throughput and exhibits good fairness performance with respect to throughput and deadline violations. The tests and simulations performed were on a 1.25 MHz CDMA channel, but such algorithm could be very useful for scheduling real-time traffic if adaptable to an OFDMA network.

A recent paper from Farrokhi et al. combines QoS scheduling and the efficient use of channel resources.[44] This builds on a previously submitted paper by the same authors.[45] It proposes to allocate specific subcarriers to specific users, thus taking advantage of frequency-selective multiuser diversity. The QoS is satisfied by allocating enough subcarriers, even if there are not enough "good" ones available. The paper assumes perfect knowledge of CSI, which is a challenge.

Multiuser multiple input multiple output

MIMO is a radio technique that transmits and/or receives over multiple antennas to exploit spatial diversity for increasing system throughput.

Consider the system in Fig. 7–6 as an example.

Different MIMO implementations exist, including single-user MIMO which uses transmit and receive diversity to increase throughput to a single user, multiuser MIMO where spatial diversity across users is used to simultaneously transmit and receive to multiple users and

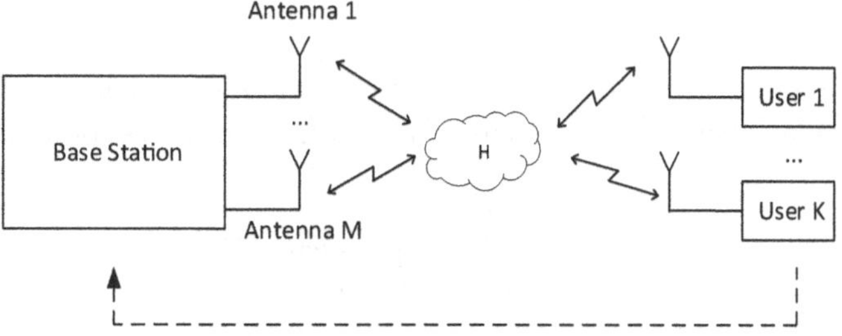

FIGURE 7–6 Multiuser MIMO system. *MIMO*, multiple input multiple output.

massive MIMO, which is a scaled-out version of multiuser MIMO to a high antenna count (say, 64 antennas).

Find a primary user using opportunistic or QoS-based scheduler algorithms and then pair the primary users with others that cause minimum interference between each other. This pairing is done by maintaining a metric of "orthogonality" between all user pairs that measures the spatial compatibility between MIMO users. Users are spatially compatible if the multiuser MIMO channels of those users can be separated in the spatial domain by means of beamforming and precoding.

Consider a broadcast channel between a single transmitter (base station) and many users,[46] like shown in Fig. 7.6:

$$y = Hx + n$$

- H is the $r \times t$ complex matrix where the (i,j)th entry defines the path between the jth transmit antenna and the ith receiver.
- y is the $r \times 1$ vector representing the received signal for the ith receiver where $I = \{1, \ldots, r\}$.
- n is the complex Gaussian noise vector.

Denote h_i as the ith row of H corresponding to the channel between the base station and the ith receiver. The channel is assumed to be memoryless and quasistatic and known to both the transmitter and receivers, which is the case in a Time Division Duplex implementation.

The number of receivers (users) is arbitrary but limited by the number of transmit antennas. The number of served users thus becomes a resource allocation problem: The scheduler selects from all active users, at any chosen time, a subset of users (and their allocated resources such as OFDMA subcarriers and transmit power) to maximize performance. Performance can be defined as sum rate across users at a given power, or similar. Intuitively, it can be understood that selected (scheduled) users need to exhibit both high SNR and good separability in the special domain (i.e., orthogonality) to ensure multiplexing is successful.

Popular algorithms are greedy, where the transmitter chooses the user with the highest channel capacity first and then finds the next user that provides orthogonality from the first user. Orthogonality can be measured as the coefficient of correlation:[47]

$$\cos\left(coeff\left(H_i, H_j\right)\right) = \frac{\left|H_i H_j^H\right|}{\left\|H_i\right\| \left\|H_j\right\|}$$

or by estimating the maximum sum rate from the remaining unselected users, which is provided by the condition number of the channel matrix from all users combined[48]:

$$cond\left(\begin{bmatrix} h_{\pi(1)} \\ h_{\pi(2)} \\ h_{\pi(i)} \end{bmatrix}\right) < threshold$$

where $cond(A) = \left\|A^{-1}\right\| \times \left\|A\right\|$. The "next user selection" continues until all r users are selected.

Architectural framework for the base station wireless scheduler algorithm

This chapter outlines a generic architecture for packet queueing and scheduling frameworks that allows different scheduling algorithms to be implemented independently of remainder MAC/Radio Link Control(RLC) processing stages.

Given the complexity of the scheduling algorithm, practical implementations are broken down into simpler stages that are executed sequentially. This breakdown sequentially reduces the dataset (user database) under consideration for scheduling. Consider Fig. 7–7.

This graph defines:

- Connected users are users that are associated with the base station. For example, mobile phones that have undergone RACH procedures and have an established control channel (bearer). These connected users may or may not have actual data available for transmission. Connected user count can be 1000 seconds.
- Active users are the subset of connected users that have actual data available for transmission and as such should be considered for scheduling air resources to. Note that active user selection can include air interface-specific limitations such as Discontinuous Reception (DRX) operation where a mobile phone is spending time in power saving mode and hence cannot be scheduled, even if there is traffic available for transmission. Active user count can be 100 seconds.
- Eligible users are the subset of active users that are considered for actual scheduling based on exhibiting "schedulability" metrics even though they may not actually end up being selected for scheduling. This is an optional stage implemented to reduce compute complexity by reducing the active user count (100 seconds) to a smaller user count of 10 seconds of users that are considered further.
- Scheduled users are the subset of eligible users that are scheduled in the specific time instance under consideration. Scheduled user count typically is in the order of 10.

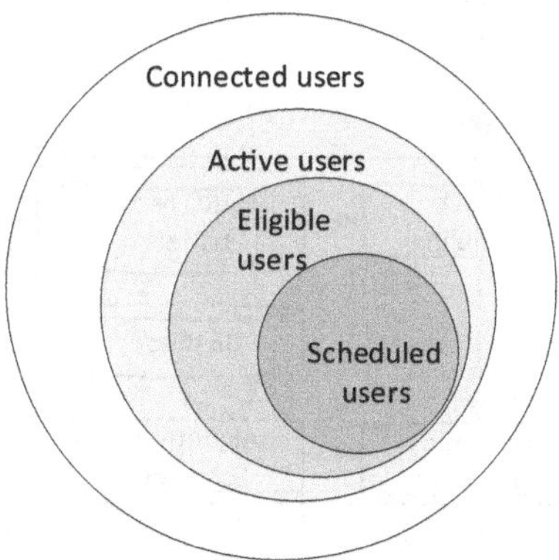

FIGURE 7–7 User database reduction for scheduler user selection.

Practical implementations breakdowns consider the following:

- Static and regular scheduling—allocate resources to prescheduled (statically scheduled) users, hybrid-ARQ (HARQ), broadcast, and other channels that are not subject to dynamic scheduling.
- Active user selection—filter out the users that are potential targets for scheduling to reduce computational load on remainder stages.
- Primary user selection—based on proportional fair (PF) scheduling to balance the competing interests of maximizing the network throughput while ensuring a minimum QoS is met. This stage is also known as "time-domain scheduling."
- Frequency resource allocation. This stage is known as "frequency-domain scheduling."
- MIMO user pairing and remainder user selection.
- Power allocation.

These stages are discussed in more detail in the following sections.

Static and regular scheduling

This scheduling stage establishes a logical grid of the air interface resources and preallocates those resources that are known in advance, such as broadcast channels, regular allocations from static resource schedulers (3GPP: semipersistent scheduling), etc. Static and regular scheduling is executed before dynamic scheduling as shown in Fig. 7–8, given that it has higher priority by design.

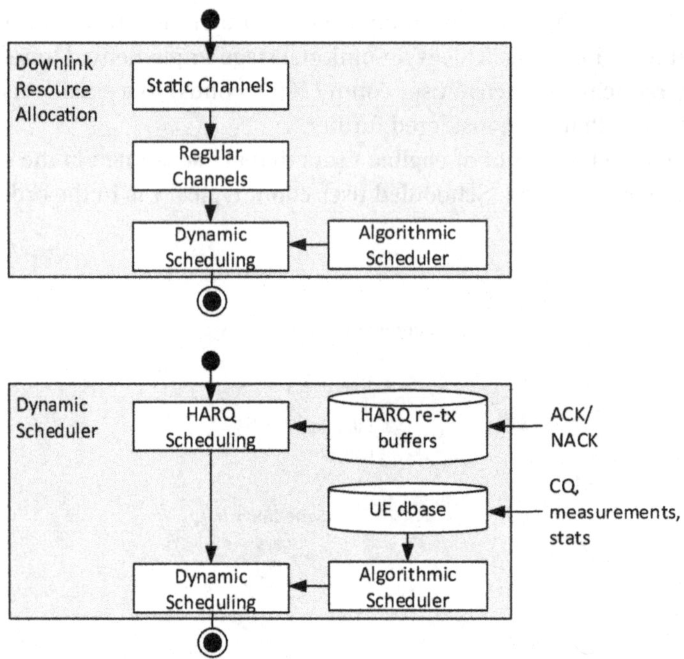

FIGURE 7–8 Breaking down the scheduler algorithm.

Active user selection

Per the definition, active users are simply defined as connected users that can technically be scheduled to and have (sufficient) data available for transmission, either in the transmit direction by a data queue transitioning from empty to nonempty, or on the receive direction by a buffer status indication. Additional limitations on eligibility for scheduling can include DRX considerations or available HARQ resources.

Algorithmically, there is no complexity in active user selection as such, but it is important to note that efficient implementation of this step in software can greatly reduce computational resources spent in the next phases through elimination of unnecessary compute resources (do not even consider scheduling a user that has no data to transmit).

Primary user selection (time-domain scheduling)

The target of a PF scheduler algorithm is to balance system spectral efficiency (and system-wide aggregate throughput) while ensuring individual user satisfaction through providing a minimum throughput to each user.

This step involves selecting which User Equipment (UEs) out of the set of UEs with data available for scheduling will be scheduled. A common approach for UE selection is to define a utility function, a term used in economics to define the level of satisfaction received by a consumer from consumption of a good or a service. A simple utility function is defined by the PF scheduling metric according to the following formula:

$$M_i = \frac{T_i^{\alpha}}{R_i^{\beta}}$$

where R_i represents the historical average rate, which is calculated as (exponential average filtering) follows:

$$R_{i,t} = T_{i,t} + ((R_{i,t-1} - T_{i,t})/T_c)$$

and T_i is the current achievable rate, according to the average signal-to-interference and noise ratio across all subchannels. T_c is the filtering time constant.

Note: Eligible UEs are those for which $M(i)$ is highest (arg max(M_i)).

Another algorithm that considers both the delay and the QoS is given by the following equation [Modified Largest Weighted Delay First (M-LWDF)]:

$$M_i = -\log(\delta)\frac{T_i^{\alpha}}{R_i^{\beta}}\frac{D_i^{\alpha}}{\Delta_i}$$

The first factor ($\log(\delta)$) represents the QoS class, the second factor represents the PF portion, and the last factor ($D_i(t)/\Delta_i$) represents the Head of Line Packet Delay/Due Delay, which ranges from 0 to 1 as the packet delay approaches the due delay.

The following is a generalization/extension of this algorithm:

$$M_i = A^{\alpha} B^{\beta} C^{\gamma}$$

With *A*, *B*, *C* presenting a scheduling factor (e.g., the QoS class, the PF Head of Line Packet Delay, and others), and α, β, and γ weights associated with each fairness factor. *A*, *B*, and *C* can each be a rate averaged metric or other.

A PF scheduler implementation that includes active UE filtering is shown in Fig. 7–9.

Frequency allocation (frequency-domain scheduling)

Frequency-domain scheduling involves the allocation of frequency-domain resources to the time-domain scheduled users (or, more precisely, a subset of the time-domain defined group of eligible users). Goal is to allocate frequency-domain resources to the different users in such a way that the users are using the subset of the frequency domain that provides most favorable channel conditions, noting that the channel conditions in the frequency domain can be different to each user as a function of where in the frequency domain that user is allocated, as shown in Fig. 7–10.

Like in time-domain scheduling, a utility function is $R_{UE,n}$ is defined, in frequency-domain scheduling as the achievable throughput of a given user as a function of the allocated slice of spectrum in the frequency domain.

The frequency-domain scheduling problem can now be formulated as:

$$\max(R) = \sum_{n=0}^{N} R_{UE,n}$$

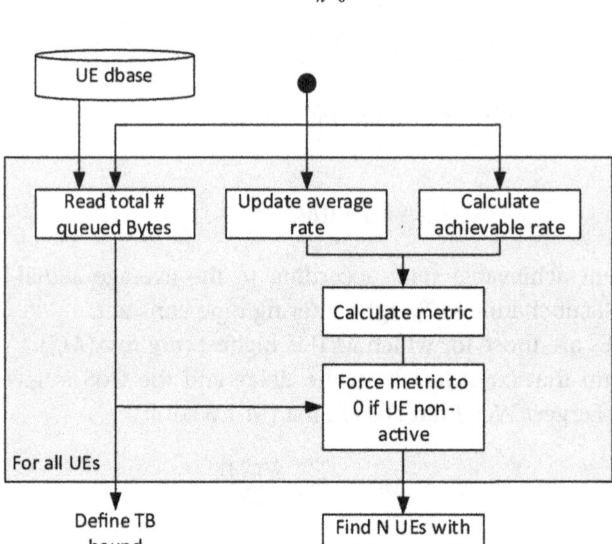

FIGURE 7–9 PF time-domain scheduling. *PF*, proportional fair.

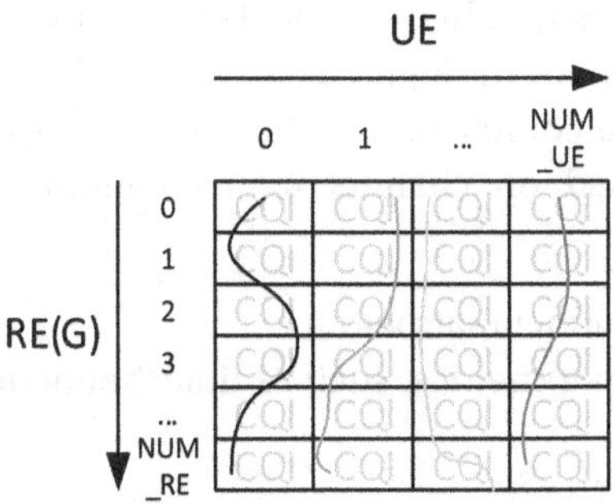

FIGURE 7–10 Frequency-domain scheduling challenge.

where n = {0, . . ., *N*) defines a slice of frequency-domain resource (i.e., RF spectrum). Note that there are allocation constraints involved—such as a requirement for users to be allocated contiguous pieces of spectrum. Typical implementations may slice the frequency-domain resources (subcarriers or resource blocks) into larger groups to simplify the algorithm.

Several algorithms for this UE to subchannel allocation can be found in literature. Using greedy allocation (best fit first) to sort to the throughput metric (in this example $R_{UE,n}$) is a common approach; but, this is known to be suboptimal in terms of finding the maximum summed rate. Linear programming can be employed to find the optimum solution to the problem, for example, by using the simplex method as described elsewhere.

The best fit first method, or greedy allocation, is among the simplest heuristic algorithms for solving the problem. The algorithm sequentially finds the highest possible utility in the matrix defined by user and n until all frequency-domain resources (n) are exhausted and/or all users are scheduled.

The best fit first method is as follows:

1. Initialize a metric value M to 0.
2. Search through every combination of user UE and frequency resource n to find the highest metric (achievable rate) for the first user, where the first user can be selected for example based on QoS priority, SNR value, or randomly.
3. Allocate the user UE and carrier ("n") by setting this carrier to nonvalid (i.e., occupied).
4. Continue steps 2 and 3 for remainder users and/or frequency resources until either group is exhausted.

In practice, this algorithm translates to a recurring search for the maximum value (achievable rate) in a two-dimensional array (users, frequency resources).

Secondary and remainder user(s) selection: multiple input multiple output scheduling

Considering a MIMO implementation, this stage involves the selection of paired users with the primary user where a paired user is selected based on channel orthogonality with the primary user as described earlier, either by calculation of the coefficient of correlation or by optimizing the selected user pool through calculation of the condition number of the combined channel matrix—or by a combination of both techniques as shown in Ref.[48]

Assuming scheduling based on channel matrix condition number, the method is as follows:

1. Initialize the combined channel matrix to 0.
2. Decide the first user on other metrics, such as time/frequency-domain scheduling as shown above.
3. Increase user count by 1.
4. Calculate the condition number associated with the combined channel vector of the already selected users and each additional candidate.
5. Select the candidate user for which the resulting condition number is lowest or below a given threshold.
6. Continue until maximum number of users is selected or threshold conditions are no longer met.

Intuition tells us that the math involved with MIMO scheduling is complex and hence compute intensive. A simplification of this algorithm is achieved by preselecting the users into groups, where the orthogonality between users is calculated offline and users are grouped together according to this calculation. In semistatic environments where channels are relatively constant (low mobility such as Wi-Fi applications), this technique can save greatly on computational cost that no longer needs to be accounted for in real time. Quasi-real-time solutions where "spare" compute resources are used opportunistically to maintain near-optimal grouped user sets are also possible.

System-level optimization

Most air interface schedule implementations focus on base station level optimization, where the base station implies colocation of MAC(/RLC/Packet Data Convergence Protocol) components with (high and low) physical layer processing in a single unit that is colocated with the antenna. Or in Open Radio Access Network language: a colocated central unit (CU)/distributed unit (DU)/ radio unit (RU) or integrated small-cell system. In this case, the scheduler (MAC layer) is colocated with the antenna for a single-cell site.

However, centralization of processing gives another degree of freedom. Let's look at the system from a very high level in Fig. 7–11.

Note the F1 interface in this picture that is defined by 3GPP in 38.475. This standard defines the low-level messages required for the *controlled* transfer of user data between CU and DU, allowing DU buffer space to remain limited and the bulk of the user buffers to be contained in the CU. This mechanism allows a single user (the middle one in the figure above) to be connected to multiple DUs at the same time, where the CU controls which DU owns the data flow to this UE. This introduces the concept of the CU scheduler algorithm that can have limited/no visibility into air interface conditions between the DU and the RU but instead has capability to route packets through different air interfaces. This CU scheduling algorithm becomes a system-level scheduler.

See Fig. 7–12 in which the RU is depicted in 3GPP language (transmit/reception point or TRP), and the DU/RU components from the previous figure are eliminated for ease of comprehension.

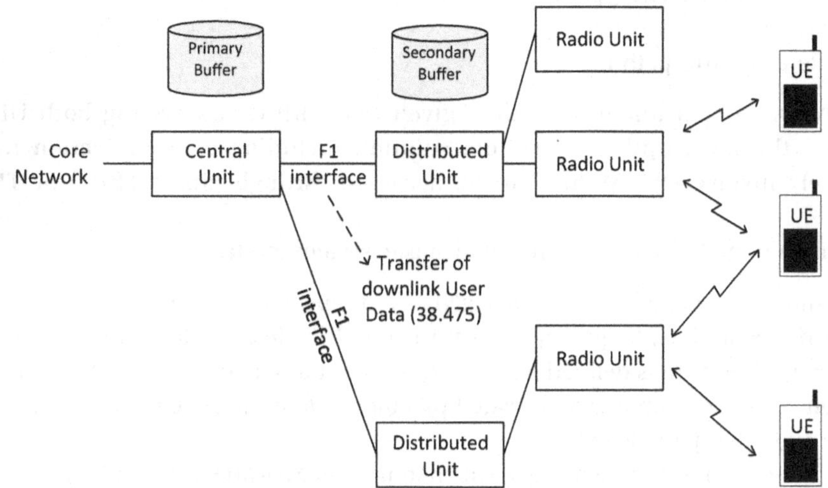

FIGURE 7–11 High-level view of an O-RAN CU/DU/RU system implementation. *CU*, control unit; *DU*, distributed unit; *O-RAN*, Open Radio Access Network; *RU*, radio unit.

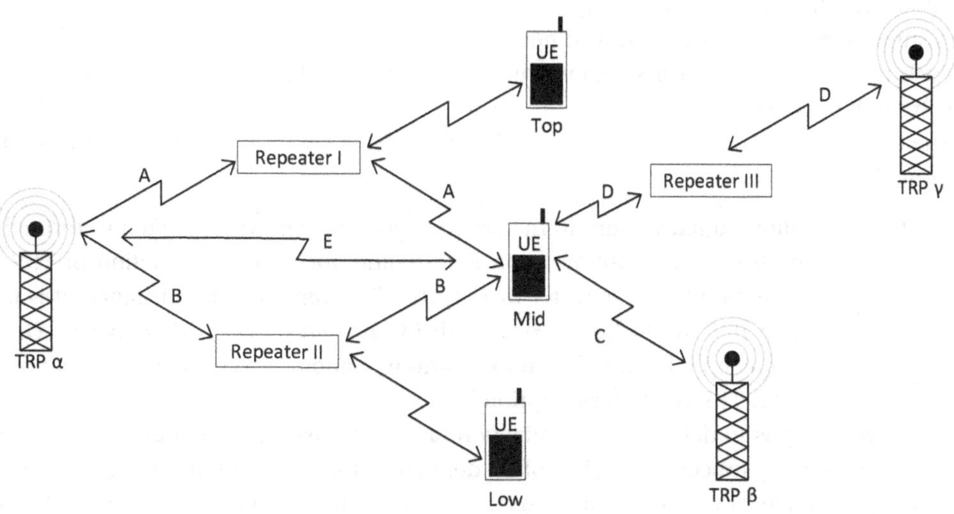

FIGURE 7–12 System-level scheduling challenge.

In this example, there are multiple paths to connect to a UE. Take the Mid(dle) UE in the figure, which can be connected through the following paths:

- from TRP α:
  - via Repeater I—path A
  - via Repeater II—path B
  - directly from the TRP—path E
- from TRP β:
  - directly from the TRP—path C
- from TRP γ:
  - via Repeater III—path D

Intuitively, we can understand that, given that TRP α is servicing both UEs "Low" and "Top," the most optimized system solution including communication to all UEs would likely involve UE "Middle" to be accessed through either TRP β or TRP γ—or both!

To analyze more formally, assume an idealistic system model:

1. CSI is known for all paths across UEs, gNBs, and repeaters.
2. The carrier width (i.e., frequency spectrum) and MCS allocated for communication between gNB and UE is defined a priori, outside of the scheduler algorithm. This decision considers path loss and related parameters to optimize channel width, transmit power, and other parameters.
3. Multiple users do not share a single subchannel—no MU-MIMO taken into consideration.
4. Control channel signaling overheads are ignored.
5. Only downlink traffic (gNB- > UE) is considered only simplify the model rather than imposing a fundamental limitation.
6. Assume only a single user is served per slot.
7. Each repeater is uniquely associated with a single TRP (uplink) but can be associated with multiple UEs.
8. The data that is to be transmitted to any UE is available centrally and can be transmitted through any TRP.

As an optimization target, we aim to maximize aggregate achieved (downlink) throughput across all UEs. The *first* step in optimization is to define, for each combination of TRP and UE the single unique path (via a repeater or not) that is optimal for performance. It is intuitive to conclude that this would be the only path to be used between that specific UE and that specific TRP for uplink and downlink communication—given that the relationship between repeater and TRP is N:1 (assumption 7).

The *second* step is to define the optimization target. We assume the goal to be defined as optimizing to a utility function, which is often defined as the aggregate throughput of the system, across multiple base stations and {user, RF channel} allocations, for each instance in time (e.g., slot or frame). This means optimization to maximize the sum rate across all allocated users:

$$\max(R) = \sum_{UE=1}^{UE=\max} R_{UE}$$

where we assume (for each UE as we defined in the first step) a known achievable rate R to that UE defined by its unique established path from the TRP to the UE.

The *third* step is to define the limitations in the system. Given assumption 6 (arguably, a bit over an over-simplification), we take it that as soon as a single TRP↔UE allocation is established, during that slot duration, that TRP can not communicate to any other UE. Enhanced versions of the algorithm can be expanded to scale to multiple UE allocations.

As we have now defined the problem and the limitations, we can think about potential solutions. Greedy allocations schemes are often thought of first:

- Out of all potential allocations between TRP and UE (across all repeater path options), find the one that gives to the first (to be allocated) UE the maximum possible throughput.
- Assign time/frequency resources to this UE and remove the TRP and repeater path to this UE from further consideration (each allocation can only be used once).
- Out of the remainder potential allocations between TRP and UE (across all remainder repeater path options) find the one that gives the next (next) highest possible throughput.
- Assign time/frequency resources to this UE and remove the TRP and repeater path to this UE from further consideration (each allocation can only be used once).
- Continue this process until no more UE can be served by an available TRP/repeater path.

This greedy scheme is often used in real applications even though it is intuitively understood not to necessarily give the best possible solution. Linear programming can be used to find a mathematically optimum solution to the problem.

Software optimization techniques

Theoretically speaking, the base station can provide time, frequency, and spatial access to a new set of users every possible scheduling instant (TTI, slot, ...). This means that the steps outlined in section "Architectural framework for the base station wireless scheduler algorithm" need to be implemented in (quasi) real time. In addition, as the number of (active) users in the base station increases and the number of scheduled users increases, the complexity of the scheduling algorithm goes up.

As a result, software optimization of algorithmic schedulers becomes important. Note that the algorithms we described earlier include specific math-centric:

- PF scheduling. Using the fact that the actual value of the UE priority metric to be sorted to is not important to the arg(max) function, rather than only the relative value of the metric compared to other evaluated UEs, this function can be simplified for real-time computation as follows: $j = \log(A\alpha B_\beta C\gamma) = \log(A\alpha) + \log(B_\beta) + \log(C\gamma) = \alpha \log(A) + \beta \log(B) + \gamma \log(C)$. This implementation is regular and highly parallelizable by performing calculations on A, B, C, ... in parallel using vectorized operations. Assuming the input parameters are structured in a single instruction multiple data (SIMD)-friendly manner, this can lead to large potential performance gains from modern CPU architectures with 128, 256, or 512-bit vector support.

- The next step is to find the set of utility metrics for which j has the highest values, which indicates the users that are selected for the next stages: time-domain and frequency-domain user allocation. This involves making an efficient implementation of the arg(max) function, which we are showing in the next section.
- Time-domain user selection and frequency-domain allocation. Both of software algorithms are in practically implemented as greedy algorithms as we discussed, where the "best" user $\leftrightarrow$ time/frequency allocation is selected and allocated after which this user is removed from the pool and the next best user is selected. This is again a variety of an arg(max) implementation and subject to performance improvement by SIMD vectorization.
- Secondary and remainder user selection/MIMO pairing which involves matrix operation (matrix multiplication, absolute value calculation, etc.) and SIMD vectorized implementations are widely discussed on, and available from the Internet.

Example: arg(max) vectorization

We saw earlier that arg(max) is one example of a commonly used function to optimize the CPU cost of a large user-count scheduler algorithm. To showcase how vectorization helps execution speed, look at this example: consider an array of values (utility metrics, SNR values, or similar), conveniently stored in a linear array in memory (32 entries shown in the example in Fig. 7–13).

The target of the algorithm is to efficiently find the highest value in this array (and, left as an exercise to the reader, the index of where in the array that value is located). Note that the target value is on the second row (value: 31).

In a vectorized implementation, we start by loading a reference vector (all zeroes), as well as the first vector of array values (the first row from the figure above), and perform a Vector Compare Greater Than operation as shown in Fig. 7–14.

0 7

0	1	2	3	4	5	6	7
8	9	10	**31**	12	13	14	15
24	25	26	27	28	29	30	11
16	17	18	19	20	21	22	23

24 31

FIGURE 7–13 Example of input array to arg(max) optimization.

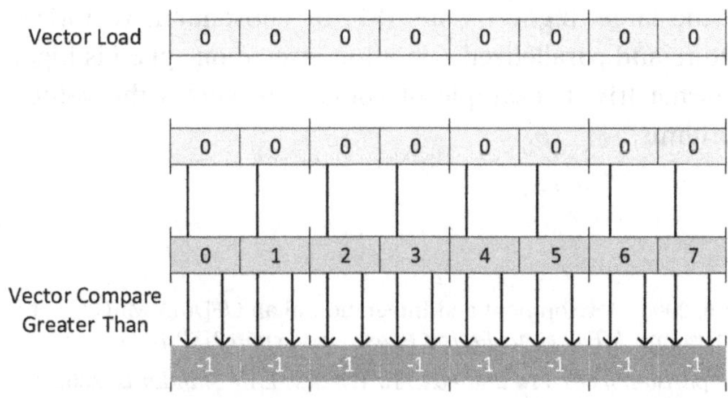

FIGURE 7–14 Vectorized comparison to find highest value (step 1).

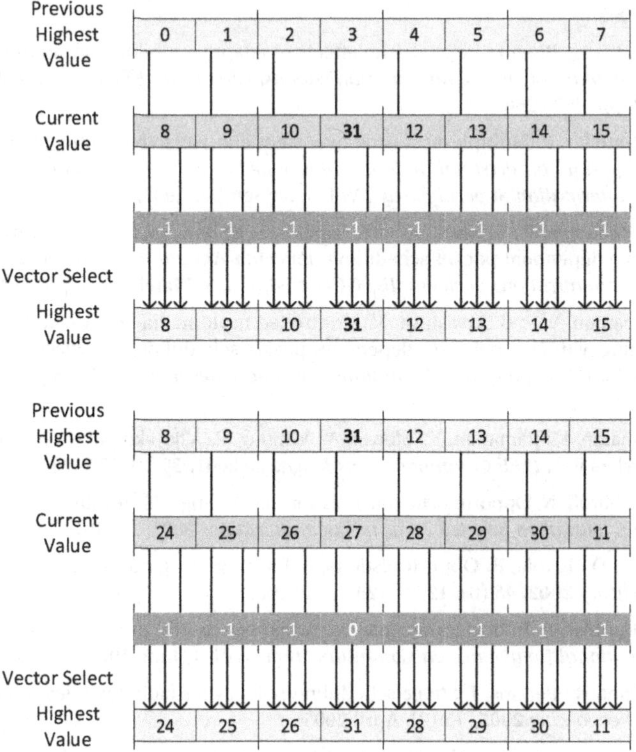

FIGURE 7–15 Vectorized comparison to find highest value (step 2 and step 3).

The output from the comparison is a selection bitmask that identifies (for each column representing the two input vectors) which row "wins" the comparison. In the second and third stages, we do this comparison between the first and second row as follows (Fig. 7–15).

This filters out the target highest value (31) by subsequent vectorized operation. Given the branch-less nature and parallelized operation, execution speed is high.

This is a (somewhat trivial) example of course but shows the value of vectorization in these types of algorithms.

References

1. Jérôme M.P. Adam, 2005, Development and Integration of an OFDMA MAC Layer Scheduling Algorithm for IEEE 802.16e Systems, *MEng Hons Project Phase One Report HSP 1835.*
2. Huotari, A. *A Comparison of 802.11a and 802.11b Wireless LAN Standards*, 2002. <http://www.linksys.com/products/images/802_11a_vs_802_11b.pdf> May 2002. (This paper is no longer available for download at this location. A copy can still be found at: <http://web.archive.org/web/20030427194830/http://www.linksys.com/products/images/802_11a_vs_802_11b.pdf>).
3. Pottie, G. System design choices in personal communications. *IEEE Personal Communication* **1995,** *2* (5), 50–67 October 1995.
4. Zou, L., Zhao, Y., Wang, B. and Liang, Q. The effects of adaptive modulation on the TCP performance. *IEEE 2002 International Conference on Communications, Circuits and Systems and West Sino Expositions*, Vol. 1, July, 2002; pp. 262–266.
5. Knopp, R. and Humblet, P. Multiple-accessing over frequency-selective fading channels. *Sixth IEEE International Symposium on Personal, Indoor and Mobile Radio Communications, PIMRC'95, 'Wireless: Merging onto the Information Superhighway'*, Vol. 3, September 1995; pp. 1326.
6. Bhagwat, P., Bhattacharya, P., Krishna, A. and Tripathi, S. Enhancing throughput over wireless LANs using channel state dependent packet scheduling. *Fifteenth Annual Joint Conference of the IEEE Computer and Communications Societies. INFOCOM'96*, Vol. 3, March 1996; pp. 1133–1140.
7. Fragouli, C., Sivaraman, V. and Srivastava, M. Controlled multimedia wireless link sharing via enhanced class-based queuing with channel-state-dependent packet scheduling. *Seventeenth Annual Joint Conference of the IEEE Computer and Communications Societies. INFOCOM'98*, Vol. 2, April 1998; pp. 572–580.
8. Andrews, M.; Kumaran, K.; Ramanan, K.; Stolyar, A.; Whiting, P.; Vijayakumar, R. Providing quality of service over a shared wireless link. *IEEE Communications Magazine* **2001,** *39* (2), 150–154 February 2001.
9. Liu, X.; Chong, K.; Shroff, N. Opportunistic transmission scheduling with resource-sharing constraints in wireless networks. *IEEE Journal on Selected Areas in Communications* **2001,** *19* (10), 2053–2064 October 2001.
10. Viswanath, P.; Tse, D.; Laroia, R. Opportunistic beamforming using dumb antennas. *IEEE Transactions on Information Theory* **2002,** *48* (6), 1277–1294 June 2002.
11. Knopp, R. and Humblet, P. Information capacity and power control in single-cell multiuser communications. *IEEE International Conference on Communications*, Vol. 1, June 1995; pp. 331–335.
12. Eryilmaz, A.; Srikant, R.; Perkins, J. Stable scheduling policies for fading wireless channels. *IEEE Transactions on Networking* **2005,** *13* (2). April 2005.
13. Zhimei, J. and Shankaranarayana, N. Channel quality dependent scheduling for flexible wireless resource management. *IEEE Global Telecommunications Conference. GLOBECOM'01*, Vol. 6, November 2001; pp. 3633–3638.
14. Guoqing, L.; Hui, L. Dynamic resource allocation with finite buffer constraint in broadband OFDMA networks. *IEEE Wireless Communications and Networking* **2003,** *2*, 1037–1042 WCNC 2003, March 2003.
15. Cheong, Y. W.; Cheng, R.; Letaief, K.; Murch, R. Multiuser OFDM with adaptive subcarrier, bit, and power allocation. *IEEE Journal on Selected Areas in Communications* **1999,** *17* (10), 1747–1758 October 1999.

16. Cheong Y. W., Cheng, R., Letaief, K. and Murch, R. Multiuser subcarrier allocation for OFDM transmission using adaptive modulation. *IEEE 49th Vehicular Technology Conference. VTC'99*, Vol. 1, May 1999; pp. 479–483.

17. Cheong Y. W., Tsui, C., Cheng, R. and Letaief, K. A real-time sub-carrier allocation scheme for multiple access downlink OFDM transmission. *IEEE 50th Vehicular Technology Conference. VTC'99-Fall*, Vol. 2, September, 1999; pp. 1124–1128.

18. Kittipiyakul, S. and Javidi, T. Subcarrier allocation in OFDMA systems: beyond water-filling. *Conference Record of the Thirty-Eighth Asilomar Conference on Signals, Systems and Computers*, Vol. 1, November 2004; pp. 334–338.

19. Kivanc, D. and Hui L. Subcarrier allocation and power control for OFDMA. *Conference Record of the Thirty-Fourth Asilomar Conference on Signals, Systems and Computers*, Vol. 1, November 2000; pp. 147–151.

20. Sushanta, D. and Mandyam, G. An efficient sub-carrier and rate allocation scheme for M-QAM modulated uplink OFDMA transmission, *Conference Record of the Thirty-Seventh Asilomar Conference on Signals, Systems and Computers*, Vol. 1, November 2003; pp. 136–140.

21. Grossglauser, M. and Tse, D. Mobility increases the capacity of ad hoc wireless networks. *Twentieth Annual Joint Conference of the IEEE Computer and Communications Societies. INFOCOM'01*, Vol. 3, April 2001; pp. 1360–1369.

22. Giroux, N.; Ganti, S. *Quality of Service in Atm Networks: State-of-the-Art Traffic Management;* Prentice Hall PTR: Upper Saddle River, NJ, 1998101–11907458, December 1998.

23. Lu, S.; Bharghavan, V.; Srikant, R. Fair scheduling in wireless packet networks. *IEEE/ACM Transactions on Networking* **1997,** *7* (4), 473–489. August 1999. Previously in Proceedings of the ACM SIGCOMM '97 conference on Applications, technologies, architectures, and protocols for computer communication. ACM SIGCOMM Computer Communication Review, Volume 27 Issue 4, pp. 63–74, September 1997.

24. Ramanathan, P. and Agrawal, P. Adapting packet fair queuing algorithms to wireless networks. *Proceedings of the 4th Annual ACM/IEEE International Conference on Mobile Computing and Networking*, Vol. 3, Issue 1, October 1998; pp. 1–9.

25. Nandagopal, T., Lu, S. and Bharghavan, V. A unified architecture for the design and evaluation of wireless fair queuing algorithms. *Proceedings of the 5th Annual ACM/IEEE International Conference on Mobile Computing and Networking, 1999. MOBICOM'99*, August 1999; pp. 132–142.

26. Bharghavan, V.; Songwu, L. U.; Nandagopal, T. Fair queuing in wireless networks: issues and approaches. *IEEE Personal Communication* **1999,** *6* (1), 44–53 February 1999.

27. Lu, S.; Nandagopal, T.; Bharghavan, V. Design and analysis of an algorithm for fair service in error-prone wireless channels. *Wireless Network* **2000,** *6* (4), 323–343 July 2000.

28. Ng, T. S. E., Stoica, I. and Zhang, H. Packet fair queuing algorithms for wireless networks with location-dependent errors. *Seventeenth Annual Joint Conference of the IEEE Computer and Communications Societies. INFOCOM'98*, Vol. 3, April 1998; pp. 1103–1111.

29. Namgi K. and Hyunsoo Y. Packet fair queuing algorithms for wireless networks with link level retransmission. *Consumer Communications and Networking Conference*, 2004. CCNC'04, January 2004; pp. 122–127.

30. Shreedhar, M. and Varghese, G. Efficient fair queuing using deficit round robin. *ACM SIGCOMM Computer Communication Review, Proceedings of the Conference on Applications, Technologies, Architectures, and Protocols for Computer Communication*, Vol. 25, Issue 4, October 1995; pp. 231–242.

31. Kim, H., Han, Y. and Yun, S. A proportional fair scheduling for multicarrier transmission systems. *IEEE 60th Vehicular Technology Conference, VTC2004-Fall*, Vol. 1, September 2004; pp. 409–413.

32. Park, T., Shin, O. and Lee, K. Proportional fair scheduling for wireless communication with multiple transmit and receive antennas. *IEEE 58th Vehicular Technology Conference, VTC2003-Fall*, Vol. 3, October 2003; pp. 1573–1577.

33. Shakkottai, S. and Srikant, R. Scheduling real-time traffic with deadlines over a wireless channel. *Proceedings of the 2nd ACM International Workshop on Wireless Mobile Multimedia*, August 1999; pp. 35–42.

34. Wang, H. S.; Moayeri, N. Finite-state Markov channel – a useful model for radio communication channels. *IEEE Transactions on Vehicular Technology* **1995,** *44* (1), 163–171 February 1995.

35. Yang, S.; Veciana, G. Enhancing both network and user performance for networks supporting best effort traffic. *IEEE/ACM Transactions on Networking* **2002,** *12* (2), 349–360 April 2002.

36. Yaxin, C., Li, V. and Zhigang, C. Scheduling delay-sensitive and best-effort traffics in wireless networks. *IEEE International Conference on Communications, 2003. ICC'03*, Vol. 3, May 2003; pp. 2208–2212.

37. Wongthavarawat, K. and Ganz, A. IEEE 802.16 based last mile broadband wireless military networks with quality of service support. *IEEE Military Communications Conference, 2003. MILCOM 2003*, Vol. 2, October 2003; pp. 779–784.

38. Guosong, C., Wang, D. and Shunliang, M. A QoS architecture for the MAC protocol of IEEE 802.16 BWA systems. *IEEE 2002 International Conference on Communications, Circuits and Systems and West Sino Expositions*, Vol. 1, July 2002; pp. 435–439.

39. Lee, H., Kwon, T. and Cho, D.-H. An efficient uplink scheduling algorithm for VoIP services in IEEE 802.16 BWA systems. *IEEE 60th Vehicular Technology Conference, VTC2004-Fall*, Vol. 5, Issue 26–29, September 2004; pp. 3070–3074.

40. Kim, D.-H.; Ryu, B.-H.; Kang, C.-G. Packet scheduling algorithm considering a minimum bit rate for non-realtime traffic in an OFDMA/FDD-based mobile internet access system. *ETRI Journal* **2004,** *26* (1), 48–52 February 2004.

41. Ergen, M.; Coleri, S.; Pravin, V. QoS aware adaptive resource allocation techniques for fair scheduling in OFDMA based broadband wireless access systems. *IEEE Transactions on Broadcasting* **2003,** *49* (4), 362–370 December 2003.

42. Wu, D.; Negi, R. Utilizing multiuser diversity for efficient support of quality of service over a fading channel. *IEEE Transactions on Vehicular Technology* **2005,** *54* (3), 1198–1206 May 2005.

43. Khattab, A. and Elsayed, K. Channel-quality dependent earliest deadline due fair scheduling schemes for wireless multimedia networks. *International Workshop on Modeling Analysis and Simulation of Wireless and Mobile Systems archive, Proceedings of the 7th ACM international symposium on Modeling, analysis and simulation of wireless and mobile systems*, May 2004; pp. 31–38.

44. Farrokhi, F., Olfat, M., Alasti, M. and Liu, K. Scheduling algorithms for quality of service aware OFDMA wireless systems. *IEEE Global Telecommunications Conference, 2004. GLOBECOM '04*, Vol. 4, December 2004; pp. 2689–2693.

45. Alasti, M., Farrokhi, F., Olfat, M. and Liu, K. Service level agreement (SLA) based scheduling algorithms for wireless networks. *IEEE 2004 International Conference on Communications*. Vol. 2, June 2004; pp. 1028–1032.

46. Tu, Z.; Blum, R. S. Multiuser diversity for a dirty paper approach. *IEEE Communications Letters* **2003,** *7* (8), 370–372.

47. Castaneda, E., 2016, An overview on resource allocation techniques for multi-user MIMO systems.

48 Lu, X.; Huang, X.; Li, W., et al. An improved smi-orthogonal user selection algorithm based on condition number for multiuser MIMO systems. *Future of Digital Communications* **2009**.

Further reading

Available at en.wikipedia.org/wiki/Assignment_problem.

Kela, P.; Puttonen, J.; Kolehmainen, N.; Ristaniemi, T.; Henttonen, T.; Moisio, M. *Dynamic Packet Scheduling Performance in UTRA Long Term Evolution Downlink;* IEEE, 2008.

Kim, I.; Lee, H.; Kim, B.; Lee, Y. *On the use of Linear Programming for Dynamic Subchannel and Bit Allocation in Multiuser OFDM;* IEEE, 2001.

Song, G.; Li, Y. Cross-layer optimization for OFDM wireless networks-part i: theoretical framework. *IEEE Transactions on Wireless Communications* **2005,** *4* (2).

Song, G.; Li, Y. Cross-layer optimization for OFDM wireless networks-part ii: algorithm development. *IEEE Transactions on Wireless Communications* **2005,** *4* (2).

Viswanath, P.; Tse, D. N. C. Sum capacity of the vector gaussian broadcast channel and uplink-downlink duality. *IEEE Transactions on Information Theory* **Aug 2003,** *49,* 1912–1921.

Wang, Y., Peng, F., Weidong, Z., Yuan, Y. MU-MIMO User Pairing Algorithm to Achieve Overhead-Throughput Tradeoff in LTE-A Systems. *2013 8th International Conference on Communications and Networking in China (CHINACOM),* 2013.

Yin, R., et al. Pricing-based interference coordination for D2D communications in cellular networks. *IEEE Transactions on Wireless Communications* **2015,** *14* (3), 1519–1532.

8

Synchronization in open radio access networks

In this chapter, we turn away from the primary data-plane functionality of open radio access networks (O-RAN) and look at all the aspects relating to synchronizing the nodes in an O-RAN network. We will start by revising the common concepts of frequency and time and discussing how networks can be synchronized and the tools made available for this. Then we look at the specific synchronization requirements for O-RAN networks and how these can be achieved in practice.

Unfortunately, the synchronization aspect of a network is often a function that is not given the consideration it deserves in system planning—often with the result that it is left until the last minute. This is bad, since without proper synchronization, the performance of a cellular network can be significantly downgraded or, in extreme cases, compromised to the point where its transmissions could violate regulatory standards.

Understanding frequency, time, syntonization, and synchronization

To understand the challenges of synchronization in an O-RAN network, or indeed any type of network, it is important to review the difference between frequency and what we call phase or time.

Frequency

Frequency is in general a well-understood concept—it is essentially just how often a signal "wiggles." Specifically, frequency is measured in Hertz, which represents the number of times the signal repeats per second. This, of course, assumes that the signal is periodic, or at least some periodicity is observable in it; otherwise the frequency would be continuously changing and would also depend over what period of time the signal was observed. However, even with these caveats, a wide range of signals have a definable frequency as shown in Fig. 8–1.

Since frequency is defined as repetitions, or cycles, per second, it can be thought of as the reciprocal of the period—the time it takes the signal to repeat—as seen in Fig. 8–1. Therefore all that is required to measure the frequency of a signal is a reference that can be used to determine the duration of a period of time over which the number of cycles is counted. In other words, we need to know what a second is. Since 1967, the second has

Open Radio Access Network (O-RAN) Systems Architecture and Design. DOI: https://doi.org/10.1016/B978-0-323-91923-4.00005-7

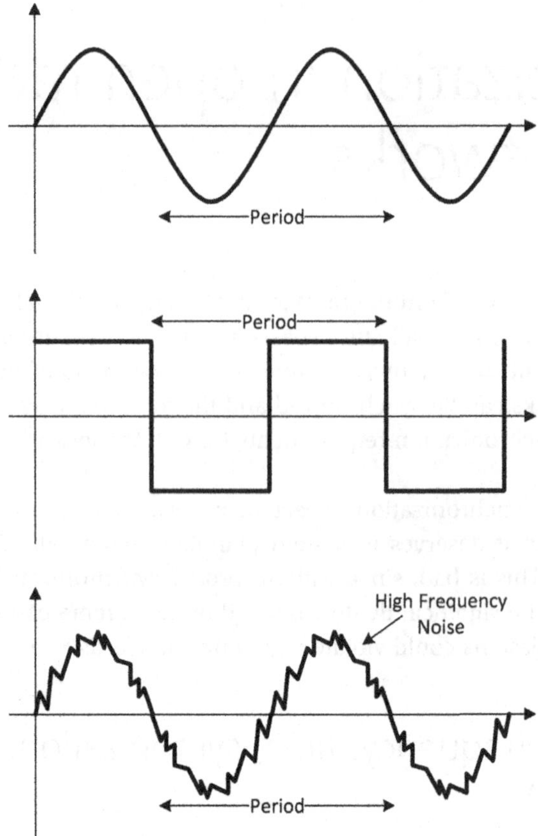

FIGURE 8–1 Definition of frequency.

been defined based on physics—specifically, the duration of 9,192,631,770 cycles of the radiation emitted as a result of an electron from a cesium-133 atom transitioning between the two hyperfine levels of the ground state under certain specified conditions. What does this all mean? The details do not really matter at all (unless you are a physicists working for a standards institute), but the key point is that this definition allows the duration of a second to be determined by anyone, at any time, in any place, just as long as they have the right equipment. The significance of this, or rather the fact that the same premise does not apply to time, will become clear later.

Of course, no one measuring frequency, or generating a signal of a certain frequency, would expect to use a cesium-based atomic clock. Instead, all practical applications use less-precise standards that can be calibrated, either during manufacture, or periodically for more accuracy, against a more precise standard, that itself can be calibrated against an even more precise one, etc., until ultimately it is calibrated against a cesium source. An example might be a quartz crystal that you buy with a datasheet accuracy of 20 ppm. This means that if used properly, it will oscillator to within 20 ppm or, 0.002%, of its nominal frequency, and this is something that the manufacturer will have ensured by testing it against a more accurate standard, maybe even a cesium standard itself, at manufacture.

Time and phase

Before moving on to look at time, it is important to distinguish the difference between two related terms—time and phase. These words are often used interchangeably and without a clear definition. However, the generally accepted definition is that "time" refers to a unique instance that can be found out from a clock and a calendar. For example, 12:31 pm on February 2, 2018 is a time. Of course, while we typically express time down to the resolution of minutes, or occasionally seconds, there is no reason that it cannot be stated considerably more precisely and, as you will see, in-network synchronization timescales at the nanosecond (one billionth of a second) are often quoted.

While some applications in a cellular network do need to know the actual, true, time, for example, call logging for an itemized bill and indeed some user applications themselves, such as a clock/calendar on a handset, in many cases, it is relative time that matters, as in what is the difference in time between two events. Additionally, at least in O-RAN applications, such events occur at a timescale well below a second—for example, the alignment of transmitted data frames for carrier aggregation. In these instances, it is more common to use the term "phase," to mean the time within a given second. For example, it will normally be enough to know that one event occurred 20% of the way through a second, whereas another occurred 55% of the way through the same second, without even actually knowing which second we're talking about.

Even though time and phase have distinct, different, meanings, for the remainder of this chapter, we will use the term time in all cases, even if what we actually care about is the phase, since the distinction between the two becomes irrelevant for most discussions.

What is time?

Time and frequency are clearly very closely related. We have already seen that frequency is defined in terms of cycles per seconds, and when we talk about time, what we really mean is the number of seconds, or parts of seconds, since something. The big difference though is that something—the starting point from which we count seconds. Formally, this was referred to as the epoch and its selection is rather arbitrary. Most of the Western World uses an epoch somewhat loosely based on events described in the Christian bible, with time measured in seconds, minutes, hours, days, months, and year from a start of 1 BCE, or more correctly, CE 1 (for Common Era). However, elsewhere in the world, different starting points are used. For example, the Hebrew calendar defines "year 1" as approximately 3760 years before the Christian calendar.

None of this is at all useful when trying to agree on a universal definition of time, so in 1960, Coordinated Universal Time, or UTC, was introduced as an internationally agreed time system that uses the Western year numbering. However, even this is totally useful for a couple of reasons. First, to keep UTC roughly in line with the rotation and orbit of the earth, it includes the concept of leap years and leap seconds. Everyone is familiar with the former, but leap seconds are less well known since they have little impact on everyday life. They are, however, used to maintain time relative to the rotation of the earth and one can be added, or removed, up to twice a year to avoid days slipping with respect to sunrise and sunset. Both leap years and leap seconds become significant when calculating the elapsed time, measured in seconds—between two events a long time apart since one must consider how many leap year days and how many leap seconds were in the period. For days added by leap years, this is merely an extra complication, since it is absolutely defined when such days are added, but for leap seconds, it becomes impossible since they are added (or removed) largely arbitrarily based on the Earth's rotational speed. Therefore it is impossible to calculate the number of seconds from one event to another, specified in UTC, over a long period without having additional information in the form of a list of leap seconds. To avoid this, an alternate timescale referred to as an International Atomic Time, or TAI, was introduced. TAI is very similar to UTC, except that it does not include leap seconds. This means that every time a leap second is added to a day TAI slips a further second behind UTC. As of early 2021, TAI is 37 seconds behind UTC. Fig. 8–2 shows how TAI slips from UTC.

The second problem, which affects both UTC and TAI, is that the division of time into years, days, hours, minutes, and seconds, is unnecessarily complex and exists only to align with the Earth's motion—something that O-RAN networks really does not care about. Even if

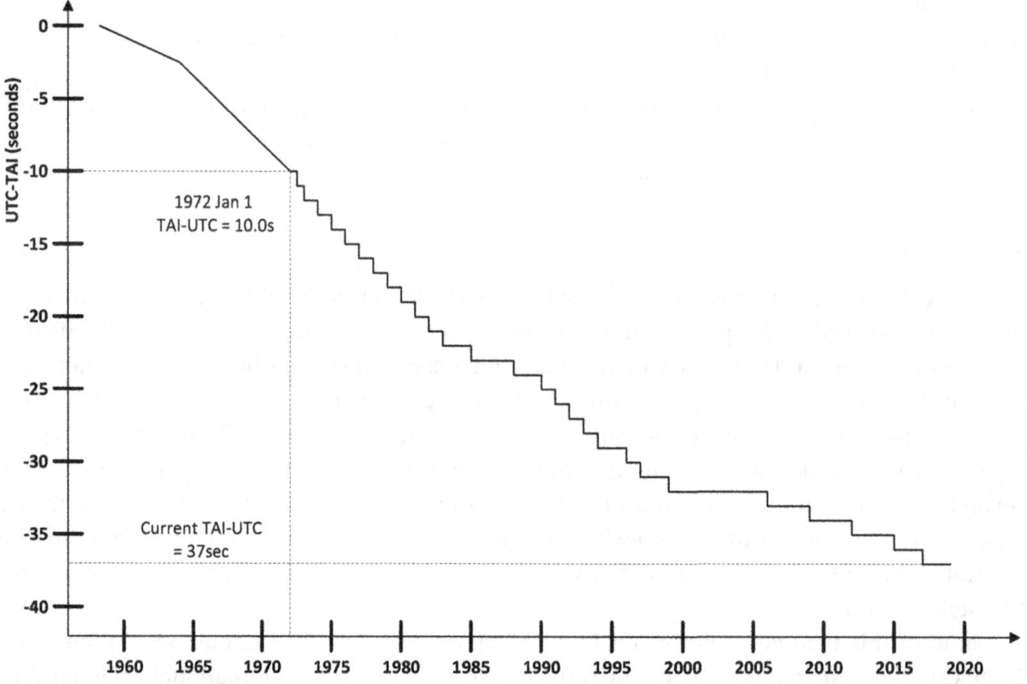

FIGURE 8–2 The relationship of TAI to UTC. *TAI*, International Atomic Time; *UTC*, Coordinated Universal Time.

we collapse all these to just counting seconds, the number of seconds since a starting point of over 2000 years ago is a large number that will result in an unnecessary amount of calculation. Therefore to make things more manageable, it is common to define a different epoch—one that is closer to the present day. There are two such epochs that are relevant to O-RAN:

- The Precision Time Protocol (PTP) epoch, as used by IEEE 1588. This counts seconds from January 1, 1970 00:00:00 TAI (which at that point was already 8 seconds ahead of UTC).
- The GPS epoch, which counts from January 6, 1980 00:00:00 TAI.

O-RAN defines frame numbering in terms of GPS time, so seconds from the GPS epoch. However, the PTP epoch is more relevant for synchronization on the network, as we will see when we look at IEEE 1588, and PTP time can easily be converted to GPS time by subtracting a constant value. Therefore when we talk about time in the following sections, we will mean PTP time.

One thing that should be clear from the preceding discussion is that compared to frequency, time is very arbitrary. Indeed, it is impossible for two people to agree on the current time without them having synchronized their clocks at some point in the past. Compare this to a frequency where, as stated earlier, any two people can agree on frequency, by virtue of agreeing on the length of a second, provided they have the right equipment—no prior synchronization is needed. This is a key consideration when it comes to time synchronization in networks, since it means that time must be transported from one place to another.

Syntonized versus synchronized

As a final piece of terminology, it is necessary to understand the difference between the terms "synchronized" and "syntonized" when applied to network applications. In common use, the term synchronized is used to refer to the general process of two events, or systems, being aligned in some way. However, in network applications, the term "synchronized" has a more specific meaning. If two nodes in the network are said to be synchronized, it means that those nodes share both a common frequency and a common time. On the other hand, if the nodes only share a common frequency, but not time, then they are said to be syntonized. The distinction is shown in Fig. 8–3.

Unfortunately, to further complicate things, the term "synchronization" is typically used to refer to the process of achieving alignment between nodes, whether the aim is for those nodes to be actually synchronized or just syntonized.

How do we get time?

As stated earlier, unlike frequency, time cannot be created locally. Therefore where do we get time from? Are there some magical things that we can "plug into" and receive time? Luckily there are—GNSS (Global Navigation by Satellite Systems) services.

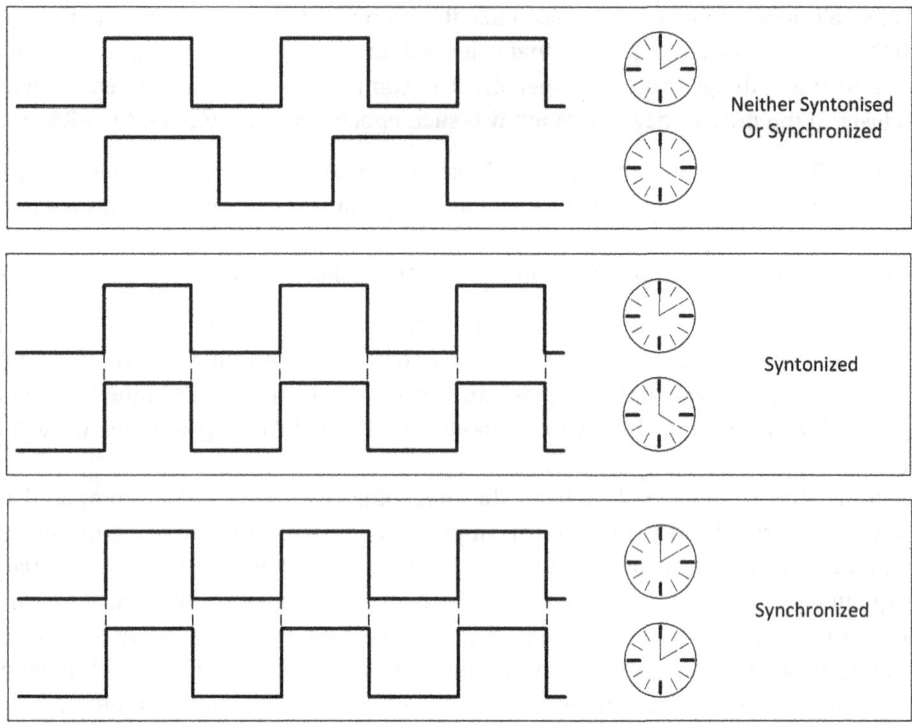

FIGURE 8–3 Syntonization versus synchronization.

Global navigation by satellite systems

Since the 1980s, governments around the World have provided "free" time as part of so-called GNSS services. The best known of these is GPS, run by the United States military, but still there are other GNSS services operating—Glonass (Russian), Beidou (Chinese), and Galileo (European). These systems are all global, meaning that they are available worldwide, even though they are owned by individual countries or consortiums (Fig. 8–4).

As the "N" in GNSS suggests, the primary purpose of these services, at least initially, was for navigation—at first for military applications but now equally, if not more, important for civilian and commercial uses—however, they very fortuitously provide precise time distribution too. All GNSS services work by using a set of orbiting satellites, referred to as a constellation, orbiting in a middle-earth orbit, meaning that the position of any satellite with reference to a point on the ground changes over time, unlike communication satellites, which appear fixed.

Each satellite in the constellation carries a very accurate clock, based around a rubidium oscillator, which is, in turn, synchronized to TAI be uplink ground stations. The satellites transmit their time to earth in a wide-area signal, as well as providing information about their orbit. GNSS receivers on the ground can receive the signals from multiple satellites

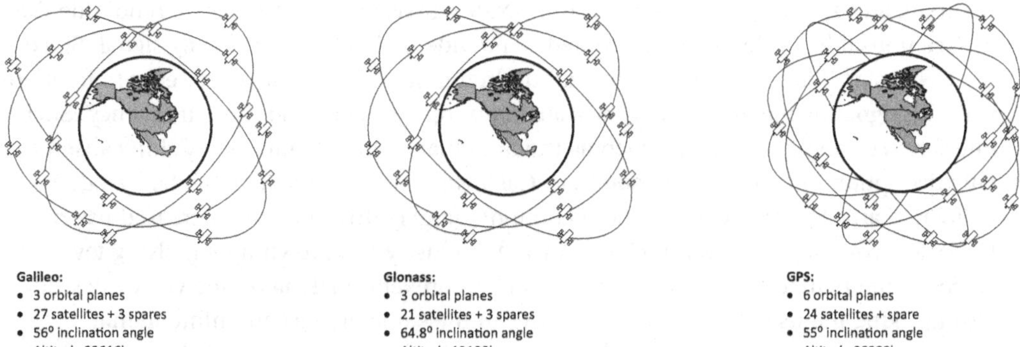

Galileo:
• 3 orbital planes
• 27 satellites + 3 spares
• 56° inclination angle
• Altitude 23616km
Glonass:
• 3 orbital planes
• 21 satellites + 3 spares
• 64.8° inclination angle
• Altitude 19100km
GPS:
• 6 orbital planes
• 24 satellites + spare
• 55° inclination angle
• Altitude 20200km

FIGURE 8–4 A GNSS constellation. *GNSS*, Global Navigation by Satellite Systems.

simultaneously and use a comparison of the transmitted times, along with the orbital parameters, to determine not only a precise location on the surface of the Earth but also a precise time. To achieve this, in general, requires reception from a minimum of four satellites, though additional ones can be used for redundancy. However, if the exact position of the receiver is already known, it is in fact possible to recover precise time using just a single satellite—something that can be used to improve reliability in fixed-location applications, such as macrocells.

Since GNSS signals are out there in the ether for anyone to receive (though the Galileo service does have the option for subscription-based encryption), it really is a free source of time for anyone to use. It is also a relatively precise source of time—modern, so-called multiband, receivers can provide a time output to within better than 30 ns of actual time (TAI). Sadly, GNSS is not without its limitations, which need to be appreciated when relying on it for a source of synchronization:

- Installation challenges and costs—To guarantee continuous visibility of the required minimum four satellites, a GNSS receiver really needs to have a properly mounted antenna, positioned so as to have a near-360 degree view of the sky. For a macrocell that is typically located in an open space on a high tower, this is largely a nonissue. However, for an in-building O-RU, it might involve provisioning of a roof-mounted antenna that can be both costly in terms of initial capex and ongoing opex related to access leases. For even harsher environments, such as an in-house femtocell, achieving reliable antenna positioning might be close to impossible.
- Urban canyons—Even with optimal antenna placement, for example, on the roof of a building—the nearby terrain and topology might be such that a guaranteed view of the necessary number of satellites is not always possible. This is often the case in the so-called urban canyons, where the sky view from even a relatively tall building is blocked by surrounding taller buildings. This is a case where sometimes the single-satellite mode can be useful—if the location of the antenna can be carefully measured at installation, only a single satellite needs to be in view to recover time from GNSS.

- Jamming and spoofing—This is when the GNSS signals are either blocked, rendering the service unusable, or deliberately spoofed to provide a plausible, but wrong signal. Since the GNSS signal is coming from several hundred miles in space, it is very easy to interfere with the signal by means of a nearby transmitter operating on the same frequency. Such interference can be accidental, such as the time that a United States navy ship caused a GPS disruption in large parts of southern California, or can be deliberate. Sadly, such deliberate attacks are becoming more common, largely driven by the increased use of GNSS for workforce location tracking. A whole industry has grown up supplying low-cost GNSS jammers that can be used, for example, to prevent tracking of one work truck but also blocking the use of GNSS for other uses in the same area as an unintentional consequence. Unfortunately, the sale and possession of such jammers is in general not illegal. Of course, attacks on GNSS can also have more sinister motives and actors. For example, spoofing of GNSS with the intention of offsetting GNSS-recovered time by few microseconds could be a very powerful technique for bringing down communication networks, with an inevitable destabilizing impact on societies.
- General availability—For any given GNSS service, there are typically only a small number of ground stations that uplink configuration and timing data to the satellites in the constellation, and it may, therefore, be a considerable time—several hours—before any given satellite can be contacted by a ground station. This means that, for example, any misconfiguration of satellites can take an extended time to correct. Just such an incident occurred a few years ago when an accidental misconfiguration of a number of GPS satellites resulted in digital radio broadcasts in Europe being disrupted for several hours as they lost the ability to synchronize via GNSS.

Alternatives to GNSS for providing synchronization do exist, such as other satellite systems using low-Earth orbit constellations. Low earth orbiting satillite based systems help mitigate against attack by virtue of having a stronger signal. Alternatively, calibrated fiber based solutions are a great source of very precise time (<1 ns with proper calibration) but unlikely to be affordable outside a small number of institutions. Therefore despite its risks and limitations, GNSS is likely to remain the main player in terms of time distribution on a global scale for many years to come.

Of course, you cannot just stick a GNSS antenna in the air and pull something representing time out of it. Instead, a GNSS receiver is needed which in itself is quite a complicated piece of hardware capable of receiving the spread-spectrum signals from the satellites, processing and decoding them, and using the extracted data to regenerate a local time source. Luckily, fully integrated GNSS receiver modules are available these days from multiple vendors for anything from a few to a few tens of US dollars depending on accuracy, features, etc. A lot of these modules can support multiconstellation operation, meaning that they can use the signals from more than one GNSS service—GPS, Glonass, etc.—to achieve improved reliability and accuracy.

The output interface of such a GNSS module typically consists of these elements—conveying frequency, phase, and time, as shown in Fig. 8–5.

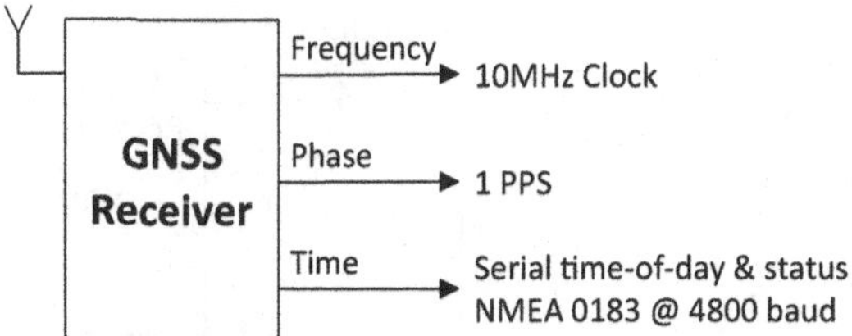

FIGURE 8–5 GNSS receiver interfaces. *GNSS*, Global Navigation by Satellite Systems.

The clock output provides a frequency that is syntonized to time from GNSS. The long-term stability of this clock should meet the requirements of what is called a Primary Reference Clock, meaning that it is effectively the same as having a local cesium frequency reference. The 1PPS, or one pulse per second, output provides phase alignment—remember, time to the subsecond level without consideration for which second. The serial interface, at least the transmit portion, is used to convey a whole host of information including time—which second the 1PPS refers to, position information, and status indications, such as whether the received information should be considered reliable. There are various protocols used for this information, some proprietary and some standard. However, by far the most common is based on a standard called NMEA-0183. Under this, the information is packaged into defined textual messages called sentences. The serial port can also be used bidirectionally for module configuration in a vendor-specific manner.

O-RAN synchronization

Based on the previous discussion, you might be forgiven for thinking that the obvious way to synchronize the nodes in an O-RAN network is to just build a GNSS receiver into each O-RU and use these to achieve time alignment between all the radios, as shown in Fig. 8–6.

Indeed, prior to the growth of heterogenous networks, back when most of the radios in a cellular network were outdoor macrocells, this was the standard method of synchronization, and as we shall see later, this is actually allowed as a method by the O-RAN specification. However, there are three good reasons not to follow this approach, all stemming from the limitations of GNSS outlined previously:

- Cost—The inclusion of a GNSS receiver and associated antenna will always have some cost impact.
- Practicality—Many O-RUs, particularly for millimeter-wave applications, are likely to be in locations where the need for a clear view of the sky for GNSS to work makes the use of GNSS simply impractical.
- Reliability—The vulnerability of GNSS to jamming and spoofing makes it a risky proposition to use directly at the radio nodes.

Therefore it is expected that the majority of O-RAN networks will instead use an alternate method of synchronization for the radios, and this takes the form of network-based synchronization.

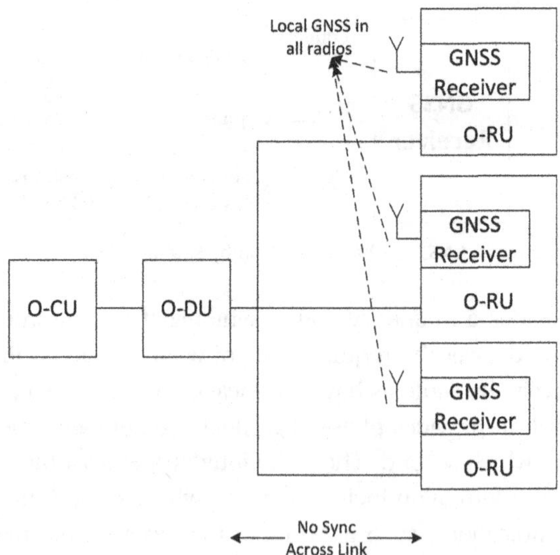

FIGURE 8–6 GNSS synchronization of O-RAN. *GNSS*, Global Navigation by Satellite Systems; *O-RAN*, open radio access networks.

Network-based synchronization

The basis of any network-based synchronization strategy is to use the network itself to distribute time and/or frequency between nodes in the network, as shown in Fig. 8–7.

Specifically, one node, commonly called the master, receives time and/or frequency from outside the network, normally through a GNSS receiver, and then transports this across the network to the other nodes—the slaves. The result is that the slaves align their time and frequency to that of the master, which in turn is aligned to actual time from the GNSS.

This technique of network-based synchronization directly addresses the limitations of relying on GNSS everywhere:

- Cost—Although providing network synchronization will increase device complexity somewhat, with a corresponding cost penalty, since the existing data transport is being used which will almost certainly be significantly less than providing GNSS receivers at each node.
- Practicality—By decoupling where time is received, in other words where the master node with the GNSS receiver is, from where time is used (the radios) it becomes possible to locate the master in a more preferable location for GNSS reception independent of the radiolocation.
- Reliability—While 100% reliability can never be achieved, placing multiple masters in the network, each with their own GNSS receiver, goes a long way to mitigating potential concerns by providing multiple, geographically distinct, sources of time. In particular, this can help against jamming and spoofing attacks since these will in general be isolated to a relatively small area.

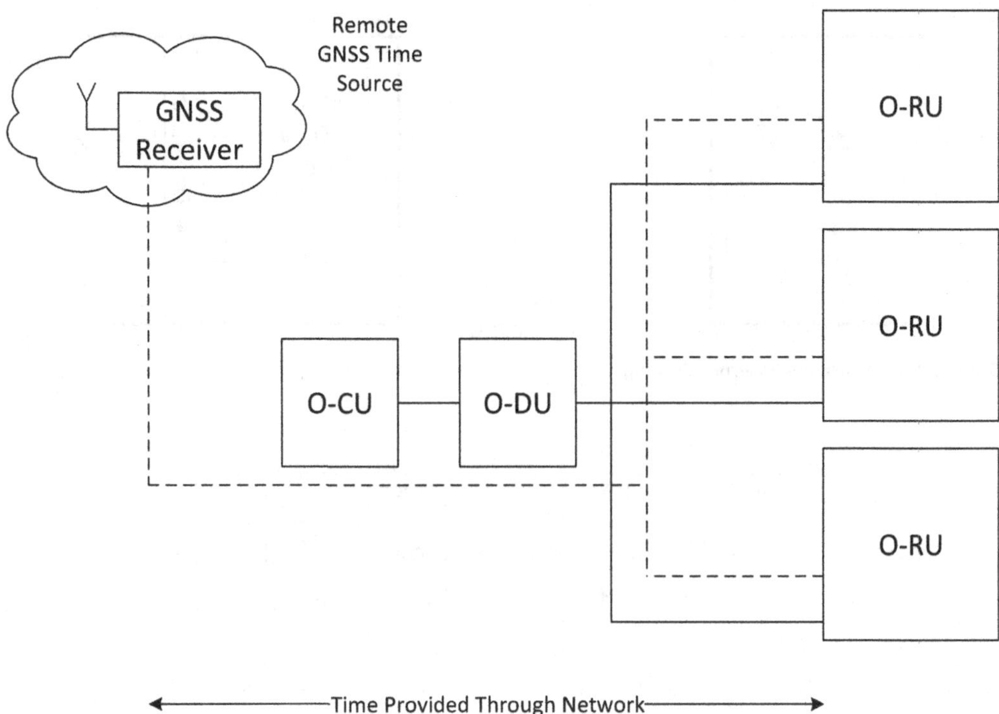

FIGURE 8–7 Network time distribution.

So, what tools do we have to transport time and frequency across a network? There are two distinct mechanisms that come into play—physical-layer transport and packet transport, and both are relevant to O-RAN.

Physical-layer transport and Synchronous Ethernet

Every network itself already has a mechanism to transport data over a wire or optical fiber and this can generally be used directly to transport a frequency from one node to another. This can be done because the transmitting side has to use a clock signal to clock the data onto the wire, whereas the receiving end typically needs to recreate a clock aligned to the bit rate of the incoming datastream to allow it to be decoded. This is shown in Fig. 8–8.

Ethernet, which is the transport most likely to be used for any O-RAN application, is what is called an asynchronous network, meaning that the nodes do not need to share a common frequency for the network itself to work—instead, each node can operate up to 50 ppm to more off nominal, allowing low-cost crystals to be used in equipment. This also means that there is not any guaranteed relationship between the frequency that was used to clock the datastream into a switch and that which is used to clock it out, as shown in Fig. 8–9. Most Ethernet switches use a store-and-forward strategy that means they will wait for a complete

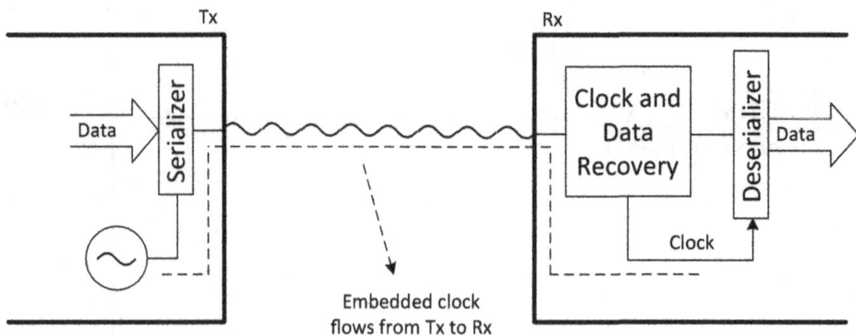

FIGURE 8–8 Asynchronous Ethernet clocking.

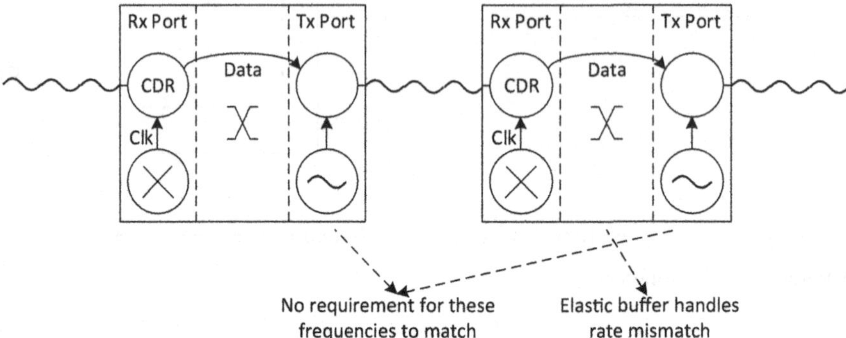

FIGURE 8–9 Ethernet rate mismatch.

data frame to be received before starting to send it out. This allows for the potential difference in transmitting and receiving rate without data loss.

However, there is nothing in the Ethernet standards that specify where the transmit clock comes from—just what error it is allowed to have. Specifically, there is nothing stopping one from taking the clock that the receiver recovers from the incoming datastream, cleaning it up with a bit of filtering, and using that as the transmit clock. This allows a frequency to propagated across the network, from the master, through a number of switches and finally to the slave nodes, as shown in Fig. 8–10.

This is the technique used by Synchronous Ethernet, though more commonly abbreviated to SyncE, which is defined by ITU-T in the recommendations (standards) G.8262 (timing characteristics of a synchronous equipment slave clock) and G.8262.1 (timing characteristics of an enhanced synchronous equipment slave clock). As the titles suggest, G.8262.1 deals with an improved, second generation, version of Synchronous Ethernet. Note that ITU-T uses the term "clock" to refer to a piece of equipment, or a function within that equipment, that is involved in the distribution and use of time or frequency—it does not imply a clock in the traditional sense of something that keeps time.

For the purpose of this discussion, the details of what the SyncE standards define are not really necessary beyond knowing that the "clean up" function shown in Fig. 8–10 is a low-pass filter with a bandwidth normally between 1 and 10 Hz. This allows each switch in the network to filter out noise on the SyncE signal, rather than allowing it to accumulate unchecked node-to-node, allowing for more switches between the master and slave.

Another aspect of SyncE, defined in the ITU-T G.8264 recommendation, is a messaging protocol that operates between nodes to indicate the quality of the clock carried on the corresponding link. After all, the receiver has no simple way of knowing if the frequency it is recovering has come from a GNSS receiver or from a cheap crystal in the upstream node. Therefore this message protocol, which is referred to as the Ethernet Synchronization Messaging Channel is used to send Synchronization Status Messages (SSMs) between nodes. A multiport switch can look at the SSMs received on each port to decide which ports carry a suitable SyncE clock and then select one of these to use as the transmit clock. This is shown in Fig. 8–11.

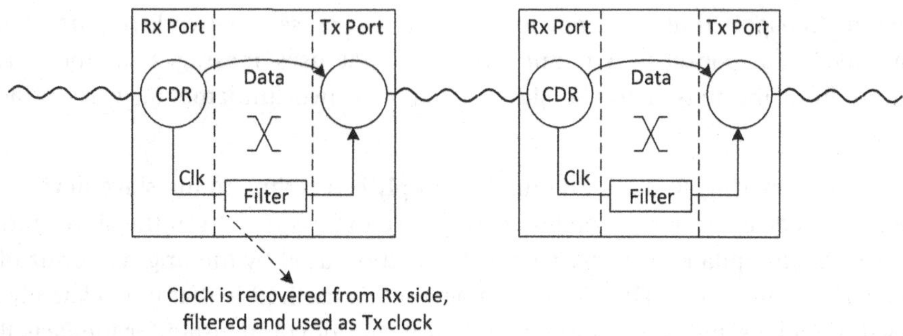

FIGURE 8–10 SyncE propagation.

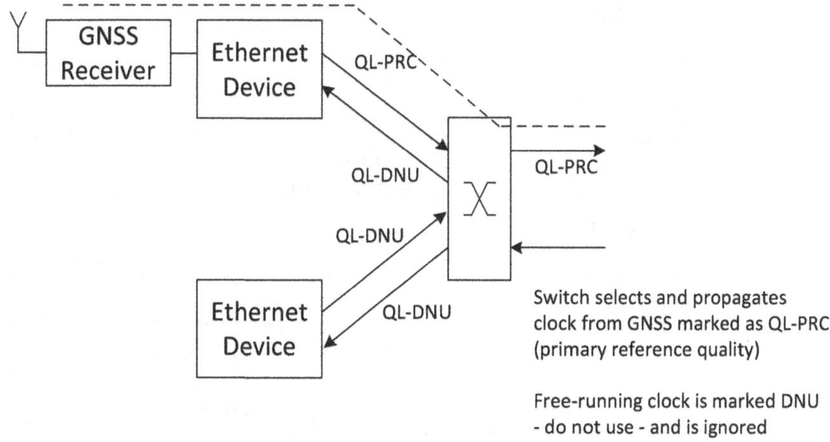

FIGURE 8–11 SyncE ESMC Messaging. *ESMC*, Ethernet synchronization messaging channel.

SyncE provides a good way of distributing frequency and can achieve high accuracies since it uses the physical layer directly, but it does have a couple of drawbacks:

- SyncE is only useable for frequency distribution. There is no way that it can be used directly to transport time.
- For SyncE to work, every switch in the network between the master and slave must support it. If any switch in the link doesn't then synchronization is lost—just like breaking a single link in a chain renders the chain useless.

Despite these limitations, we shall see that SyncE has a role to play even in time distribution and since the incremental cost of including SyncE capability in new networking equipment is relatively low most new-build networks now being deployed include it. However, the majority of legacy networks do not support SyncE because of the fact that doing so would require new equipment to be deployed throughout the network.

Packet time transport and precision time protocol

To be able to transport time, and not just frequency, across a network requires a different tactic—specifically, a packet-based time transport that uses messages on the network to transfer time from the master to the slaves. There are two fundamental ways of achieving this:

- One-way protocol (Fig. 8–12)—The master simply sends time to the slave devices, either through a broadcast or one-by-one or in response to a request from the slave, and the slave uses this to update its clock. This is the method used by the original forms of Network Time Protocol (NTP). The advantage of this method is that it is relatively simple. However, it had a significant drawback in that it cannot compensate for the facts that the messages carrying the time information experience a finite delay between the master and

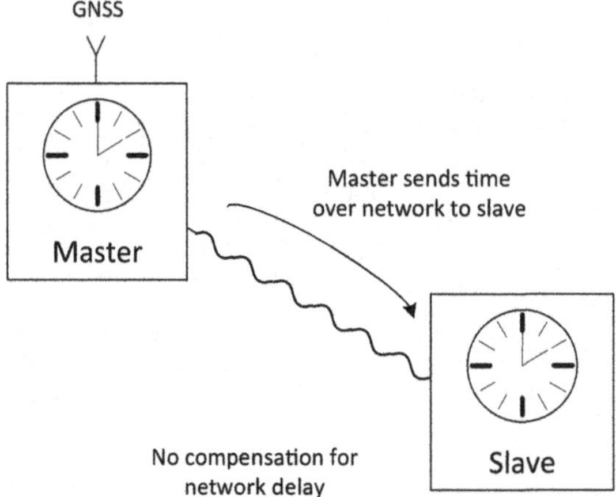

FIGURE 8–12 One-way time distribution.

slave. In other words, when the slave receives the time message it only knows what the time was at some point before now, rather than what it actually is now. Whether this is a problem or not depends on the required accuracy and the message delay through a network. For example, setting the clock of a PC using NTP over a network that has an average latency of 50 ms is probably okay for the majority of applications, whereas trying to achieve submicrosecond level time alignment, as is needed for O-RAN, over a network even a hundred times better would be nonstarter.

- Two-way protocol (Fig. 8–13)—The inaccuracy caused by the finite delays in a network can be largely eliminated by using a two-way protocol, in which messages flow in both directions between the master and slave. These methods combine one-way time delivery with a ranging mechanism that is used the gauge the message delay between the master and the slave, which can then be used by the slave to correct the received time. By far the most common two-way time transport protocol, and the one of relevance to O-RAN, is PTP, as defined by the IEEE 1588–2008 standard—hence it is alternate name of just 1588. The following section gives an overview of PTP.

An introduction to PTP

PTP was originally designed for time synchronization across small, localized, networks for industrial automation applications such as the synchronization of robots on a production line. However, it was quickly appreciated that the underlying concept, though not the particular initial implementation, could be used for a wide range of applications across both large and small networks. Therefore a second, incompatible, version of PTP was defined and ratified in the 2008 release of the IEEE 1588 standard (the original dating from 2002). A subsequent third release of the IEEE 1588 standard in 2019 added some additional functionality

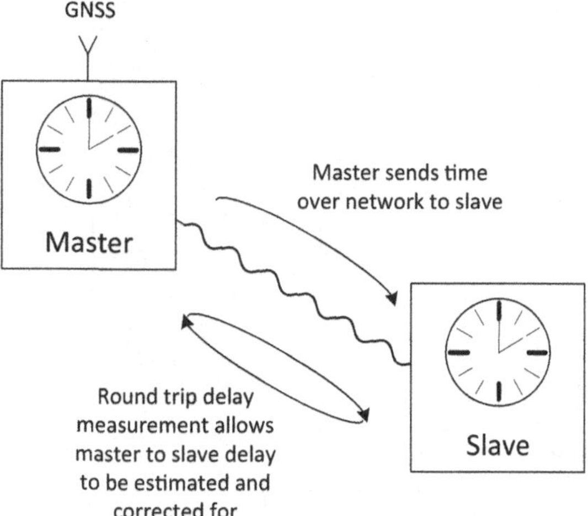

FIGURE 8–13 Two-way time distribution.

and clarification but is fully backward compatible with the 2008 version, and none of the new changes are directly relevant to O-RAN.

The IEEE 1588 standard is architected in such a way as to provide a framework that is flexible enough to be used for many applications with different synchronization requirements and different network types, including non-Ethernet networks. However, this flexibility is a double-edged sword as it makes designing equipment that is fully compliant to the standard unnecessarily complex. Therefore the standard allows for a layer of customization to be applied in the form of a PTP profile. The idea of a profile is that it defines an acceptable subset of the flexibility offered by PTP targeted at a specific application. IEEE itself, or another standards body, or indeed any other interested party, can create such a profile and a number of profiles have been defined to date. However, only two of these are relevant to O-RAN, both defined by ITU-T:

- Full-timing support (FTS)—Defined in G.8275.1 and designed to support the highest accuracy time transport over specially architected networks
- Partial-timing support (PTS)—Defined in G.8275.2 and designed to transport less-precise time (but still of the microsecond level) over more general network

The difference between these two profiles and their uses will be addressed shortly.

The operational details of PTP are very extensive, and it would take too long to give a full explanation, so what follows is an overview that is sufficient to understand how PTP can be used for O-RAN synchronization.

Fig. 8–14 shows the general topology of a PTP network.

There are one or more Grandmasters, which take time from, typically, GNSS and provide this to the network. The end nodes—the devices that require time, such as the individual

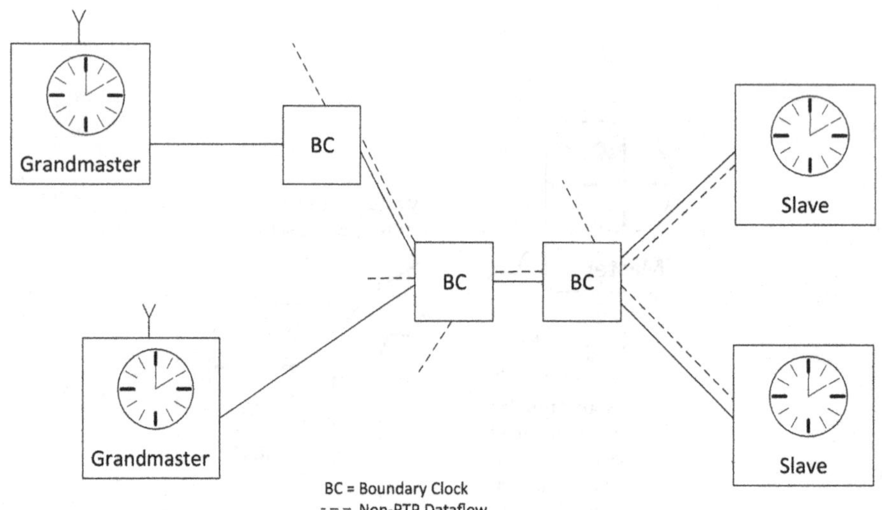

FIGURE 8–14 PTP network topology. *PTP*, Precision Time Protocol.

O-RUs—are the slave and ultimately receive time from the Grandmasters. In between, there could be a number of so-called boundary clocks, which can be thought of PTP repeaters. These have two or more ports and receive time on one port and retransmit it on the other ports. In general, a boundary clock is a specialized Ethernet switch or router, so each PTP port corresponds to a data port and the device also forwards non-PTP data traffic between ports. In the FTS and PTS profiles, these three types of nodes are referred to as:

- T-GM—Telecom grandmaster
- T-BC or T-BC-P—Telecom boundary clock or telecom boundary clock-partial, depending on if it is used for FTS or PTS
- T-TSC or T-TSC-A—Telecom time slave clock, again depending on the profile

There are a number of different messages used by PTP, which depending on the profile are sent either as raw Ethernet frames, with their own EtherType value, or as the payload of UPD/IPv4 or UDP IPv6 datagrams, and all of which have a common header format followed by message-specific information. These messages can be divided into two basic types:

- Event messages—These are the messages that are actually used to transport time across the network and therefore are considered time-sensitive. These messages are timestamped at each end, meaning that the sender captures the exact time the message is sent using its clock and the receiver does the same with its local clock.
- General messages—Messages that are used by PTP but those are not time-sensitive, so do not get timestamped.

The first step in PTP synchronization is to decide which node is going to be master in the network. It might seem obvious that this will be the one with a GNSS receiver attached to it. However, as mentioned earlier, it is possible that there might be multiple such candidates to provide a degree of redundancy. Although any one of these could do the job of Grandmaster for the network, only one can in general have that role at once. To decide which one, each candidate for the role broadcasts a general message, called an Announce message, that outlines the quality of the clock the node can offer. Every slave in the network receives all of these Announce messages and selects the best candidate to listen to. The actual comparison of who is better is performed using a comparison process called the Best Master Clock Algorithm, which differs slightly between PTP profiles but always with the intention of selecting what will ultimately be the most accurate source of time. Likewise, each master candidate itself can see the Announce messages from other candidates, and if there is a better option out there, a given candidate will enter a passive state where they do not become Grandmaster since another node could do a better job. The details of all of these selection processes are handled by a standardized state machine in each node and therefore every node ultimately reaches the same decisions. Fig. 8–15 illustrates this selection process.

This whole selection process is dynamic and continues to run after the initial selection. This means that if a better potential master suddenly appears on the network, then it can take over as the Grandmaster role, and likewise, if the current master disappears for any reason or loses its accuracy (e.g., as the result of GNSS jamming attack affecting it), then a better candidate that's been lurking in the background can step up and take the Grandmaster role. It is this master selection mechanism that allows PTP to provide robust synchronization.

Once everyone is agreed on whom the Grandmaster is going to be, which in turn determines how PTP will flow through any boundary clocks, the nodes can actually start synchronization—as in alignment with their local clocks and tick rate to that of the Grandmaster. This is a two-stage process and involves repetitive messages exchanges as shown in Fig. 8–16 to initially syntonize the slave's clock—make it tick at the right rate—and then synchronize it—make it tell the correct time.

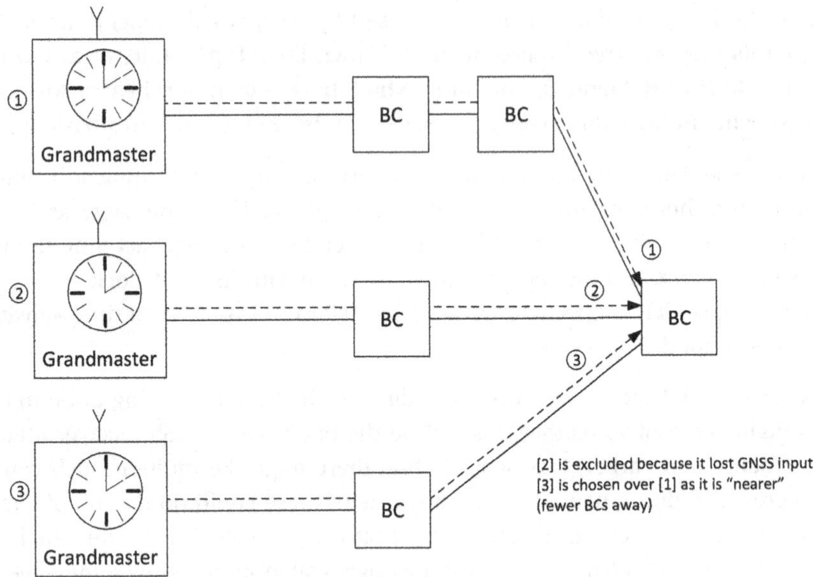

FIGURE 8–15 PTP master selection process. *PTP*, Precision Time Protocol.

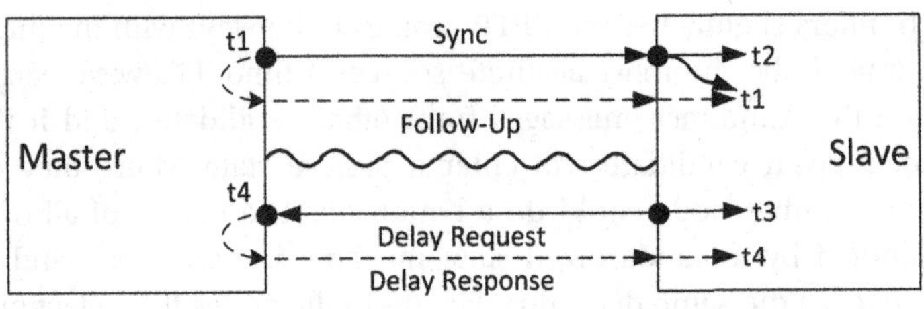

● Timestamp captured as message passes this point
t1 timestamp is sent to slave either in sync message itself or in separate follow-up message

FIGURE 8–16 PTP synchronization message flow. *PTP*, Precision Time Protocol.

As can be seen in Fig. 8–16, the Sync and Delay Request messages are event messages—they get timestamped at each end—whereas the Delay Response and Follow-Up messages are general messages and are not time-critical. The flow of messages is such:

- The master sends a Sync message to the slaves and captures the exact instant this was sent (the t1 timestamp). This timestamp is either included in the Sync message as it is sent, or sent in a related Follow-Up message if the master's hardware cannot handle retime modifying of a message. Either way is equivalent in performance and all slaves are required to support both methods.
- The time at which the slave receives the Sync message is captured as t2 using the slave's local clock.
- The time at which the slave sends the Delay Request message is captured as t3, again using the slave's local clock.
- Finally, the time at which the master receives the Delay Request message is captured against actual time as t4.
- The Delay Response message does not involve any timestamping but instead is used to pass the final t4 timestamp back to the slave, since it cannot piggyback on another message as t1 does with the Sync message.

Once all these stages have been completed, the slave has all four timestamps t1–t4 ready for processing. The information in these timestamps can be used by the slave to achieve both syntonization and synchronization to the master.

Syntonization is achieved by comparing the difference between t1 timestamps across many messages with the difference between the t2 timestamps. This is shown in Fig. 8–17.

In Fig. 8–17, it can be seen that the master sends the Sync messages with a certain gap between each message. Ideally, this gap will be constant, but in reality, it will change a bit because of factors such as software scheduling. However, the slave can always determine the gap between the sending of consecutive Sync messages through the difference between the t1 timestamps of those messages. On the slave side, the perceived gap between each message will be different, since the timestamping clock is not currently running at the correct tick rate. By comparing the perceived gap (from the t2 timestamps) with the actual gap

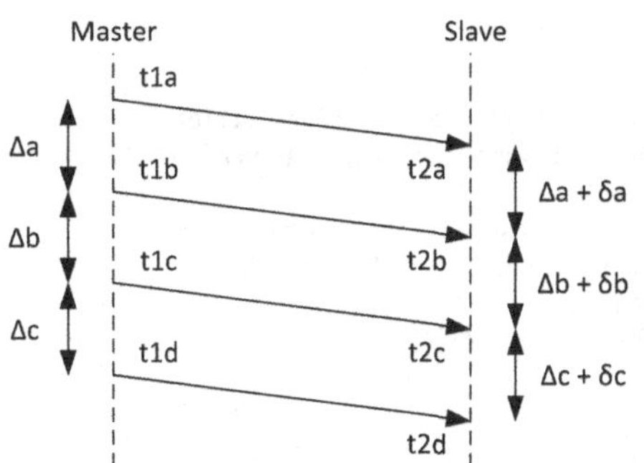

Slave uses t1 and t2 timestamp information to calculate δn values. Average of δn represents the slave's "tick rate" error. Slave adjusts local oscillator to make δn zero.

FIGURE 8–17 PTP syntonization process. *PTP*, Precision Time Protocol.

(from the t1 timestamps), and averaging over many cycles to mitigate the effects of network variations, the error in the local tick rate can be determined and the rate adjusted.

The first step of the slave synchronizing to the master involves the calculation of the round-trip delay, which is then divided by two to give the one-way (master to slave) delay. From Fig. 8–16, this round-trip delay is the time between the master sending the Sync message (t1) and the time it receives the Delay Request message (t4) minus the time it took the slave to send the Delay Request message after receiving the Sync message, which is t3 – t2. Therefore,

round-trip delay = (t4 – t1) – (t3 – t2)
  or, by rearranging:
round-trip delay = (t2 – t1) + (t4 – t3)
  and the one-way delay is half of this, so:
one-way delay = [(t2 – t1) + (t4 – t3)]/2

An important consideration in these calculations is that because the slave clock is not necessarily synchronized to the master clock yet—after all, that is what this whole process is being done to achieve—the value of t2 relative to t1, and of t4 relative to t3, could be way different to the actual network delays, because the current value of the slave's clock (used for t3 and t4) could literally be years wrong compared to actual time (used for t1 and t2). However, this does not matter since it is only time difference that matter, (t4 – t3), and not absolute time at this stage. Therefore the calculation stated before is always valid. However, what is important is that the slave's clock is ticking at the correct rate; otherwise, the (t4 – t3) calculation would not be accurate. This is why the previously discussed syntonization step is important.

Once the one-way delay from master to slave is known, the slave can use this to offset the time received from the master in the Sync messages (t1) and set its clock accordingly as shown in Fig. 8–18. At this point, both syntonization and synchronization have been achieved.

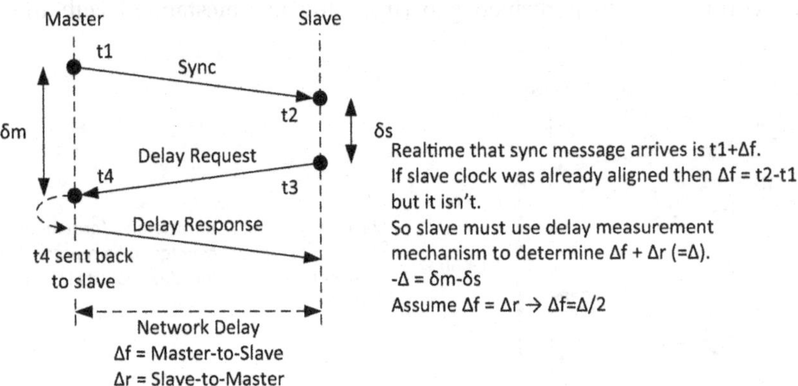

FIGURE 8–18 PTP synchronization process. *PTP*, Precision Time Protocol.

This synchronization process is then repeated continuously, with several such message exchanges every second normally, to maintain synchronization of the slave clock to read time from the master. This ongoing activity is necessary since the syntonization from the initial step can never be perfect, meaning that the slave clock will be ticking with a fractional error. Such an error can quickly accumulate if the clock is not regularly resynchronized.

Of course, nothing is ever this simple in reality, and in real networks there a number of factors that can impact the ability of a two-way time transfer to transport time:

- Timestamping accuracy. The previous example assumes that we can timestamp messages as they are sent and received with arbitrary accuracy and precision. In reality, this is not the case. The time counter that we use to maintain time, and which is captured to generate the timestamp, has a finite resolution—typically in the order of a few nanoseconds. Additionally, hardware limitations generally mean that there is some variation packet-to-packet as to what is defined as the "exact moment" of sending or receiving a message. Both of these factors effectively add noise to the timestamps which can impact the accuracy with which synchronization can be achieved.
- The above calculations made the assumption that the network is symmetrical, at least in terms of the delays experienced by the timing packets. Specifically, it assumed that the one-way delay for master to slave (which is what we need) is exactly half the round-trip delay of master–slave–master (which is what we can measure). In a perfect network, this may be a valid assumption, but in real networks, there will in general be asymmetries. These can be caused by the network infrastructure itself, such as different fiber lengths in each direction when separate transmit and receive fibers are used, or different optical pathlengths if two wavelengths are used on the same fiber, or can also be caused within the nodes themselves by factors such as where in the interface circuitry the timestamping.
- A much bigger problem is what is called Packet Delay Variation (PDV)—or jitter as it is commonly known to network engineers. This refers to the fact that each message sent between the master and slave likely would not take the same time to complete the journey, even though the physical path is unchanged. This results from delays in any switches or routers between the master and slave caused by other traffic on the network. Ethernet and the overlying protocols all support the concept of some form of quality of service to allow messages that are considered time-sensitive, such as the synchronization ones, to be prioritized over other packets, but none of these mechanisms can get around the issue of head of line blocking, as shown in Fig. 8–19.

As can be seen in Fig. 8–19, a high priority message arrives in a queue to leave a switch just as a lower priority message has started being sent. Even though the new message has a higher priority it has to wait for the earlier one to complete since Ethernet does not provide any mechanism for preempting transmissions once started. The effects of this type of blocking can be large. As an example, Fig. 8–20 shows the measured variation for Sync messages (dark) and Delay Request messages (grey) going through a network of 10 switches that are also carrying other traffic at a constant rate. It can be seen that in this particular test, which

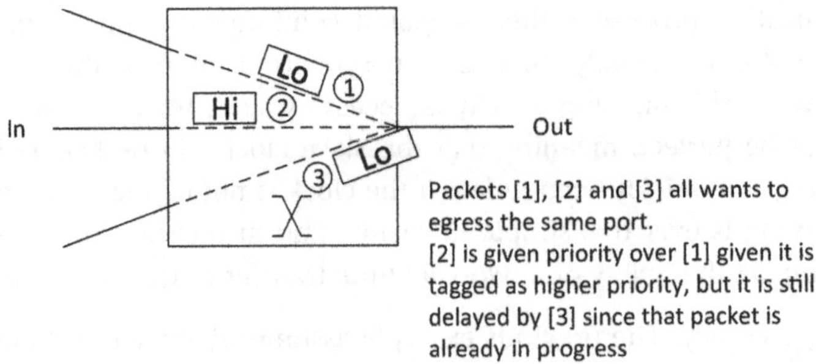
Lo
①
Hi
②
In
Out
③
Lo
Packets [1], [2] and [3] all wants to egress the same port.
[2] is given priority over [1] given it is tagged as higher priority, but it is still delayed by [3] since that packet is already in progress

FIGURE 8–19 Head of line blocking.

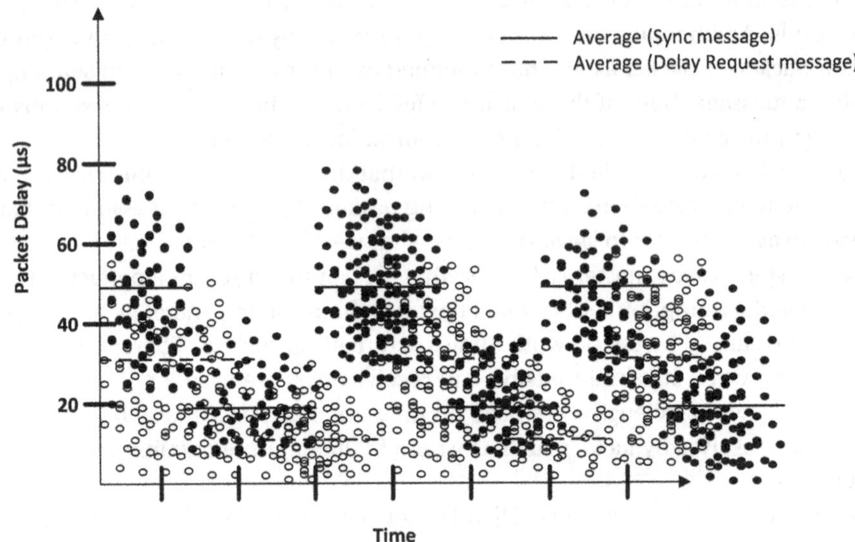
Average (Sync message)
Average (Delay Request message)
100
80
60
40
20
Packet Delay (µs)
Time

FIGURE 8–20 Example network PDV. *PDV*, Packet Delay Variation.

comes from an older ITU-T standard (G.8261) and is probably rather more draconian than would ever be seen in a real network, the PDV very closely matches the traffic loading on the network.

The effect of PDV on the syntonization and synchronization calculations is that each time the calculations are done there will be a different result. For example, during the syntonization phase, it is meaningless to compare the t2 and t1 timestamp differences between two consecutive Sync messages if those messages actually took different times to get through the network.

In order for a PTP slave to provide useful time recovery, each one of these factors needs to be considered and mitigated in some way. Looking at each one in turn:

- Timestamping accuracy—The key to reducing errors related to the timestamping of the event messages is to design the solution such that the messages are timestamping as

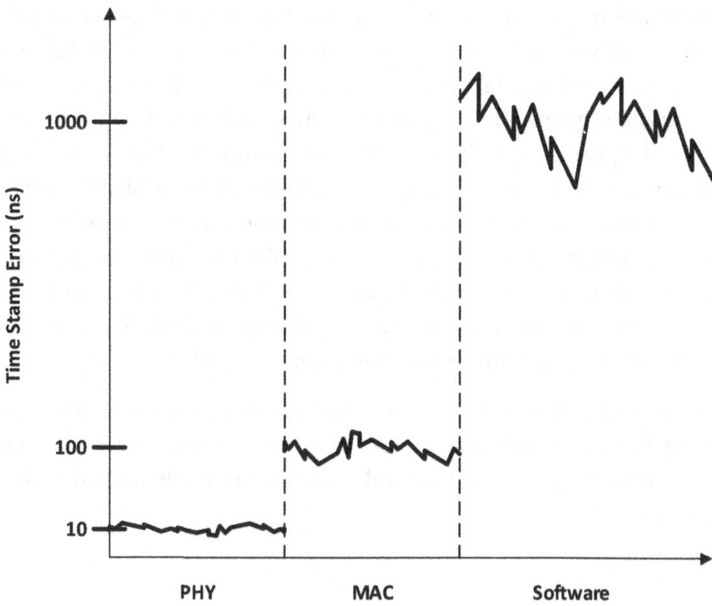

FIGURE 8–21 Effect of timestamping point.

close to the physical network connection as possible. As soon as the timestamping point is pushed away from the actual cable into the bowels of the hardware there will be FIFOs, clock domain crossings, etc., between it will add errors to the timestamping accuracy. If the hardware has no timestamping capability and it is instead handled by software then the accuracy will be even worse given the nondeterministic nature of most operating systems. This is illustrated in Fig. 8–21. We will come back to where timestamping realistically needs to be done in an O-RAN application later.

- Network asymmetry—if there is a true, static network asymmetry then the only way to eliminate the effect of this on the accuracy of the synchronized time is to manually program the error into each device based on either knowing the asymmetry through design analysis or measuring it in some way. For example, if a slave is connected to the master by a pair of fibers that are not length matched, with the result that there is a 1 μs difference in the time it takes data to go in each direction then, without any correction, this will cause an offset in the recovered time of exactly half this—500 ns. However, if the network installer tells us about the 1 μs difference, we can program the slave to always offset its time by 500 ns to null out the error. Alternatively, when installing the slave, we could temporarily hook up a local GNSS receiver to the slave and compare the time from that with the time we derive from PTP and program the difference as a known offset.
- Network PDV—This is the tricky one, and the problem that can make or break the usefulness of PTP. There are two ways that PDV can be handled. The first is to accept

that the network will have PDV and force the slave to somehow filter this out. In order to do this, the clock recovery servo in the slave—the software function that processes the timestamps to achieve syntonization and synchronization—has to use smart techniques that allow it to try and gauge the PDV and the effect it has had on each message received and use advanced algorithms to filter out the effects of PDV. How well this can be done depends on a number of factors—internal ones to the slave, basically how smart the boffins who developed the servo are, and external ones such as both the amount and nature of the PDV. Often, a leading-edge servo implementation can achieve PDV filtering of a couple of orders of magnitude. In other words, if the PDV experience by the event messages is 100 μs, the recovered time will be accurate to 1 μs. As we shall see, this might be acceptable for some applications but not others.

The second way of handling PDV is to do whatever can be done to eliminate it. If there is little PDV affecting the event messages, there is not so much to have to filter out. This is where we come back to FTS, PTS, and the value of boundary clocks. Consider the three networks shown in Fig. 8–22.

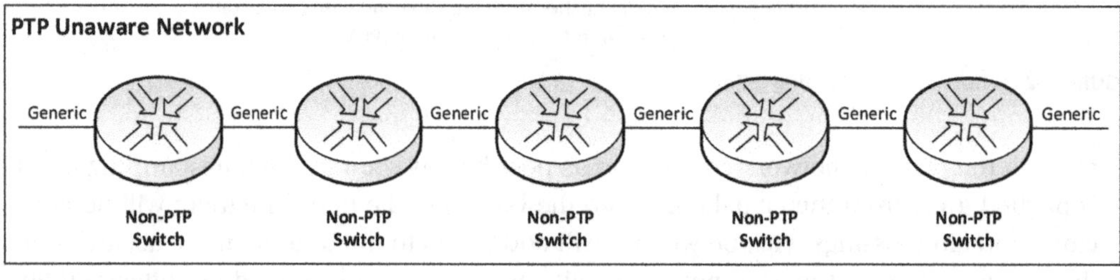

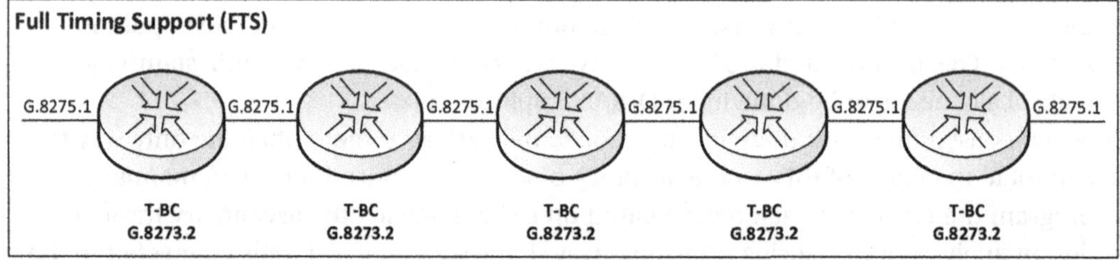

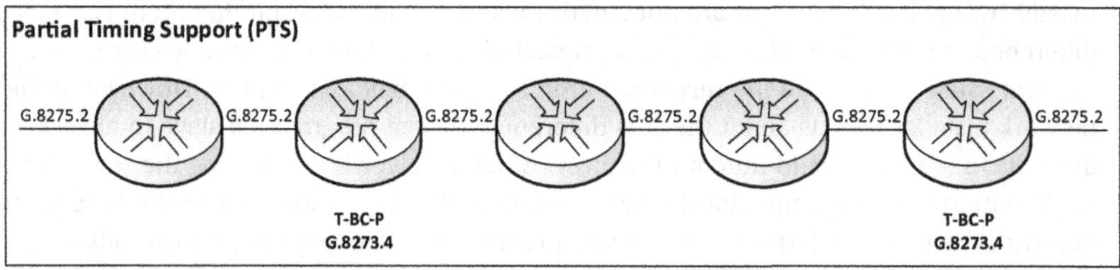

FIGURE 8–22 PTP network types. *PTP*, Precision Time Protocol.

Each of these networks has a PTP master providing time to a PTP slave through a network of five switches. This difference between these three is whether those switches also act as PTP boundary clocks. In the first network, none of the switches act as boundary clocks, meaning that the PTP event messages have to pass through each switch in exactly the same way as any other data packet would—this is a PTP-unaware network, since the switches are unaware of what PTP is. The second network, on the other hand, has every switch also function as a boundary clock. A boundary clock can be thought of as a PTP slave and one or more masters all rolled into one. Rather than just transferring PTP packets as a switch does, it actually uses PTP itself to maintain a local clock. Whichever port the Grandmaster is attached to, either directly or through other boundary clocks or switches, acts as a PTP slave and recovers time in just the same way as a slave would. However, rather than using that time for some local function (such as synchronizing a cellular radio), it uses the time to become a local master on all its other ports—providing that time to slaves further downstream (potentially through more boundary clocks or switches). This network is called a FTS network since every switch in it is also a PTP boundary clock—in other words, every switch understands and supports PTP. The final network is a cross between these two—a PTS network, in which some but not all of the switches are boundary clocks.

So, what benefit does adding boundary clocks have? Well, remember that the biggest source of PDV is head of a line blocking happening to the PTP event messages as they pass through the switches in the network. However, if the switch is replaced by a boundary clock, no event messages actually pass through the switching function itself to get blocked. In other words, replacing a switch with a boundary clock eliminates the portion of overall PDV that would have come from that switch. Obviously, the more switches that are replaced by boundary clocks, the lower the overall PDV between the master and slave.

It should now be clear why we said earlier that FTS is designed to provide the best time accuracy but over specifically architected networks, whereas PTS gives reduced accuracy, but over more general networks. The improved accuracy comes from virtually eliminating PDV altogether in an FTS network, but at the expense of needing to build the network with boundary clocks everywhere.

Putting PTP and SyncE together

Having just learned about PTP and how it can be used to achieve both syntonization and synchronization—in other words, provide time and frequency to the slave—you might be forgiven for thinking that SyncE as a frequency-only transport does not have any role to play. After all, PTP can do it all. However, in practice, it turns out that there are significant benefits of using both SyncE and PTP in conjunction to achieve the best overall solution.

First, as we will see in the next section, O-RAN defines both frequency and time alignment requirements. As already said, being a physical-layer technique, SyncE can provide very stable frequency to the slave. Therefore although PTP could be used to provide both frequency and time to an O-RU, it makes perfect sense to use SyncE for frequency aspect if it is available in the network itself.

The second benefit of SyncE comes into play in determining the performance of PTP itself. The key to being able to achieve accurate time synchronization with PTP is to first achieve frequency syntonization. If the frequency that forms the tick rate of the local clock in the slave device is itself unstable, it makes being able to synchronize time that much harder. Syntonizing the slave tick rate is effectively a necessary steppingstone to synchronizing the actual time. However, if you already have a stable frequency from SyncE then why not bypass the PTP syntonization step and go straight to synchronization? The actual details are somewhat more involved, since although the frequency from SyncE is likely to be more stable than the frequency recovered from PTP, accumulated noise in the network can mean that the frequencies are not perfectly aligned, which would, in turn, cause more problems with synchronization and, therefore, has to be allowed for. However, for the purposes of this discussion, the message to take away is that PTP + SyncE will always be better than PTP on its own.

Cellular network synchronization requirements

We now know all we need to know about how PTP and SyncE can be used to distribute time and frequency around a network. Now to look at how this applies to O-RAN and, in particular, the synchronization of the fronthaul network between the Distributed Units (DUs) and associated Radio Units (RUs).

To understand the role of synchronization in O-RAN, we need to take a step out and look at the requirements of the cellular networks that O-RUs are serving.

The requirements for these networks are laid down by 3GPP in their Release 15, and beyond, standards. Specifically, the document of interest is 3GPP TS 38.104 (Base Station radio transmission and reception). All of these requirements are defined in terms of the transmit signal measured at the antenna port of the radio device, so the O-RU output in this context.

Frequency accuracy and stability

TS 38.104 defines separate frequency accuracy limits for macrocells and pico/small cells:

- For a macrocell, the frequency must remain within ± 50 ppb (parts per billion) of its nominal value.
- For a pico or small cell, the frequency must remain within ± 100 ppb of its nominal value.

The relaxation of the accuracy requirements for smaller cells is possible because their signal is much more local and, therefore, less likely to cause interference and is desirable since it could allow the use of lower cost components.

Expecting the instantaneous frequency to fall within these limits at all times is both unrealistic and unnecessary, so 3GPP further defines that the stability is observed over a 1 MS window. This means that if one was to continuously measure the frequency error over a period of time and then select any 1 ms period within that measurement time, the average frequency within that period has to be within 50 or 100 ppb of nominal. This definition allows short-term frequency errors outside of the limits, while ensuring that over longer periods, the limits are met.

Time accuracy

For radios that cooperate in a MIMO or carrier aggregation environment, TS 38.104 defines different time accuracy requirements depending on the type of carrier aggregation supported and the supported bands (sub-6 GHz or millimeter wave). Table 8–1 shows the time alignment required between radios. In an O-RAN environment, such radios will all be connected to the same O-DU.

Additionally, 3GPP TS 38.133 (requirements for support of radio resource management) defines time alignment requirements between any radios that are within transmission range of each other to support TDD operation. These limits are 3 μs for small radius cells (<3 km) and 10 μs for wider coverage macrocells (>3 km).

For any given radio, both the TS 38.104 and TS 38.133 time alignment requirements must be met. This is shown in Fig. 8–23.

Table 8–1 3GPP carrier aggregation sync requirements.

Carrier aggregation type	Sub-6 GHz	Millimeter wave
MIMO	65 ns	65 ns
Intraband contiguous	260 ns	130 ns
Intraband noncontiguous	3 μs	260 ns
Interband	3 μs	3 μs

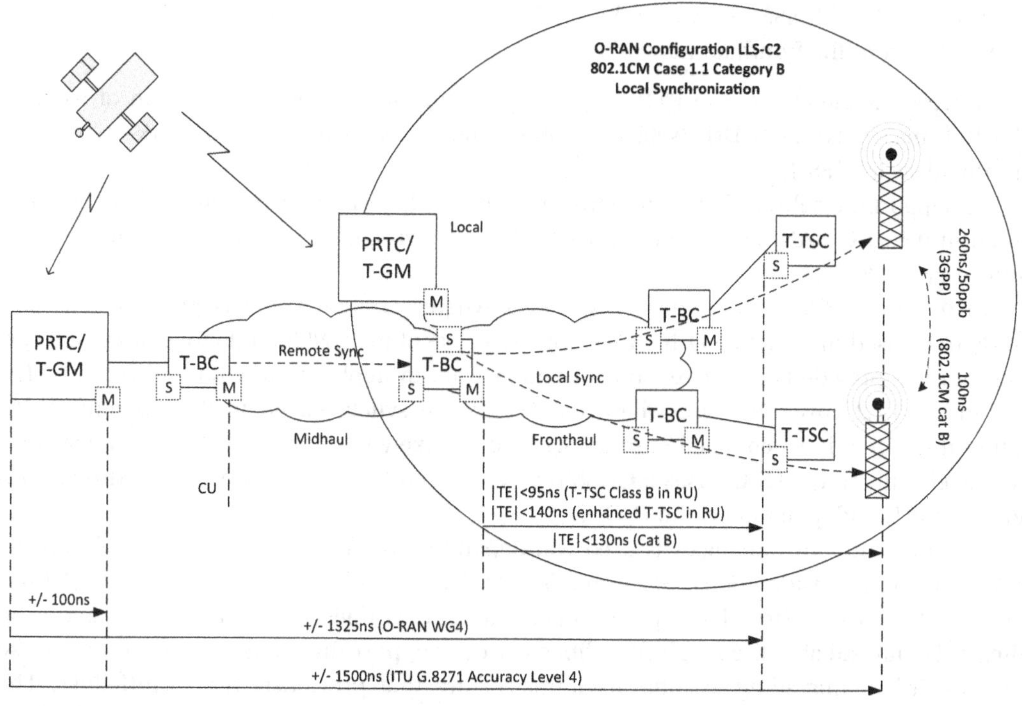

FIGURE 8–23 3GPP synchronization requirements.

Synchronization in O-RAN

Now that we know how to synchronize a network and know what the synchronization requirements at the O-RU outputs are, we can finally get to the crux of the matter—how do we achieve what is required in practice? This will be split into two discussions. The first looks at O-RAN synchronization at the network level, whereas the second covers what is required of the equipment itself and the implementation details.

O-RAN network-level synchronization

In order to meet the air interface synchronization requirements of both 3GPP TS 38.104 and TS 38.133 in an O-RAN network, it is necessary to synchronize the O-DUs to each other (to meet TS 38.133) and to synchronize the O-RUs to their parent O-DU (to meet TS 38.104).

Since the O-DUs are connected to both the backhaul and fronthaul network, there are three possible ways they can receive time synchronization:

- through a local GNSS receiver in the O-DU,
- via PTP from the backhaul network, and
- via PTP from the fronthaul network.

The O-RUs on the other hand connect to just the fronthaul network so only have two options:

- through a local GNSS receiver in the O-RU and
- via PTP from the fronthaul network.

However, if the O-RUs use PTP for their synchronization, there is an option of whether the PTP master is the O-DU itself, or another node on the fronthaul network—typically a fronthaul switch (FSH).

To support the subset of these options that a network designer would most likely be interested in using, O-RAN defines four so-called lower layer split control-plane architectures as shown in Fig. 8–24.

Of these, LLS-C4 is the legacy "GNSS everywhere" approach and is probably the least likely to be used in an actual O-RAN network, because of the GNSS limitations discussed earlier, but it is provided for in case someone does want to deploy it for whatever reason. LLS-C1 and LLS-C2 cover the case where the O-RUs are synchronized to their parent O-DU, either through a direct connection or a FSH, and are likely to be the architectures most commonly encountered. LLS-C3 uses a GNSS receiver within the FSH to provide synchronization to the O-RUs and, potentially, the O-DU.

Which of the four topologies is used will depend on a number of factors, not all directly related to sync considerations. For example, LLS-C1 can only be used when the O-DU has enough downstream (fronthaul) ports to allow a direct connection to each O-RU, and even then it is only suitable if enough fiber links exist to support this. This may, for example, be the case when upgrading existing 4G infrastructure that previously used CPRI links. The

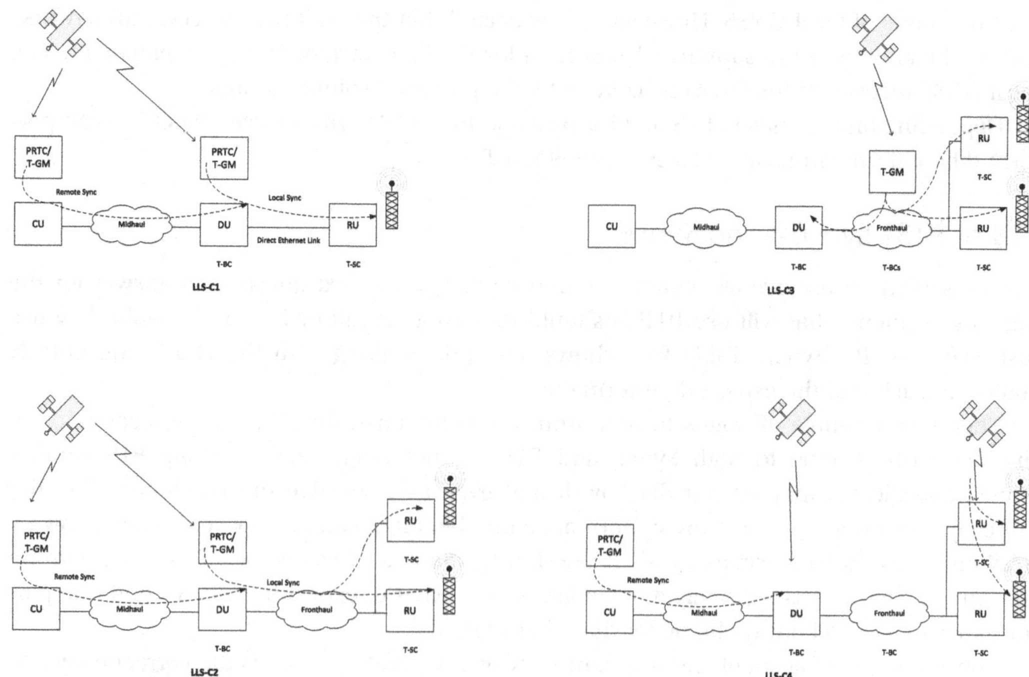

FIGURE 8–24 O-RAN Low Level Split (LLS) Architectures. *O-RAN*, Open radio access networks.

LLS-C4 topology on the other hand would likely only be preferred when the fronthaul network is not good enough to support IEEE 1588 operation, since the deployment costs of local GNSS receivers are likely to be higher and the availability will be compromised by the previously discussed limitations of GNSS. An example of such a case could be when the O-RUs are connected to an existing, unmanaged, network such as that could be found in a large office building. However, in this case, the overall performance of the network in terms of latency and jitter would also need to be considered to ensure it met the radio data requirements. The choice between LLS-C2 and LLS-C3 will typically depend on what capabilities the fronthaul network provides. If an operator is deploying their own network, they can make this choice based on factors such as which location—FSH or O-DU(s)—works best for a GNSS receiver. If the fronthaul network is provided by a third party, the decision will likely be driven by what the network provider can supply. If they offer time-as-a-service, LLS-C3 may be the more cost-effective route, whereas if the network does not provide sync services, LLS-C2 will be needed.

In Fig. 8–24, for each LLS architecture, there are a number of options as to where the O-DU gets its synchronization from. Whether this is achieved using a local GNSS receiver or PTP (generally from the backhaul network, but potentially from the fronthaul in the case of LLS-C3) will frequently be determined by the capabilities of the backhaul network since for many O-RAN installations, this will be existing network and likely provided by a third party. If the backhaul network already supports PTP, it probably makes sense to use this to avoid the limitations of local GNSS. However, it is expected that the majority of backhaul networks will not be architected to support PTP, at least for the first wave of O-RAN deployment, so a local GNSS receiver at the O-DU is likely to be the preferred solution initially.

Depending on the chosen LLS architecture and the O-DU sync source, the O-DU can perform different synchronization roles as shown in Table 8–2.

FTS or PTS? SyncE or no SyncE

Having settled on the overall synchronization topology, the next question to answer for the network segments that will use PTP is should they support full or PTS and should they use just PTP or PTP + SyncE. Table 8–3 shows the options along with the ITU-T and O-RAN stance on each and the expected performance.

There are a couple of points to note from the table. First, that ITU-T, the creators of all the standards relating to both SyncE and FTS, do not really address using FTS without SyncE. Specifically, they do not disallow it and even acknowledge the possibility of it, but critically they do not provide any performance numbers that can be expected. This leaves O-RAN on its own a bit since, as we shall see shortly, there is a clear reason for wanting to use this option, yet there is not really any guidance as to what to expect from it. Second, the performance of FTS will always be better than that of PTS.

Considering the backhaul and fronthaul networks separately, what is the correct choice to make for each? For the backhaul network, as was the case for whether PTP was even an option or not, the choice is likely moot since a legacy network will be used and all it depends on the services that network can support. Typically, in such a case, unless the backhaul

Table 8–2 O-DU sync functions.

LLS architecture	O-DU sync source	O-DU sync functionality as seen by fronthaul network
LLS-C1 and LLS-C2	Local GNSS	Grandmaster
LLS-C1 and LLS-C2	Backhaul PTP	Boundary clock
LLS-C3	Local GNSS or backhaul PTP	Passive
LLS-C3	Fronthaul PTP	Slave
LLS-C4	Local GNSS	N/A

PTP, Precision Time Protocol; *GNSS*, Global Navigation by Satellite Systems.

Table 8–3 SyncE and Precision Time Protocol (PTP) profile combinations.

PTP profile	SyncE?	Expected performance	ITU-T stance	O-RAN stance
FTS	Yes	Best	Supported	Allowed
FTS	No	Good	Mentioned but not characterized	Allowed
PTS	Yes	Not so good	Supported	Allowed but with caveats
PTS	No	Worst	Supported	Allowed but with caveats

FTS, Full-timing support; *O-RAN*, open radio access networks; *PTS*, partial-timing support.

network is operated by the O-RAN provider then PTP, if available, is likely to be provided in the form of "sync as a service." In other words, the network operator says our network will give you time via PTP, but we get to decide which profiles we provide and whether SyncE is available or not.

For the fronthaul network, the choice is likely to be more flexible since this network will often be designed from scratch just for the specific application. Clearly, FTS with SyncE is the gold-standard and should always be the option of choice if there is such a choice. However, one stumbling block to this is what seems to be a growing interest in implementing the O-DU function on standard, data center-type, hardware, such as rack-mounted servers with plug-in NIC cards to provide the network connectivity. The majority of such equipment available today does not include SyncE as an option, meaning that it is likely that a significant number of O-RAN deployments will be required to operate without SyncE. We shall see later what additional requirements this imposes on the actual equipment design.

What about using PTS in the fronthaul network? Please don't. Just don't! It is not clear why O-RAN even entertains the option of PTS for fronthaul since it is very unlikely to be able to achieve the performance levels needed for many applications. Specifically, when ITU-T defined PTS, they did it with a view of providing microsecond level accuracy across a network, which clearly would not meet the requirements of intraband carrier aggregation outlined earlier. One exception to this warning would be in the case of an LLS-C1 or LLS-C3 architecture with no other switches in the fronthaul network. In such cases, the physical network is actually identical between full and PTS since there are no intermediate nodes between the master and slave that may or may not be boundary clocks. In this case, the only difference between the two profiles will not be performance, but rather the format of the PTP messages. Details of these differences can be found in the ITU-T definitions of the two profiles, but, for example, FTS uses raw Ethernet frames, whereas PTS s used UDP/IP datagrams.

The previous discussion begs a question. What happens in an LLS-C1 or LLS-C2 configuration that uses PTP to synchronize both the O-DU and the O-RUs, but in the specific configuration, the backhaul and fronthaul networks use different PTP profiles—typically PTS on the backhaul side, since that is what the network operator provides, and FTS on the fronthaul side, since as we've shown that will offer the best performance? This scenario is shown in Fig. 8–25.

The issue here is, what sync role is the O-DU now performing? It might appear to still be a boundary clock, because it is certainly performing the same function as if it was, say, FTS on both sides—taking time from the backhaul and passing it on to the fronthaul network, as well as using it locally for its own synchronization. However, there is subtlety here, in that all the ports of a boundary clock have to use the same profile. The reason for this is that the IEEE 1588 standard views a boundary clock as a component in a sync network rather than a bridge between two networks, and all the nodes in a given network have to operate with the same profile to be able to interwork. O-RAN on the other hand views the backhaul and fronthaul networks as separate—as indeed they are in terms of

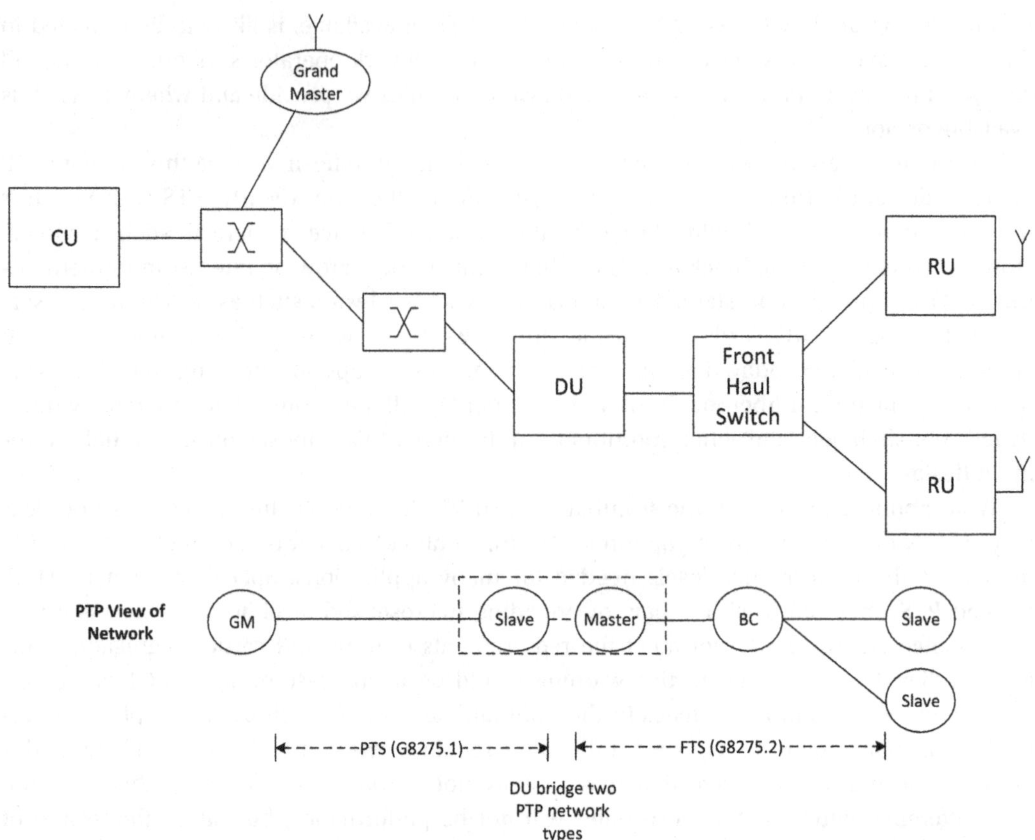

FIGURE 8–25 Mixed PTP profiles. *PTP*, Precision time protocol.

dataflow—and, therefore, there is no requirement for PTP running on the backhaul network synchronizing the O-DU to operate in the same way as PTP on the fronthaul network from the O-DU to the O-RUs, even though time itself is flowing from one network to the other. Instead, the O-DU in this case is actually operating as what is called a gateway clock or Interworking Function by ITU-T. Again, this is something that the ITU-T standards do not currently address beyond acknowledging the existence of, and is another case of the where the need-driven requirements of a standard such as O-RAN precede the more formal adoption of an idea by ITU-T.

O-RAN sync equipment requirements

We now need to turn our attention away from the network as a whole and concentrate on the requirements of the individual O-RU and O-DU equipment when it comes to synchronization. For the purposes, this discussion we will concentrate on a fronthaul network in an

LLS-C1 or LLS-C2 configuration—in other words, a network in which the O-RUs are all synchronized using PTP to their associated O-DU, either through a point-to-point fronthaul link (LLS-C1) or through one or more FSHs of routers (LLS-C2).

The O-RAN specification is currently designed to consider two use cases—networks supporting contiguous carrier aggregation (Category A) or networks only supporting noncontiguous (or no) carrier aggregation (Category B). From Table 8–1, this imposes O-RU to O-RU air interface synchronization requirements of 130 and 260 ns, respectively. Now, this is a relative requirement between the two O-RUs, but each O-RU does not synchronize to the other O-RUs—instead, they all synchronize to the associated O-DU. Therefore the synchronization limit between the O-DU and any one O-RU is half of this number, since one O-RU could be off by a certain amount in one direction, while another one is off by the same amount in the opposite direction, meaning that the error between the two is twice whatever the error of each one is, as shown in Fig. 8–26.

However, this does not mean that the network connection between the O-DU and O-RU can use up all of this synchronization budget because the O-RU itself will introduce additional errors between its input, the fronthaul network connection, and its output, the antenna connector. Therefore some of the sync budgets have to be assigned to this error, eating into the error that the network itself can add. What causes this additional error? Well, there can be a number of sources but the biggest factor is how well the PTP slave function in the O-RU can recover time from the network, in other words, the error between the packets on the network a subsequent physical clock output, such as a 1PPS signal, that drives the radio functionality of the O-RU. Additionally, the radio itself will add some further errors due to data buffering, clock domain crossing, etc. In reality, O-RAN assumes that the PTP slave function is fully integrated into the O-RU so it will often be impossible to identify how much of the overall error can be attributed to each function.

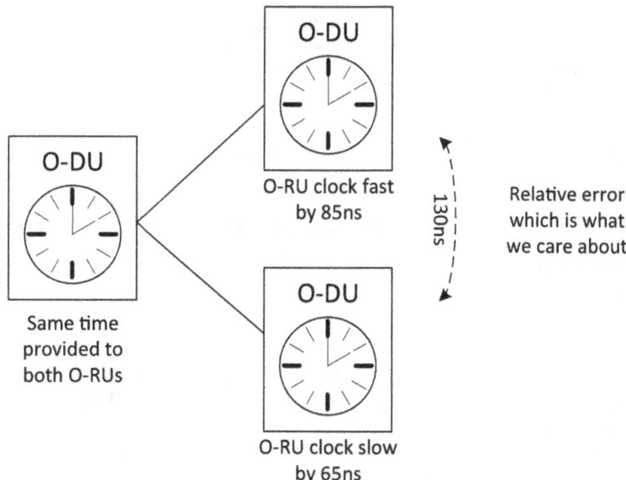

FIGURE 8–26 O-DU to O-RU limits.

In order to allow some flexibility in simplifying system design in architectures that do not need to work in a contiguous carrier aggregation environment, with its more stringent sync requirements, O-RAN defines two classes of O-RU with regards to synchronization capability:

- A standard RU—This allows for the O-RU itself to introduce up to 80 ns of additional error between the network connection and the antenna connector. This roughly corresponds to what ITU-T refers to as a Class B slave clock.
- An enhanced eRU—This reduces the O-RU error budget down to 35 ns. This performance level is more comparable to an ITU-T Class B slave clock.

Fig. 8–27 shows how these numbers work into the overall budget for the different use categories. Note that no Category B (noncontiguous CA) operation, either an RU or and eRU can be used. However, to support Category A (contiguous CA) operation, only an eRU has the performance needed to meet the requirements. Therefore when designing an O-RU consideration needs to be given as to what type of application it will be used for.

Sync solution implementation

So, within an O-DU or O-DU, what does the actual sync subsystem implementation look like?

Frequency-only systems

In the days when only frequency synchronization was a requirement, the key component to achieve this was the phase-locked loop (PLL), as shown in Fig. 8–28.

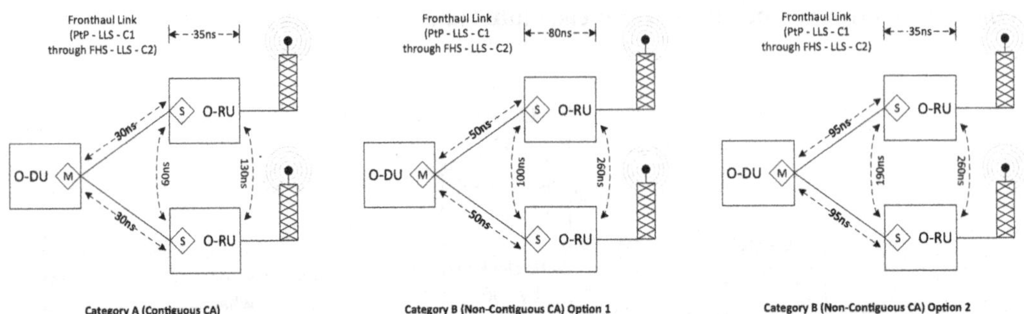

FIGURE 8–27 Overall fronthaul budgets.

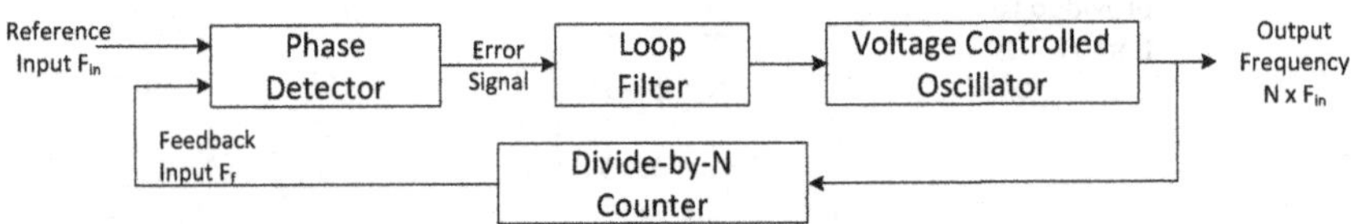

FIGURE 8–28 A PLL. *PLL*, Phase-locked loop.

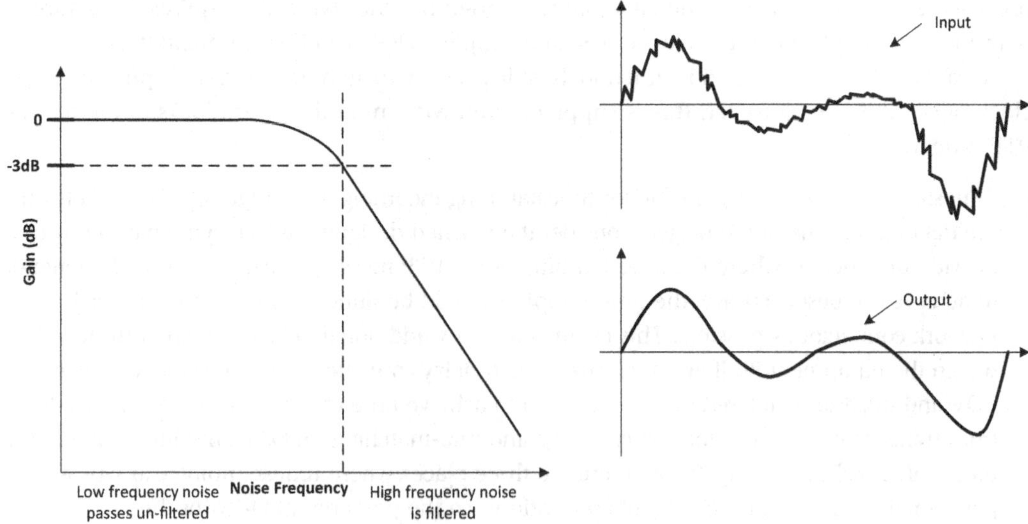

FIGURE 8–29 The filtering effect of a PLL. *PLL*, Phase-locked loop.

Conceptually, a PLL is relatively simple. It takes a clock from the wider network, for example, a clock recovered from a SyncE or SONET link and low-pass filters it to produce a cleaner output signal. The low-pass filtering function essentially removes any noise in the input signal above a certain frequency—the PLL bandwidth. In this way, any noise introduced to the signal by the network is removed. In other words, the recovered clock is cleaned up. How much noise to remove depends on the bandwidth of the PLL—the lower the bandwidth, the more noise is removed. An example of this can be seen in Fig. 8–29.

However, this filtering does not come without a price. The PLL is not really removing the input noise, but rather replacing it with noise from the local clock, which will typically be some form of quartz-based oscillator. Therefore the lower the bandwidth of the PLL, the higher the stability of the local oscillator needs to be, otherwise the noise of the oscillator itself will have a significant impact on the filtered clock. For this reason, PLLs that are used for filtering network noise, as opposed to, for example, noise arising between components within a piece of equipment, are typically used with a temperature compensated or stabilized oscillator, such as a TCXO or OCXO. These oscillators not only have much better stability (and hence lower noise) than a simple crystal oscillator but also have a price many times higher—several tens of dollars in the case of an OCXO.

Time-synchronization systems

As we move toward time synchronization using PTP, potentially in conjunction with SyncE, the architecture of the sync subsystem becomes considerably more complex. This complexity arises from the fact that PTP uses packets transported over the data-plane for synchronization, as opposed to physical signals, meaning that there is now the requirement to add data processing capability to the sync subsystem—something that typically requires a mixture of both hardware and software. Xoxo shows an example PTP + SyncE implementation.

It can be seen that the PLL function is still present to handle the SyncE physical-layer clock noise filtering. However, this is supplemented with multiple other blocks to handle the PTP function:

- Timestamper—This is responsible for timestamping incoming and outgoing PTP packets on the data-plane. One of the biggest considerations when designing a PTP sync solution, with or without SyncE, is where the timestamping of the PTP message packets is done. In general, to achieve the best accuracy, the timestamping should be done as close to the physical network connector as possible. This ensures that any additional delays that are introduced within the equipment itself are minimized. Such delays are effectively the same as network PDV and need to be filtered out by the servo to achieve time recovery. These errors result from things such as clock domain crossings and rate-matching FIFOs within the design of the data path. As shown in Fig. 8–30, there are three places where timestamping can typically be performed. In general, for fronthaul applications, so the ports on an O-RU or the downstream ports on an O-DU, the timestamping needs to be performed at the PHY level to achieve the required accuracy. For backhaul connections, so the upstream ports on an O-DU or the ports on an integrated small cell, MAC level timestamping will be acceptable. Software-based timestamping in the network driver really should be avoided for any telecom-grade application—the accuracy will in general simply not be good enough.
- Network driver—A software function that is responsible for sending and receiving PTP packets on the network. This is typically part of the same driver that handles other, nondata plane, packets such as the actual O-DU to O-RU data transport. In many cases, this driver will be part of CPU operating systems, for example, Linux.
- PTP stack—Another software function that implements the crux of the PTP solution. This portion of the software generates and processes the PTP packets and handles all the details such as selecting which PTP master to use on the network. The local side output from the PTP stack is a series of t_1-t_4 timestamps extracted through the PTP message flow as outlined previously.

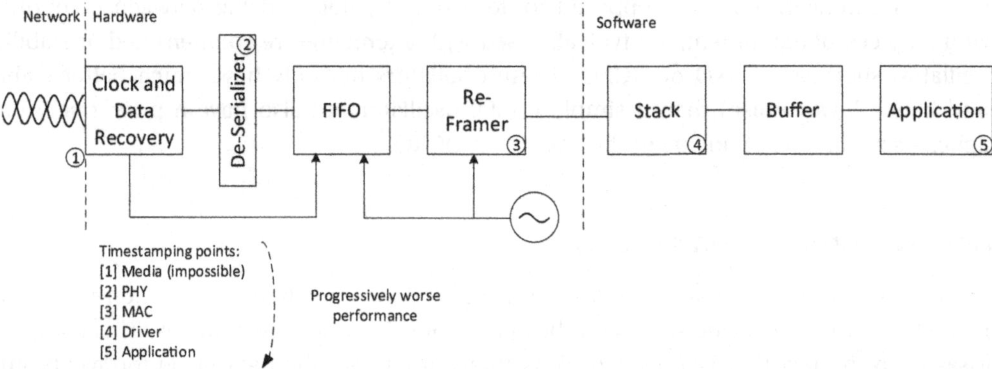

FIGURE 8–30 Timestamping points.

- Time recover servo—Yet another software function, the importance of which is often overlooked. The function of the servo is to filter the time information contained within the flow of timestamps from the stack and use this to control the alignment of the local physical time counter. How this is done, and how complex the servo needs to be, depends on the amount of PDV on the network that must be filtered out. For FTS networks with very low PDV, the servo is little more than a low-pass filter—it can be thought of as a software PLL, implementing a low-pass filtering function for data in much the same way a PLL device does for a physical clock. However, for PTS networks, which will, in general, have much higher PDV, a much more sophisticated servo implementation is needed. This class of servo often employs advanced proprietary (and closely guarded by their vendors) packet-selection algorithms to pick and choose dynamically which timestamps to process and which to ignore. A good servo of this type can often achieve better than two orders of magnitude filtering. For example, if 100 μs of PDV was present (which is a realistic figure for some PTS networks), the servo would still be able to align the time within the local device to within 1 μs. Fig. 8–31 shows the vastly different performance between a simple servo (in this case from the Linux ptp4l package) and a commercial proprietary servo, when used in a PTS network.
- Digitally or numerically controlled oscillator (DCO or NCO)—A hardware component that converts the output from the software servo into an actual physical clock ticking at the rate of the PTP master. Various implementations are possible, but the net result is the same—a clock that can be tuned up or down with a very fine resolution (often better than parts per billion) to align the local clock rate to real time, as received from the PTP master.
- Time-of-day counter—The final hardware component is the time-of-day counter, which is responsible for storing the actual time-of-day information, typically in the form of the number of seconds and nanoseconds since the epoch. In many cases, phase is more

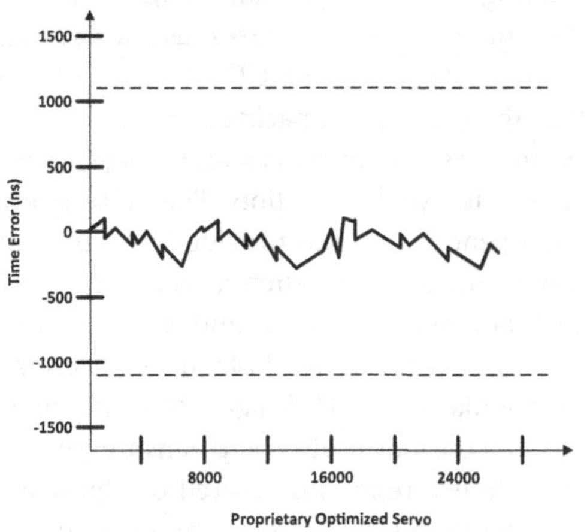

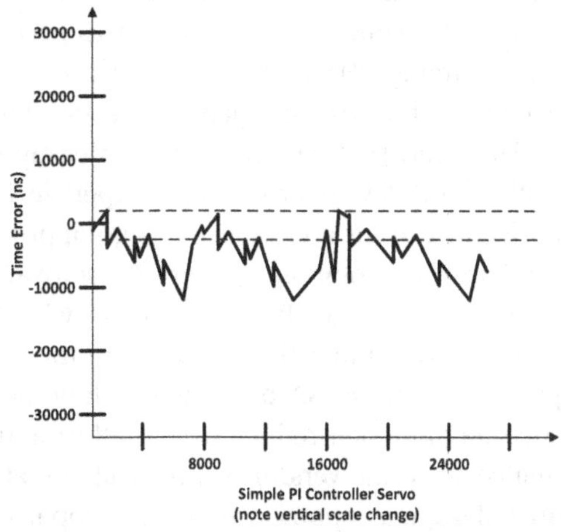

FIGURE 8–31 Servo performance comparison.

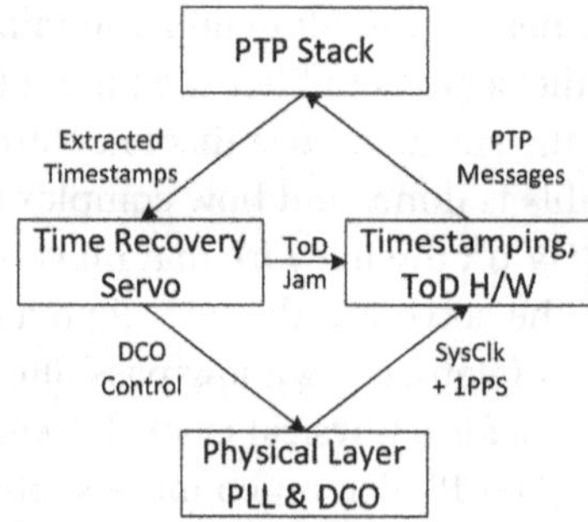

FIGURE 8–32 PTP + SyncE solution flow. *PTP*, Precision time protocol.

relevant than absolute time, in which case a so-called 1PPS signal can be generated from the time-of-day counter. This is a signal that pulses once per second and is aligned to the time-of-day counter, typically with its rising edge corresponding to the start of each second. This 1PPS signal can be used to phase align the air interface, for example, without having to worry about the details of actual time in terms of date, hours, minutes, and seconds.

The interaction of these hardware and software components is shown in Fig. 8–32.

Designing a PTP + SyncE solution

It should be clear from the preceding section that quite a lot of design thought and effort is needed in designing a PTP + SyncE sync subsystem, especially when compared, for example, to the frequency-only equivalent. This is largely due to the involvement of software elements as well as hardware, and the cross-over between the sync and data planes.

It is possible to take a traditional building block approach to this and for an O-RAN equipment designer to develop the solution themselves. For example, they could use a standard PLL with a DCO capability for the SyncE filtering and DCO portions, implement the time-of-day counter in an FPGA or as part of an Ethernet PHY in some cases, use an existing timestamping PHY and network driver, and either attempt to write the PTP stack and servo from scratch or use an open-source solution, such as the Linux ptp4l package.

However, such an approach can easily result in unforeseen challenges and delays, especially if the developer's area of expertise is not rooted in synchronization. Therefore many designers opt for a reference design approach from a vendor that can provide both some of the hardware components and the software functions. For example, such as solution might include a hardware PLL/DCO along with PTP stack and servo software and the necessary time-of-day counter IP to implement the complete solution on a defined platform, for example, one of the FPGA-based SoC solutions that are popular in O-RU designs. Not only does such an approach reduce design effort and time to market, but it also helps ensure performance since the vendor will typically be able to provide test reports measured on the reference design, as opposed to the developer needing to set up a test environment and conduct their own testing to gain confidence in the performance.

The effects of timestamping location and resolution

Oscillator selection and holdover

One consideration that is often significant in the design is the sync subsystem is the selection of the local oscillator, since this can easily become the single most expensive sync-related component. There are two factors to consider when deciding on the oscillator requirements:

- the stability needed for normal operation and
- the stability needed when the network clock source is lost.

For normal operation, we have already mentioned the need for a stable oscillator for the physical layer, SyncE, filtering. The easiest way to understand this is to recognize that a PLL acts as a low-pass filter for its input signal and a high pass filter for its local oscillator. In other words, the output noise of the PLL, which is the value that needs to meet the limits in the various standards for robust performance, is a mix of the input noise and the local oscillator noise. The lower the PLL bandwidth, the more the local oscillator noise comes into play and vice versa. Now, a PLL used for a local function such as jitter cleaning may have a bandwidth of 100 kHz + , meaning that long-term stability of the local oscillator really is not an issue. However, for a SyncE PLL that has to filter network noise, the bandwidth is usually in the 1–10 Hz range, meaning that the long-term stability of the local oscillator comes into play. Typically, as already stated, a SyncE PLL will require a minimum of a TCXO to provide the necessary stability and could require an OCXO if support is needed for legacy networks that require a bandwidth as low as 100 mHz, though this is unlikely to be a factor in O-RAN.

However, it is also necessary to consider the local oscillator stability needed for the PTP filtering, since that too has an "effective bandwidth," which is related to how aggressive the servo is at filtering out network PDV. This is normally considerably lower than the SyncE bandwidth—between 50 and 100 mHz for FTS and as low as 1 mHz or less for PTS. And this is where the value of SyncE comes into play. If PTP was used on its own, this would be the bandwidth that would have to be considered when selecting the local oscillator. In other words, an OCXO would likely be needed even for FTS. However, when SyncE is used in conjunction with PTP the physical-layer clock recovered from SyncE is itself a stable clock source that supplements the local oscillator. This means that the local clock source to the PTP DCO is coming from the filtered SyncE clock rather than directly from the local oscillator and, therefore, it might be possible to use a lower stability oscillator. This is shown in Fig. 8–33.

Table 8–4 summarizes the typical oscillator requirements that might be expected for an O-RAN application.

The second part of the equation is what is required of the local oscillator when the network time source itself is lost. There are a couple of reasons that such a loss could occur:

- The time source on the network, for example, the PTP Grandmaster or the GNSS receiver driving it, could go offline.
- The network connections between the time source and the slave device could fail.

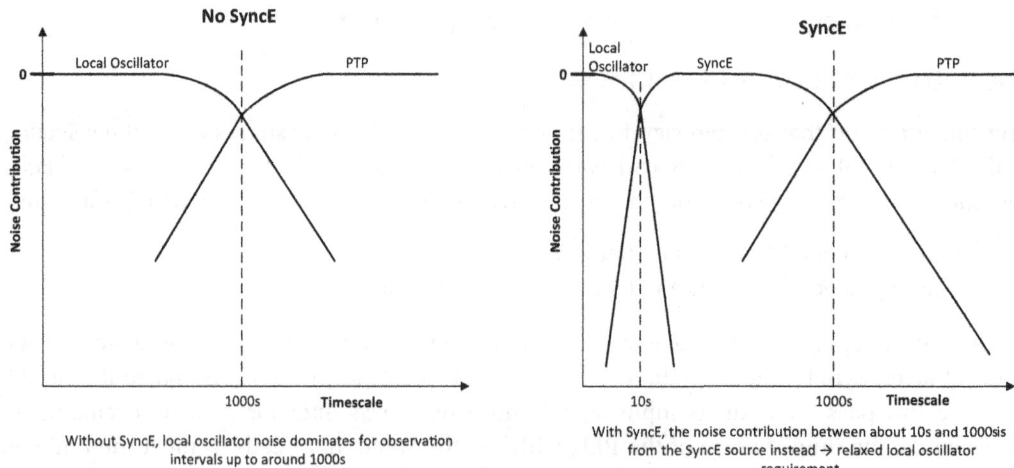

FIGURE 8–33 SyncE + PTP oscillator requirements. *PTP*, Precision Time Protocol.

Table 8–4 Required oscillator stability.

PTP profile	SyncE?	PTP servo bandwidth (mHz)	SyncE bandwidth (Hz)	Typical local oscillator
FTS	Yes	50–100	1–10	TCXO
FTS	No	50–100	n/a	OCXO
PTS	Yes	1	1–10	TCXO
PTS	No	1	n/a	OCXO

FTS, Full-timing support; *PTP*, Precision Time Protocol; *PTS*, partial-timing support.

Traditionally, in legacy frequency-only systems such as SONET, which relied on synchronization of all nodes for the network itself to work, it was customary to require each individual node to remain within the synchronization limits for 24, or even 72, hours in the event of a failure. This, of course, requires a very stable local clock source, since there is no longer anything to lock to so it is free-running, but 24 hours holdover could be achieved with a TCXO normally, with an OCXO only being needed for the extended 72 hour requirement. However, in a time-synchronization system, the situation is considerably different since any error in the local oscillator frequency will accumulate in the time domain. It is not necessary to go into the details of the calculations here, but the conclusion is that to maintain time alignment even within the network-wide limit of 1.5 μs for an hour requires an OCXO, and if 24 hours is needed then you might need a super high stability, double-oven, OCXO—available, costing over $200!

Therefore few O-RAN equipment vendors will be willing to put anything close to a 24-hour holdover capability into their equipment, though it is often listed as a requirement early in the design phase before reality sets in. However, does O-RAN really need such a holdover level? The answer is almost certainly not, for a number of reasons:

- If the network time source itself is so important, why not have a backup in case it fails? PTP allows for multiple Grandmasters and it is not unreasonable to expect such a backup to be available. Therefore it might be that protection against the time source failing is already available.
- If the network time source fails and I also have SyncE, will that fail too? In a typical network it would not, since the node driving the SyncE link will have its own frequency holdover, meaning that the SyncE clock might degrade but it will still in general be more stable than the local oscillator over an extended period. This allows a PTP + SyncE slave to use something called time holdover—where it no longer receives continual time information over PTP, but still has a good enough physical-layer clock from SyncE to keep its local time accurate for long enough without a super stable local oscillator.
- If the sync failure is due to the network connection to the device itself failing, for example, the fronthaul link between the O-DU and O-RU, is the fact that sync has been lost really a problem? The sheer fact that the data link itself has failed presumably means that the O-RU can no longer perform a meaningful function, with or without sync. It may as well just turn off the radio portion until service is restored. Hopefully, a nearby O-RU in this case can absorb the extra demands on it without degrading the performance of the cellular network too much during the outage.

Therefore in reality, there tends to be a disconnect between the holdover capabilities that are initially stated as a "must have" requirement and what typically gets deployed once a more detailed analysis of design complexity and cost has been completed.

Unraveling the standards spaghetti

No discussion of synchronization would be complete without a look at the various standards bodies that are involved in defining the standards that govern the network and equipment requirements, and how these all interact.

When it comes to O-RAN synchronization, there are four parties defining these requirements:

- 3GPP—The owners of the overriding standards controlling each generation of cellular equipment. They are the ones responsible for ensuring that end equipment, such as handsets, from any vendor can communicate on networks that are themselves made up of equipment from a number of different suppliers. 3GPP's focus is primarily on defining the air interface between the network and the user equipment, and these definitions include the synchronization requirements as seen at the air interface.
- IEEE—The Institute of Electrical and Electronics Engineers handles standardization across a broad range of topics. Two of these standards are particularly relevant to O-RAN. One of them is, of course, IEEE 1588, which defines the PTP itself. As already discussed, when

O-RAN talks about "1588" it is typically referring to the IEEE 1588–2008 version of the standard, which is not backward compatible with the earlier IEEE 1588–2002 flavor. The newer IEEE 1588–2019 version is, however, backward compatible with the 2008 variant and it is expected that over the coming years, more features from the 2019 version will be added to PTP implementations, and at some point, it might be that a future version of O-RAN calls for some of these features. The second IEEE standard that is relevant to O-RAN is IEEE 802.1CM-2018, which defines the synchronization requirements for fronthaul networks. This standard effectively takes the air-interface numbers from 3GPP and translates those into network and equipment limits for a fronthaul network serving such an air interface. It is from the IEEE 802.1CM limits that O-RAN derives its sync requirements.

- ITU-T—The standardization branch of ITU, which itself is an agency of the United Nations. ITU-T has been around for well over 100 years and traditionally has been responsible for defining the operation of telecom networks so that multiple networks from many operators spread across the globe can move data between each other. The structure and terminology (they use "Recommendation" rather than "Standard" so as to not be seen to try and tell any country what to do) are rather convoluted, but one small portion of it—a group referred to as Study Group 15, Question 13—is responsible for defining synchronization in telecom networks, which today means things like SyncE in networks that could be used for fronthaul applications.
- The O-RAN Alliance—As the name suggests, this pseudo-standards body (as in they are more needs-driven and do not necessarily have the oversight that the other bodies have) owns the O-RAN specification and all aspects of it, including synchronization.

Unfortunately, the different focus areas and histories of the various standards bodies mean that they are not always in agreement. In particular, ITU-T often takes a more cautious, and hence less reactionary, approach to standardization, meaning that it frequently appears to be playing catch-up. An example of this is the characterization of FTS networks without SyncE—something that is awaiting further study by ITU-T, but that is already being required by multiple O-RAN operators. Because of these differences, occasionally discrepancies arise between standards. In general, when designing an O-RAN network, the requirements and allowances documented in the O-RAN standards themselves should be the over-arching driver.

Further reading

The chapter aims to provide an introduction to synchronization for nonexperts and an overview of the sync requirements and implementation for O-RAN. It is impossible in a single chapter to even really scratch the surface of this subject and hence a large amount of detail has been intentionally overlooked. To get broader, and more in-depth, information on the topic, there are a number of sources that can be explored:

- Solution vendors whitepapers and presentations—Many players in the space, both the semiconductor and software companies supplying the building blocks, such as PLLs, and the system level solution providers, selling the likes to PTP Grandmasters, have libraries of material available. Some of this will be very product-specific, but much of it also contains more widely applicable material.

- Synchronization conferences—There are a number of annual conferences held that specialize in just synchronization. Two of these—WSTS (Workshop on Synchronization and Timing Systems) held in the United States, and ITSF (The International Timing and Sync Forum) in Europe—specialize in providing presentations targeting real-world, topical, applications such as O-RAN synchronization. Other conferences, such as ISPCS, are more technology-focused in their scope.

Software performance

If there ever was a "how long is a piece of string" question, it would be "how good will my software perform on this system." Software performance greatly depends on the amount of time spent both on architecting an efficiency environment in which software is set up to run fast (say, optimizing the processing flow for data locality, providing user-space drivers, and so on), and software application stack optimization itself. Many developers tend to treat performance as an afterthought and respond with a "just use more cores" answer if performance is not at the desired level. We try to counter this with up-front analysis of what performance should be achievable and a tracking mechanism to make sure that reality matches to theory—or that appropriate mitigation measures can be taken. Let's build a mindset of performance-aware design and system architecture. This includes planning measurement experiments to ensure that results are both representative and reproducible. Software also needs to be instrumented to facilitate data collection. Finally, once the performance-critical components of the software are identified, they are measured early and often to validate the models that have been built and to verify earlier predictions.

Note that for the purpose of such benchmarks, it is useful to identify a few "key" use cases (centering around bitrate, packet rate, user count, etc.). Performance metrics are divided into three categories, each of them tackling separate areas of the performance picture, which are as follows:

- cycle counts per SW component or Processing Cycle Budget Analysis
- Central Processing Unit (CPU) loading summaries
- System-on-Chip (SoC) performance counters

We can also break performance aspects down to use-case components:

- L2/stack performance—including Enhanced Common Public Radio Interface (eCPRI) as an example of an packetization protocol potentially running under Linux
- L1/Physical Layer performance
- accelerator/offload-associated overheads

Each of these components are broken down in more detail later in this chapter. We also discuss performance mitigation techniques: what can we do when achieved performance does not match to the target and we need to deploy a product into the market anyway?

All components of the solution we are describing are handled assuming a software-centric implementation. It is obvious that when offloaded to hardware acceleration, the loading does not exist.

Open Radio Access Network (O-RAN) Systems Architecture and Design. DOI: https://doi.org/10.1016/B978-0-323-91923-4.00012-4

Packet processing cycle budget analysis

The concept of "Processing Cycle Budget Analysis" comes from the digital signal processing (DSP) world, where dimensioning is historically done by dividing the available multiply–accumulate capacity of the device (measured often as available CPU clock cycles–number of DSP/CPU cores multiplied by the frequency of operation) between the jobs that need processing. This concept can be carried over to the packet processing world where we define a "cycle budget" as:

$$\text{Cycle budget} = \frac{\text{Core frequency} \cdot \text{number of CPU cores}}{\text{Packets per second needed to be processed}}$$

This defines an available number of clock cycles (the budget) that presents the compute horsepower available to spend on packet processing. For packet forwarding applications, this budget is both well understood/documented and very low. Consider[1] which advertises 233 Gbps (or 347 million packets per second) of L3 forwarding at 64-byte packet sizes on a 28-thread, 2.3 GHz machine, or 28 × 2.3 G/347 M = 186 clocks/packet. Note that such forwarding applications are a very trivial example and do not represent a base station stack. Nonetheless, it establishes a methodology and a categorization of processing complexity:

- <1000 clocks/packet for packet forwarding/routing applications
- <10,000 clocks/packet for complex packet operations that include payload manipulation (e.g., Packet Data Convergence Protocol and Radio Link Control-type applications)
- <100,000 clocks/packet for applications where the payload needs to be interpreted (control plane applications)

Again, these are extremely high-level approximations and obviously depend on the processor architecture as much as on software efficiency and details of the application.

The next step is obviously cycle count measurement (and tracking) of the actual application to establish a more accurate set of metrics. Per-component [and this could be an Application Programming Interface (API) call, a set of API calls or entry/exit points from/to network interfaces] cycle count tracking is implemented by reading the CPU timebase (stopwatch) before and after execution of each SW module and logging the cycle count readings in a logging database. This database is extracted to a host for offline analysis.

Cycle counting is identifying which SW components (as such, their respective owners) are causing excessive CPU loading. As such, cycle counting is the baseline for focusing optimization effort and a useful tool in tracking optimization "actual" versus "target" numbers. Iteratively, the results of cycle counting in a product are input into the modeling baseline for the next generation, as such increasing accuracy and trust in the modeling HW/SW codesign approach.

In case cycle counts are not readily available, we can use rule-of-thumb math:

- Write C-code, compile, and count additional lines of assembly.
- Assume a metric (say, 0.7–1) lines of assembly/CPU clock cycle.

- Write pseudocode and estimate 7 lines of Reduced Instruction Set Computer (RISC) assembly lines per simple line of C-code.
- Execute code on a core/SoC simulator, even if cycle accuracy is not guaranteed but only estimated.

Example: Enhanced Common Public Radio Interface complexity analysis

As an example of packet processing complexity analysis, let's look at eCPRI termination CPU cost. Performance requirements are defined by U-plane IQ termination and associated eCPRI (or equivalent) processing. This is broken down to the following:

1. eCPRI frame receive/transmit operation. This is everything that needs to happen to get the software to the start of the eCPRI header and typically handled by Data Plane Development Kit (DPDK) drivers and/or hardware components in the Network Interface Card. This includes the overhead in Linux to wake up and handle the eCPRI-received frames—even though this latency can be brought down to near-zero by means of moving toward a polling mode implementation or similar techniques.
2. eCPRI/Open Radio Access Network (O-RAN) FH C/U framing/deframing: Assumed implemented in software rather than hardware for this discussion, this is a (DPDK) software routine that generates eCPRI headers in the transmit direction and/or parses them in the receive direction. In this regard, eCPRI is no different from any other fast path packet processing implementation that is implemented in networking stacks and all optimization techniques apply. DSP-like functions are explicitly outside of the scope of this function: this is strictly about (un)packing IQ samples in Ethernet frames.
3. The IQ extraction and (optional) decompression: this is the building/consumption of the IQ buffer that the DSP algorithms use in the receive direction and generate in the transmit direction. These functions involve a memory copy operation (read and write of IQ samples) as well as lightweight signal processing to perform the (de)compression activity.

Topics 1 and 2 are benchmarked like other software benchmark analysis operations and broken down into a clock cycles/function as we did for other packet processing components. The cycle count/packet is then converted into CPU loading by multiplication with the eCPRI packet rate.

We take the example from earlier with the following modem configuration:

- 5G/NR eMBB DL/UL OFDMA modem
- Time Division Duplex (TDD)
  - downlink slot: 2 symbols PDCCH, 1 symbol Demodulation Reference Signal (DMRS), and 11 symbols Transmit
  - uplink slot: 2 symbols PDCCH, 1 symbol DMRS, 10 symbols Receive, and 1 symbol Rx/Tx switch
- up to 100 MHz single-carrier RF bandwidth
- 30-KHz subcarrier spacing, 0.5-ms TTI. 1 × DMRS/slot
- 8 antennas, 4 layers
- 16 UE/TTI

- downlink/uplink throughputs of ~2.5 Gbps at Medium Access Control (MAC)/PHY level

We calculate theoretical IQ sample rate as: 3276 × 14 symbols (peak) × 32b (IQ) × 2000 slot/second = 2.93 Gbps. Assuming 10% packetization overhead and (roughly) 1-KB eCPRI frames, this is ~400 K frames/second. Assuming ~1000 CPU clocks/packet (benchmark to be sourced from benchmarking), this equates to 400 M clock cycles/second/100 MHz carrier/layer. At 4 layers, this is 1.6 G cycles/second or roughly 1 × 2 GHz core/100 MHz carrier.

Topic 3 is implemented like any other DSP algorithm, using Single Instruction Multiple Data (SIMD) or similar operations. We estimate the overall CPU loading as done typically in DSP algorithms by the number of operations—as we are doing next for Physical Layer complexity analysis. Each sample (say, 9 bit decompression) requires:

- an overhead of about 20 clocks/RE to extract the exponent value
  - or ~1.7 clocks/sample with 12 samples in 1 RE
- load operations to load the compressed vector into a register
  - assuming a 128-bit SIMD engine, assume ~7 I + Q samples/operation or 0.14 clocks/IQ sample
  - and an additional ~7 I + Q samples/operation or 0.14 clocks/IQ sample to load an offset value of the compressed as we need for combining
- two shift operations per I or Q value to align the 2 components of the 9-bit element
  - assuming a 128-bit SIMD engine and operation on 8-bit values (16 operations/clock), 0.25 clocks/IQ sample
- an OR/XOR operation per I or Q value to combine the two components
  - assuming a 128-bit SIMD engine and operation on 16-bit values (8 operations/clock), 0.25 clocks/IQ sample
- an expansion operator per I or Q value to place the combined 9-bit value into a 16-bit register
  - assuming a 128-bit SIMD engine and operation on 16-bit values (8 operations/clock), 0.25 clocks/IQ sample
- a multiply (or shift) operation per I or Q value to apply the exponential
  - assuming a 128-bit SIMD engine and operation on 16-bit values (8 operations/clock), 0.25 clocks/IQ sample
- a store operation to store the resulting 16-bit sample in memory
  - assuming a 128-bit SIMD engine and operation on 16-bit values (8 operations/clock), 0.25 clocks/IQ sample

Aggregate performance is added as ~3 clocks/sample in theory of (assuming 50% utilization of the vector unit for intrinsics based code) ~6 clocks/sample in practice. Given 3276 × 14 symbols (peak) × 1 (IQ sample) × 2000 slot/second = 92MSPS, we count for 550 MHz/layer or (roughly) 1 core/4 layers.

From the aggregate of topics 1/2 and 3, we can estimate performance target of 2 CPU cores/4 × 100 MHz layer for this operation. While not to be interpreted as a hard performance target, we can use this number to track achieved performance versus target performance and quickly identify major misses.

Physical Layer complexity analysis

We show a deployment example of the NXP LX2160 multicore Arm processor in a 5G Physical Layer processing scenario as per Fig. 9–1, assuming a high-level functional mapping as shown in the figure, with the shaded areas offloaded to an accelerator:

A generalized use case is defined through the following key parameters:

- 5G/NR eMBB DL/UL OFDMA modem
- TDD
  - Downlink slot: 2 symbols PDCCH, 1 symbol DMRS, 11 symbols Transmit
  - Uplink slot: 2 symbols PDCCH, 1 symbol DMRS, 10 symbols Receive, 1 symbol Rx/Tx switch
- up to 100 MHz single carrier RF bandwidth
- 30-KHz subcarrier spacing, 0.5 ms TTI. 1 × DMRS/slot
- 8 antennas, 4 layers
- 16 UE/TTI
- downlink/uplink throughputs of ~2.5 Gbps at MAC/PHY level

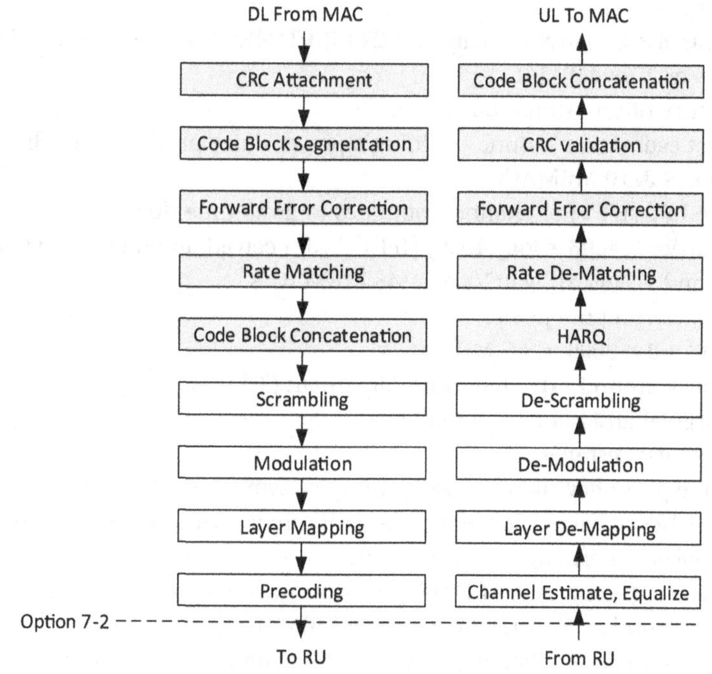

FIGURE 9–1 L1 processing in DU for Shared Channel complexity analysis. *DU*, distributed unit.

Phase 1: theoretical analysis

We do a performance analysis for the eMBB receive operation by breaking down the processing chain into low-level operations that can be counted in a spreadsheet calculator, for example:

- Channel estimation. The received eCPRI channel carries post-fast Fourier transform (frequency domain) IQ samples. In the DMRS symbol, extract all subcarriers which contain DMRS pilot sequence. The frequency-domain channel estimation function implies multiplying each subcarrier with the known transmitted DMRS sequence. That will give a channel estimate per subcarrier. Optionally, this channel estimate is then passed through a smoothing filter to improve estimation quality. From a complexity perspective that means that once per slot (2 K slot/second) a complex MAC (CMAC) is applied to all subcarriers in a layer. Computational complexity scaled with $N_{carrier} \times N_{Receive\ Antenna} \times N_{layer}$. Optionally, this step can be implemented in the antenna processor/digital front end, but this would add additional antenna processor complexity as well as additional communication overhead between distributed unit (DU) and antenna processor.
  - 1 carrier × 3276 subcarriers/received symbol × 0.5 DMRS sequence RE occupancy × 8 received antennas × 4 layers × 2000 slot/second = 105 MCMAC/s or 420 MRMAC/s.
- Equalization Matrix Compute. Equalization can be done in different ways, we assume Minimum Mean Square Error, a commonly used algorithm (MMSE) on a resource block (RB) granularity.
  - 1280 CMAC/8 × 8 matrix × 1 carrier × 273 RB/DMRS × 2000 slot/second = 699 MCMAC/s or 2796 MRMAC/s.
- Time–frequency offset estimation.
  - Time offset estimation is done by doing a power delay profile of the channel estimation: ~3210 MRMAC/s.
    - Inverse FFT (IFFT) of channel estimates to go to time domain.
      - 1 carrier × 4096 × $\log_2(4096)$ (IFFT) × 8 received antennas × 2000 slot/second = 786 MCMAC/s or 3145 MRMAC/s.
    - $I^2 + Q^2$ to compute power.
      - 4096 × 8 × 2000 = 65 MRMAC/s.
    - Find peak around expected Time Alignment (TA).
      - Marginal and not accounted here.
  - Frequency offset estimate.
    - Compute phase deviations across consecutive symbols over Phase Tracking Reference Signal (PTRS), Channel State Information Reference Symbol (CSI-RS), and Demodulation Reference Symbol (DMRS).
      - Assuming worst case of DMRS: 3276 subcarriers × 8 received antennas × 2000 slot/second = 52 MCMAC/s or 210 MRMAC/s.
    - If phase noise is significant, an average over many symbols is required to remove phase noise.

- Channel equalization. This is the application of the MMSE weights so comes as a matrix multiply between received samples over Rx antennas and MMSE weights. As such, computational complexity per symbol is estimated as $N_{carrier} \times N_{antenna} \times N_{layer}$.
  - 1 carrier × 3276 subcarriers/received symbol × 8 received antennas × 4 layers × 2000 slot/second × 10 symbols = 2097 MCMAC/s or 8388 MRMAC/s.
- Demodulation. In this step, the output of the channel equalization phase is converted to Log Likelihood Ratios (LLRs), where the number of LLRs equals the $\log_2$ of modulation order. The complexity depends on the modulation order and scales with $N_{carrier} \times N_{layer}$. Complexity per LLR assumed to be 3 real operations (alpha × N − |Li|) as per 3GPP standards.
  - 1 carrier × 3276 subcarriers × 4 layers × 8 LLR [256 quadrature amplitude modulation (QAM)] × 3 real operations × 2000 slot/second × 10 symbols = 6290 MROPS/s.
- PRACH is assumed to follow a regular frequency domain correlation processing flow for short sequence. Symbol FFT is assumed to be performed in Remote Radio Head (RRH) as explained earlier. Rest of the processing is assumed to be done in DU following breakdown given next, leading to a total of 1876 MROPS. It is also assumed that PRACH is processed within a slot, but larger latency might be acceptable leading to lower average complexity.
  - Frequency Domain Root Sequence Correlation: 1 carrier × 8 received antennas × 2 PRACH preambles × 6 PRACH occasions × 1 PRACH slot per subframe × 139 sequence length × 2000 slot/second = 27 MCMAC/s or 107 MRMAC/s
  - IFFT: 1 carrier × 8 received antennas × 2 PRACH preambles × 6 PRACH occasions × 1 PRACH slot per subframe × 256 $\log_2(256)$ IFFT size × 2000 slot/second = 393 MCMAC/s or 1573 MRMAC/s
  - preamble coherent combining: 1 carrier × 8 received antennas × 2 PRACH preambles × 6 PRACH occasions × 1 PRACH slot per subframe × 256 IFFT size × 2000 slot/second = 49 MCADD/s or 98 MRADD/s
  - power computation and antenna noncoherent combining: 1 carrier × 8 received antennas × 6 PRACH occasions × 1 PRACH slot per subframe × 256 IFFT size × 2 operations ($I^2 + Q^2$) × 2000 slot/second = 49 MRMAC/s
  - preamble detection/peak search: 1 carrier × 8 received antennas × 6 PRACH occasions × 1 PRACH slot per subframe × 256 IFFT size × 2 operations (noise averaging and max search) × 2000 slot/second = 49 MROPS/s
- Sounding Reference Signal (SRS) is assumed to follow a regular frequency-domain channel estimation given the narrowband allocation. Time-domain channel estimation would provide performance improvement for a wideband configuration at the expense of larger complexity. Symbol FFT is assumed to be performed in RRH as explained before. Rest of the processing is assumed to be done in DU following breakdown, leading to a total of MROPS. It is also assumed that SRS is processed within a slot, but larger latency might be acceptable leading to lower average complexity.
  - 1 carrier × 48 RB/comb × 4 combs × 12 cyclic shifts × 4 symbol × 2000 slot/second = 18 MCMAC/s or 74 MRMAC/s

Aggregate compute cost (excluding PRACH/SRS, who are either implemented in the radio unit (RU) or in background tasks that are not considered performance critical) is 21GOPS/second. Now, this is a strictly theoretical number that will not be achieved for two reasons:

1. Even when implemented by means of highly optimized software kernels, general-purpose processors rarely achieve even half of the theoretical performance due to overheads associated with load/store operations, data conversion, cache misses, dependencies in the processor pipeline, and other inefficiencies. Based on benchmarking, we see 50% effective utilization of an SIMD engine to be a realistic maximum, even when operating from warm cache memory and an otherwise optimized environment.
2. The environment described above does not consider cache misses, overheads for data conversion and similar operations, and so on. We assume another factor of 2 as a first approximation to account for the delta between low-level kernel performance and system performance when multiple kernels are integrated together to form a complete L1 chain.

In aggregate, we come to $21 \times 4 = 84$ GOPS/second required processing. Assuming a lower end 128 SIMD-capable CPU core at 2.2 GHz operation, each core can do $128/32 = 4$ multiply–accumulate or 8 (multiply and accumulate are 2 separate operations) operations or roughly 18 GOPS/second. Similarly, when operating in 16-bit (fixed point), the number of GOPS/second doubles to 36.

As a first pass, we conclude that Physical Layer processing per carrier takes roughly (84/36 to 84/16 =) 3 to 5 CPU cores.

Phase 2: performance proof points

A very high-level analysis has shown above. In the next stage, we need to prove that the assumed cycle counts for each individual software kernel is achievable in real life. This requires actual software development—building proof points of the actual application or acquiring performance relevant software components from the device or other vendors.

This stage allows us to eliminate guesswork around topic 1 abovementioned (low-level kernel inefficiencies). We still have overheads associated with integration inefficiency (topic 2) but we start eliminating risk in the development cycle. As an example, let's assume we need to validate performance for QAM demodulation (LLR generation) in software. 3GPP 38.212 defines QAM mapping for 16QAM as follows: quadruplets of bits, b(4i), b(4i + 1), b(4i + 2), b(4i + 3), are mapped to complex-valued modulation symbols d(i) according to:

$$d(i) = \frac{1}{\sqrt{10}}\{(1-2b(4i))[2-(1-2b(4i+2))]+j(1-2b(4i+1))[2-(1-2b(4i+3))]\}$$

or, shown graphically in Fig. 9–2.

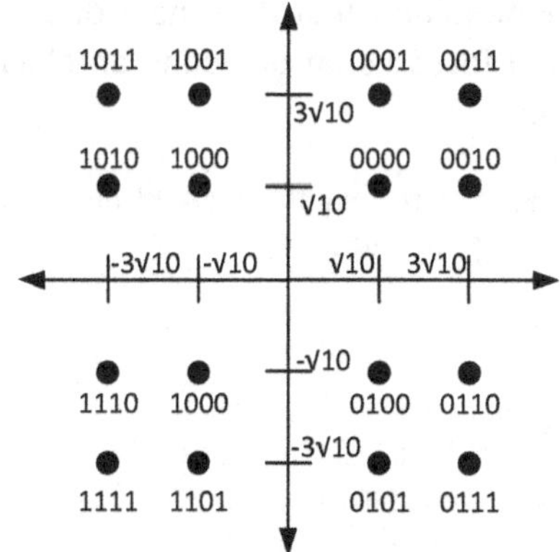

FIGURE 9–2 QAM demodulation. *QAM*, quadrature amplitude modulation.

Assuming an signal-noise-ratio γ and a complex (R(eal) + I(imaginary)) data subcarrier output (from channel estimator) d(n) for each symbol, LLR calculations are shown in the following:

$$L0 = -\gamma R(d)$$

$$L1 = \left(\frac{2}{\sqrt{10}}\gamma - |L0|\right)$$

$$L2 = -\gamma I(d)$$

$$L3 = \left(\frac{2}{\sqrt{10}}\gamma - |L2|\right)$$

Assume the input data format to be an I/Q sequence of 16-bit values. The LLR outputs have to be quantized to 8 bit to be passed on to the descrambler and decoder units.

Vectorized QAM modulation is implemented in a two-step process:

1. LLR generation. In this step, input samples are converter to LLRs. This step is implemented by executing the math as outlined above for L0, L1, L2, L3 in parallel on a set of 4x {I + Q} samples. The LLR generation process is not implemented by means of table lookups, but by directly implementing the required math.
2. LLR storage to memory. In the previous step, each output vector that is generated holds 8 × either {L0, L1, L2, L3}. These vectors are required to be stored in an interleaved manner, and with a limited LLR size (assume here 8 bit). Interleaving can be implemented using mask/shift/operations to truncate and combine the individual LLRs, or by using the *<vsri_n>* instruction that does a vector shift/merge operation combining these instructions.

We do an inner-loop implementation that does not concern with the precision analysis and quantization boundaries but focusses on depicting an efficient implementation of basic math functions in the algorithm:

```
void llr16(const int16_t* input, int bytes, uint8_t* output) {
    //uint16_t* out = (uint16_t*)output;
    uint8_t* out = output;
    int i, j;
    // Gamma
    const int16x8_t gamma = vdupq_n_s16(32000);
    // Two over square root ten multiplied by gamma
    const int16x8_t two_square_ten_gamma = vdupq_n_s16(0xefff);
    int16x8_t in;
    int16x8_t LzeroLtwo, LoneLthree;
    for (i = 0; i < bytes; i + = 16*1) {
        // load 1 x 16B at a time
        in = vld1q_s16(input + i);
        // multiply with gamma
        LzeroLtwo = vqdmulhq_s16(in, gamma);
        LoneLthree = vsubq_s16(two_square_ten_gamma, vabsq_s16(LzeroLtwo));
        // build 8 bit LLRs anr combine them, store to memory
        vst1q_u8(out, vreinterpretq_u8_s16(vsriq_n_s16(LoneLthree, LzeroLtwo, 8)));
        // next set of input values
        out + = 16;
    }
}
```

This inner loop is quickly developed and can be used as an input for productized kernel development later by the software team. It is also an excellent benchmarking tool that can be used to compare device vendor performance.

In this example, benchmarking shows 16 LLRs are generated in $\sim$6 CPU clocks on an Arm A72 core. Compare this to the theoretical expectation set earlier ("... $\times$ 3 real operations ..." per LLR) without any overheads and the later assumption of 50% overhead for low-level kernel implementation.

It is to be noted that 6 CPU clocks in an Arm A72 with 128-bit vector support gives $6 \times 8 \times 2 = 96$ operations for generation of 16 LLRs or 3 operations per LLR. This great match between theory and practice may be rare but the point is the same nonetheless: over time, we learn how to map theoretical calculations to practical implementations well.

Phase 3: stack development

The last stage in the analysis cycle is actual software implementation where we compare subsystem (transmit chain, receive chain, etc.) level performance against the initial theoretical projections to find out where performance bottlenecks in the software implementations exist and optimization is needed.

Central Processing Unit loading summary

Although the per-component cycle counting provides good insight into which components need to be optimized, it is not always the best tool to identify where the cycles are consumed. For this purpose, a CPU loading profile, at a very fine granularity (i.e., per symbol name) is more useful. An advantage of using Linux as a development environment is that such profile can be generated with open-source tools such as Oprofile (http://oprofile.sourceforge.net/about/). Oprofile is a system-wide profile for Linux systems that helps identify high-level hotspots and areas to delve deeper into for optimization. Oprofile tools are commonly used and included popular commercial Linux distributions including Red Hat Enterprise Linux and SUSE.

System-on-Chip performance counters

Both software cycle counting and Oprofile do not provide insight into the SoC-level performance metrics that we established before, such as:

- Target Instructions Per Cycle/Clock (IPC)
  - Including associated L1/L2 ICache and DCache hit/miss ratios.
  - Gives insight into level of optimization of the software application. As a rule of thumb, an IPC of below 0.7 is an indication of limited use of capabilities of the processor and room for optimization efforts that improve cache utilization.
  - The counters are typically associated with the core vendors (e.g., Arm, Intel, or AMD x86) and not with the SoC manufacturer.
- Bus utilization
  - Most SoCs have a vendor specific infrastructure to measure bus utilization. Alternatively, a static (paper) analysis can provide insight into bus utilization limits. Hitting internal bus limitations present a "hard stop" on system performance optimization that dictates a move to a new dataflow through the system.
- Double data rate (DDR) memory utilization
  - Most SoCs have a vendor specific infrastructure to measure bus utilization. As we explained before, a DDR utilization above 50% implies increased memory latency and associated suboptimum performance.

Such metrics are taken from SoC HW counters on a running system. These counters are implemented in each IP block in the device, with the specific purpose of being able to track system-level performance. They can include timestamping and "packets or operations/second" type metrics.

SoC-level performance counters are mainly useful to verify that system-level parameters have enough headroom (e.g., bus/DDR loading). They also help identify SW architectural issues such as low cache hit rates due to code/data locality, etc.

Life-of-a-packet double data rate utilization analysis

DDR utilization is difficult to estimate up-front and is often forgotten in a performance analysis scenario. As we indicate earlier, DDR bound systems show up as performance bound implementations where "the system simply cannot do better, whatever we throw at it." To help with DDR bandwidth dimensioning before using SoC or other counters for field measurements, consider using a life-of-a-packet analysis to trace the use of DDR memory bandwidth as the packet goes through the system. Say, the example of an IPSec networking flow that uses a lookaside accelerator for the actual ciphering operation:

- Ingress from Ethernet through networking subsystem write to system memory ($1 \times W$).
- Read and write operation from/to DDR for IPSec (de)cipher operation ($1 \times R + 1 \times W$). Note that this path is optional—in a low-latency operation, the packet payload will be accessed from on-chip cache.
- Egress through Ethernet read from system memory ($1 \times R$).
- 50% overhead to all to accommodate for packet context store/retrieval, including cipher keys, etc.

Aggregate DDR bandwidth consumption for 50 Gbps operation: $4 \times$ throughput ($2 \times R$, $2 \times W$) or $4 \times 50 = 200$ Gbps. Including 50% overhead, this becomes 300 Gbps. Assuming a multicore SoC equipped with dual 3.2GTPS 64-bit interfaces an aggregate of 410 Gbps throughput, we deduce a consumption estimate of 73%, or in other words a DDR controller utilization that is marginal for worst case operation (but likely better in real life where platform caches alleviate bandwidth).

Mitigation techniques: what if the product does not meet performance targets?

3GPP offers several mechanisms to reduce the computational load on the system allowing to make a softer boundary between high-level product specifications and software/systems performance achieved. Particularly in multisector applications where a DU is serving more (ideally) many sectors, averaging and pooling are in fact intentional techniques rather than mitigations.

We explore a few of these techniques here, while noting that the software-centric nature of O-RAN products implies that performance can be yet one more feature that gets upgraded over time. Early deployments that target few users and limited throughput may need less performance than mature products.

- Bandwidth partitioning (BWP). We explained the concept of "bandwidth partitioning" mainly from the perspective of the client rather than the base station. However, given that BWP provides a fine granularity of control over uplink and downlink air resources (individually!), BWP can be a good tool to explicitly bound both fronthaul bandwidth (reduced allocated air interface bandwidth reduces fronthaul bandwidth in a well-defined manner) as well as computational and accelerator/memory IO bandwidth.

- Other reductions in Physical Layer specifications such as used multiple input multiple output layers, supported modulation type (e.g., 256QAM uplink support), etc. help reduce computational load in a similar manner while maintaining 3GPP compliance of the end-product.
- Performance-aware scheduling implies making the MAC layer aware of Physical Layer constraints. Consider the example of limited Hybrid Automatic Repeat Request (HARQ) memory or DDR bandwidth where 5G asynchronous HARQ allows the scheduler algorithm to schedule HARQ (re)transmissions based on the knowledge of system limitations rather than only air interface resource availability.
- Deadline supervision is a technique where a software component does not run to its logical completion point but has intermediate "cutoff" points where it checks for compliance with a time deadline. Consider MAC delivery of Functional Application Platform Interface information to the Physical Layer or RU delivery of an OFDM symbol over the air as an example—in both cases, it is preferred to send a partially empty (dummy) frame instead of late delivery of a complete frame.

Development environment optimization

Contrary to the "implement first, optimize later" mantra, we feel that it is imperative to "ground-up" architect the solution to be implemented in an architecturally optimized way. This means, for example, a memcpy-free U-plane, and other optimization techniques shown in this book. Development teams that are under pressure to add features do often not have time to change an existing implementation, so the up-front architecture work is a worthwhile investment! At the same time, it is important not to miss any low-hanging fruit and ensure that hardware, development environment, and system software are set up for achieving maximum performance:

- Compiler and flags. First, even though GNU Compiler Collection (GCC) is the most used compiler, for reasons of familiarity and availability, use of a proprietary compiler can be a very easy way to get a double-digit percentage performance bumping with minimum effort. Also, within the GCC family, we suggest using the latest compiler version. For example, some of our internal benchmarking showed around 5% performance improvement moving from a very old GCC 3.4.x to GCC 4.4.x, and another 5% moving to GCC4.7.2.
- Optimized libraries for key functions. The practice of pushing for optimized basic libraries such as memset/memcpy is well established in the embedded industry. Often, other libraries (e.g., math) are used sporadically in a software stack, but still often enough to have a significant performance impact. Implementing such libraries in an optimized matter (optimization to the use case, precision, etc.) can offer significant performance gains.

- Prefetching. Both instruction and data prefetching can be useful. The base station application uses multiple processes/threads on multiple cores. Given the large footprint of the codebase and the user contexts, this means that the achieved performance in terms of Instructions Per Cycle/Clock (IPC) is often relatively low. Performance improvement thus comes from prefetching both code and data.
- Cache locking. Like prefetching, cache locking can provide for an improved IPC. System software should provide appropriate APIs.
- Large memory pages limit the software load associated with Memory Management Unit (MMU) misses by allocating packet buffers from a single large MMU entry rather than from standard Linux 4-KB pages.
- Interrupt coalescing/NAPI ("New API"). The overhead associated with interrupts and (associated) task switches is significant. As such, most networking stacks provide the option to tune the tradeoff between response time (latency) and throughput.
- Optimized hardware configuration. The default device configuration as provided by the Board Support Package typically is not optimized for final hardware. For example, DDR timing settings may be less strict than supported by the actual hardware. Review of such settings between hardware and software teams can provide an overall performance gain.
- Limit context switching. Remove unnecessary context switches. Examples include regular message queue checking using a separate process (could be triggered on message send), or excessive process-count where tasks could better be managed from within a SW process.
- Removal of system calls from the real-time path and replace necessary functionality with optimized, user-space functions. Use of timers from the user space is a great example of code that can efficiently be replaced with single-instruction timebase-read functionality.

Software optimization techniques

It is never possible to provide an exhaustive list of software optimization techniques. Based on our experience, we can define a checklist of commonly used strategies to get the optimum performance out of application software:

- Use the optimizer. Some programmers avoid compiler optimization for various reasons. Historically, some optimizers have caused logic problems. Performance-centric applications should all be compiled with the –O3 optimization level unless there is an explainable reason why not to. We recommend that this level of optimization be used for almost all circumstances.
- Avoid branches where possible. Branches cause instruction memory stalls depending on the target address. The programmer may know in advance which code path is the more likely to occur. In that case, the code can be structured to execute that path first, thus eliminating some run-time testing and branching. A common example where you can eliminate dynamic branches is in cascaded if-then-else statements. For example:

```
if (bar = = 0)
  almost never get here
else if (bar < 0)
  sometimes get here
else
  almost always get here
```

is better rewritten as:

```
if (bar >0)
  almost always get here
else if (bar <0)
  sometimes get here
else
  almost never get here.
```

To keep the code more readable even with complex if-then-else statements and without having to refactor or retest code, use the likely/unlikely compiler hints. In fact, this makes the code more readable because the most likely path is visible to the observer.

- Avoid branches by making one code path common and only test one condition. So, the following code snippet:

```
if (condition)
 x = 1;
else
 x = 2;
```

Could be rewritten more efficiently as:

```
x = 2;
if (condition) x = 1;
```

Avoid switch/case constructs like the following, especially for small code segments. Although the source is clear, it generates numerous tests and branches that make it inefficient. It also does not distinguish the most likely, or "hot" path from others.

```
switch (var) {
    case 0:
        x = 1; break;
    case 1:
        x = 2; break;
    case 4:
        x = 9; break;
    default:
        x = 12; break;
}
```

- Consolidate common code paths to reduce testing. This suggestion may seem like it is too obvious to mention, but we have found that as an application goes through a series of revisions, additional tests and branches can be introduced. If an application has undergone updates to the logic, you should review the code path to be sure that unnecessary tests and branches have not been introduced. What you may find in an application is as sequence like this:

```
if (cond){
  common path here
  x=1;
 }
 else
  x++;
 if (x==1)
  do something
```

  This code fragment can be consolidated to:

```
 if (cond){
  common path here
  x=1;
  do something
 }
 else
  x++;
```

- Avoid direct control register bit masking. Use optimized macros for bit operations such as BitIsSet and BitIsClear, even if you need to custom define them. These macros can be optimized to produce efficient code. Also, using these macros will isolate your application from later implementation changes. If more efficient implementations of these macros are devised, or bit positions need to shift for some reason, the applications will not need to change to take advantage of the improvements.
- Avoid use of global variables, file statics, and pointers wherever possible. A local variable is a lot easier for the compiler to disambiguate, so it usually gets promoted to a register for the lifetime of the routine. This can save cycles for each reference to the variable by avoiding load and store operations. In some cases, it may be beneficial to copy some of these variables to locals. For example:

```
// Compiler can't tell if someOtherGlobal and *p might overlap so it generates
// a load *p every time through the loop.
  foo (int* p)
  for (expensive loop){
    if (someOtherGlobal==*p){
      do something
    }
    someOtherGlobal=something.
  }
```

```
// If you know that they never overlap, copy *p before entering the loop.
// Here compiler knows local can never intefere someOtherGlobal and should
// put local into a register avoiding a load each time through the loop.
foo (int* p)
int local = *p;
for (expensive loop){
  if (someOtherGlobal = = local){
    do something
  }
  someOtherGlobal = something.
}
```

- Make conditionals as specific as possible. If a variable "x" is an unsigned integer (int32u) and it has to be checked for the value being less than a variable "n", the conditional test could be coded as: "if (x < n)." The conditional does not need to test that (x > = 0) since that condition should be met by the definition of the variable. In an example we came across, on examination of the assembly language, it was found that by encoding the conditional as "if (x > = 0 && x < n)," the optimizer was able generate code that was 4 cycles faster than the "if (x < n)" implementation.
- Avoid volatile declarations wherever possible. Use the "volatile" construct only on control, I/O and other addresses that change asynchronously by hardware or have side effects when read. A volatile variable or field cannot be promoted to a register, cannot be eliminated, cannot be part of a common subexpression, etc. As a result, every reference to a volatile variable generates a load or a store. If you load a volatile field and know that it cannot change because of some other semaphore (i.e., "avail" bit says you own it) copy the field into a local variable and reference the local copy instead.
- Avoid function calls where possible. Function calls require cycles to set up the arguments and the stack so should be avoided when possible. This can be done by putting the code inline, or by using the "inline" directive for functions to have the compiler, insert a copy of the code. Care should be taken with this, however, since it can increase the amount of instruction memory used. Function calls can also cause shared instruction memory stalls, further slowing performance.
- Avoid function calls with many arguments. Famously, in the MIPS/Arm processor architectures, the first four arguments are passed in registers and can be referenced directly. For x86–64 Linux, the number is six. All arguments after this have to be marshaled into memory and are accessed indirectly through a pointer in the called routine. This takes additional cycles and reduces performance.
- Pass function arguments by value instead of by reference. The compiler can store parameters passed by value in a register for faster access than it could if it had to dereference a pointer and fetch the value.
- Create a global variable for values that must be referenced by many routines. This avoids the overhead of repeatedly pushing the value on the stack. But care must be taken to prevent contention issues when using this approach.

- Use variable sizes that are intrinsic to the processor (typically 32 or 64 bit) when possible. Declaring variables as byte or half-word types can cause the compiler to generate additional code when it is referenced, or when operations are performed on that variable. Additional cycles are needed to mask off bits when it is loaded to a register, or to properly calculate overflow in math operations with other data types.
- Avoid serializing tasks where possible. Try to identify independent tasks that can be performed concurrently and initiate them without waiting for them to complete. The completion logic can be more complex because it must accept one of several resulting events, but this can significantly increase performance.
- Avoid superfluous context switches. Do not automatically put a context switch at termination of a block of processing, such as at the end of a receive or transmit function. Only place them where there is no further processing that can be performed. Consider not doing a context switch during accelerator/DMA calls that have relatively fixed latencies. Instead, try to perform a block of processing that matches that latency period.
- Do as much work as possible prior to checking for the completion of operations. Get as much processing done before performing a check on the completion of an asynchronous request, which, if not completed, would result in a context switch. This would include operations such as Protocol Data Unit (PDU) payload completed, queue operation token availability check, etc.
- Move nonessential logic out of the critical path. All blocks of logic in the critical path should be evaluated to see if they could be performed elsewhere by a less critical resource. So, for example, if an application was maintaining statistics or table values in the critical path, it might be possible to offload this processing from the realtime path to a background operation that is not time critical. It might even be possible for the processor to push the relevant data up to the cloud and have processing be performed there.
- Avoid shared and global memory accesses. Shared accesses within a multicore environment take more clock cycles. Accessing memory by one core from the other core memory goes across the coherent interconnect within the SoC and can take 10 or 100 seconds extra clock cycles in the worst case. It can be more efficient to distribute data among cores (e.g., in statistics) and doing aggregation in a separate (background) process.
- Tune use of interrupts. In an realtime operating system but more so in high-level operating system such as Linux, interrupts incur a large overhead in saving and restoring context information. Instead, your application should not set up the interrupt vector for that event but could add a test in the normal code (polling mode operation) to handle the event or exception condition without interrupts.
- Tune table configurations. For example, always make sure that a hash table has at least two times the number of keys that will be active at any given time. If possible, reduce the amount of payload returned by any table lookup to fit in a small and sequential amount of cache lines.
- Avoid simultaneous requests for the same resource. The sequencing of events in most multicore processor applications will produce a random distribution of requests for

resources by the various cores in the processor. It can happen in some instances, however, that an application will tend to generate multiple, simultaneous requests to a given resource within the chip—such as an IO device or an accelerator.

- Review the code generated by the compiler. As a final step in your optimization process, you could have the compiler output a listing of the assembly code it generated. There are some infrequent sets of circumstances that can prevent the optimizer from generating the most efficient code possible. Examine the assembly listing for extra moves and branches that could have been eliminated by the compiler but were not. In these rare cases, you may want to manually inline or try other implementations that might be more easily optimized. For example, it was found that by simply changing the return type of foo() from unsigned char to int reduced the number of instructions from 19 to 14 (see code example next). Different sized operands appear to have prevented the optimizer from producing more efficient code.

```
int ext;
void halt();
void printf(char *format, ...);
static __inline__
signed char foo()
{
  if (ext > 0)
    return 1;
  else
    return 0;
}
void start()
{
  if (foo()){
    printf("pass");
  }
  else{
    printf("fail");
  }
  halt();
}
```

Reference

[1] Intel Data Plane Development Kit (DPDK), https://www.intel.com/content/www/us/en/communications/data-plane-development-kit.html

Interoperability and test

This chapter gives an overview of steps involved in test (first) and interoperability (secondary) efforts for telecommunications equipment in general and Open Radio Access Network (O-RAN) products specifically. We focus on testing of central unit (CU), distributed unit (DU), and radio unit (RU), assuming split options 2 and 7.2.

We cannot dictate an overall test strategy rather than to first provide a broad overview of common-sense testing strategies and then highlight key components in the area of testing that need to be covered for wireless telecommunications and/or O-RAN products.

Testing is broadly separated between development testing which mostly owned by the development team and implemented during the software implementation phase to ensure that (small) software components adhere to the required performance. Development testing is often done by the software developer as part of the coding process (think about usage of printf() during software development to ensure the code does what was intended) but often not formalized and hence not repeatable.

System tests represent a type of testing in which application software stacks, enablement software [board support package (BSP) including operating system], and hardware components are combined and tested as a system to confirm that they will interact as per their requirements. This typically carried by System Validation Test (SVT) team.

Development testing

Testing during the development phase is always under pressure, given that testing takes a lot of effort and each project is under time pressure. However, per the famous book "Mythical Man Month,"[1] and as known to most experienced engineers and managers, it is found that the cost of finding a bug increases with the time, which is proven many times through various studies. Basic testing needs to be incorporated into the design flow for any selected software development methodology. "Agile development" is not an excuse for eliminating test efforts!

Static analysis

Static code analysis is done to identify two areas of code improvement:

- Adherence to coding rules or quality metrics such as code complexity and number of lines per function call.
- Finding bugs. Think about indexing into an array beyond its size or other provable evidence that the code will fail.

Open Radio Access Network (O-RAN) Systems Architecture and Design. DOI: https://doi.org/10.1016/B978-0-323-91923-4.00015-X

- Security issues. The classic example is buffer copying outside of intended boundaries. A good security approach requires security aspects to be considered on every level.

Static analysis can be done by multiple means, ideally all at the same time:

- Peer review where software engineers review each other's code. Potentially implemented in a hierarchical manner where senior engineers provide sign-off on junior contributions before they can be accepted into the main development branch.
- Code analysis tools (such as Coverity) which is an automated form of the peer review and can provide additional metrics to evaluate code complexity such as Knots and Cyclomatic Complexity that help pinpoint to potential areas to target code simplification or additional test focus to ensure correct behavior. This can be done as part of an automated build environment with regular (nightly) regression tests.

Functional

Unit testing is software verification and validation method in which a development engineer verifies whether a certain piece of source code in isolation is working as per the desired behavior or not in the development phase. Its more applicable to development team and should mandatorily be done by developer on the modified code before code freeze.

Unit testing is typically done by building a harness around a developed component/code snippet. The test harness exposes the target code to defined test vectors and behavior is compared to expected behavior. Unit testing tests not only the code but also the compiler/linker and the target hardware (or simulated hardware). Unit testing is an inherent part of any code development.

In networking products in general, and wireless specifically, unit testing tools include pcap file analysis to provide automated interpretation of Ethernet traffic, specifically, DU/RU testing uses, defined through (MATLAB-generated) known orthogonal frequency division multiplexing (OFDM) frames that can be compared against the digital version of the transmitted signal in a regression test environment.

Tool Command Language (TCL) scripting is often used to automate testing. Note that unit and functional tests often can be automated and made part of an automated regression test suite that runs overnight or on code check-in, ensuring a product functionality is not broken as features are added or other bugs are fixed.

Functional testing is divided into subcomponents to test (Fig. 10–1):

- Adherence to 3GPP or other relevant specifications at algorithmic function level. Examples include fast Fourier transform, forward error correction encode/decode functions, etc.
- Integrated uplink (UL)/downlink (DL) chain tests that compare against 3GPP standards such as 38.211/38.212 (Physical Layer) or 38.321/38.322/38.323 [Medium Access Control (MAC)/Radio Link Control (RLC)/Packet Data Convergence Protocol (PDCP) Layers]. At this point, conformance matrices against 3GPP features can be built.
- Integration with a control plane interface that includes functionality such as defined in 38.213/38.214 (Physical Layer) or 38.331 [Radio Resource Control (RRC) Layer].

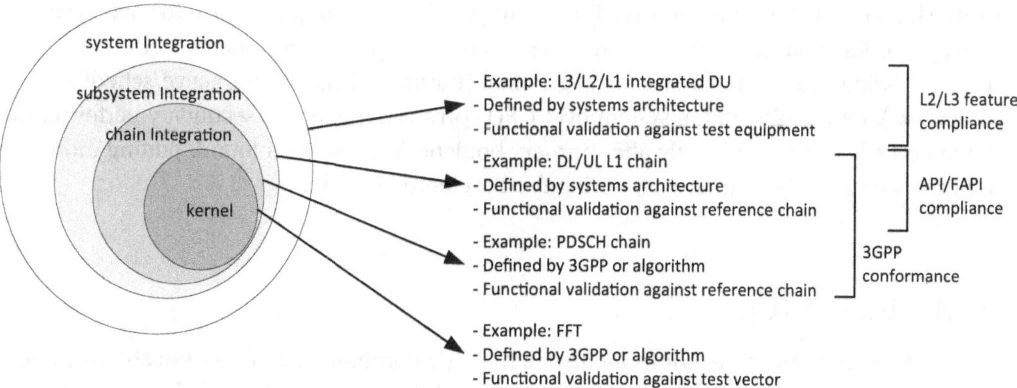

FIGURE 10–1 Various functional validation scopes.

A well-defined set of functional tests includes negative testing which is defined as testing against known invalid inputs. Consider invalid packet header formats, invalid timing, or incorrectly sequenced control plane messages as well as malicious packets that are intended to cause undesirable system behavior including application crashes.

Performance

Retroactive software optimization is a painful and slow task. For this reason, we want to make performance benchmarking an up-front exercise—we elaborated on this topic in Chapter 9, Software Performance. At unit level, software components can be benchmarked for performance by the developer team and compared against performance targets as set during the system architecture phase. Consider aspects such as:

- *I/O termination*: Performance is counted in Gbps or number of carriers/antennas and the associated clock cycle count/Central Processing Unit (CPU) loading associated with this throughput indication. Worst case scenarios are defined up-front for fronthaul, based on compression, fragmentation, beamforming, etc. These numbers should include validation against latency requirements as outlined for the DU and RU. Latency can be measured by using General Purpose Input/Output (GPIO) toggles triggering an oscilloscope or other hardware measurement tool or by timestamping in a CPU core for a software-centric implementation.
- *Signal processing chain:* Signal processing performance is measured both in terms of (1) algorithmic performance as compared to an MATLAB reference using waterfall charts and similar tools and (2) cycle count performance against the targets as defined during systems architecture phase.
- *Stack/packet processing:* Performance is measured using cycle counting (as for signal processing performance), throughput benchmarking as for I/O termination performance, or a combination of the two. Stack performance testing is highly dependent on the parameters

with which the device under test (DUT) is configured, including packet size (64 byte, IMIX/390 byte or large packet/1024 byte); stack featured (PDCP encryption mode, RLC retransmission mode); and system configuration (number of connected/active/scheduled users, etc.). Partitioning of the system over CPU cores can mean that secondary performance bottlenecks become visible only after primary bottlenecks have been found, putting more emphasis on early benchmarking, during the development cycle.

System test setup

System tests include multiple components such as the combination of a Physical Layer and a MAC/RLC/RRC stack in the case of a DU system and thus include the interface between the two (in this example, the L2/L1 FAPI interface). The system test setups deployed should emulate the typical 5G Standalone (SA) and Non-Standalone (NSA) deployments as shown in Fig. 10–2.

The complexity of the test environment means a high potential associated cost with establishing the test environment itself including cost of (emulated) core network, CU, DU, RU, and User Equipment (UE) components. Multiple well-established vendors sell dedicated test equipment, but for cost reduction reasons, test equipment can be replaced with off-the-shelf units, for example, RU, UE, and MeNB at the cost of reduced visibility into the internal workings of the equipment. Tradeoffs between cost and performance need to be made by the SVT. Test equipment also includes the following:

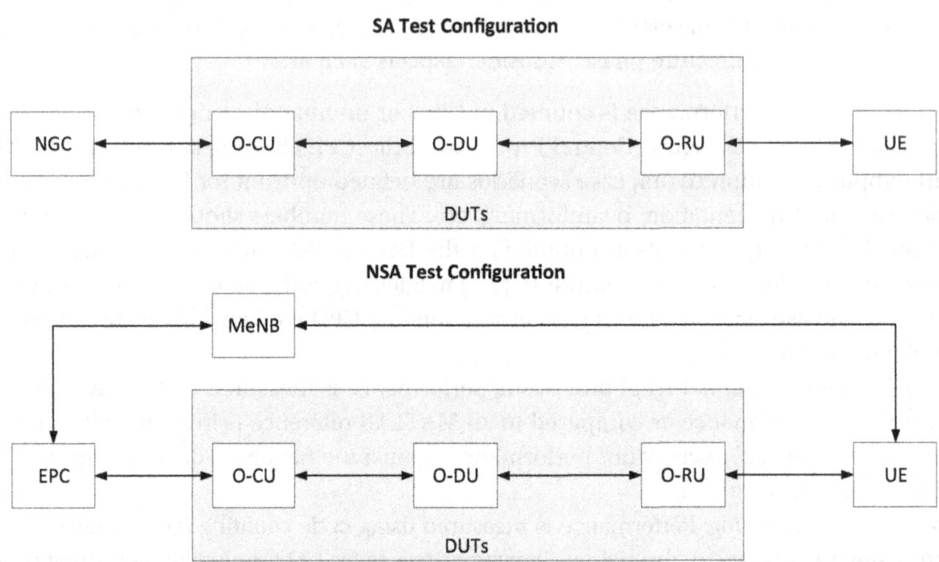

FIGURE 10–2 Stand Alone (SA, top) and Non-Stand Alone (NSA, bottom) system test environments.

- Traffic generators. Traffic sources/sinks that can generate or process deterministic traffic patterns (throughput, packet size, protocol, etc.) as well as deliver test reports to capture key performance indicators (KPIs), including received throughput, latency, packet loss, etc.
- Protocol analyzers are typically connected through a spoofing port that makes their existence invisible to the equipment communicating over the network. Protocol analyzers can analyze control plane and user plane traffic.
- RF and spectrum analyzer.

Feature testing

Most software stack and/or equipment vendors maintain a "compliance matrix" where each feature of the 3GPP standard that is under development is individually accounted for (Supported, Not Supported, Under Development). This is a requirements traceability matrix typically built by taking the 3GPP standard (note to track which version of the standard!) and copying a chapter/subchapter and/or paragraph in this standard into a table that is check-marked for compliance. Keeping to the 3GPP standards definition of feature name and sequence in which compliance is noted allows for different teams (and customers) to quickly familiarize themselves with the compliance matrix (Fig. 10–3).

Unit testing can be implemented at this feature-by-feature level in a system test setup—even when some features are more easily tested in a unit test environment. These tests are mainly targeting CU and DU systems which is where the bulk of 3GPP defined features are hosted. In addition, system-level tests can be targeted to the traceability matrix to ensure that the complete system has been tested. Note that this system-level testing can be highly complex and require in-depth knowledge of 3GPP standards and complex (third-party test) environments.

Assuming abovementioned example for PDCP-level feature testing, including pointed scenarios for the following examples (and more!):

- ciphering and integrity using AES32, SNOW3G, and ZUC algorithm for user plane and control plane data
- header compression and decompression for user plane data using different Robust Header Compression profiles
- 3GPP NSA and SA deployment options
- in-sequence delivery of upper layer PDUs at reestablishment of lower layers
- duplicate elimination of lower layer service data unit (SDUs) at reestablishment of lower layers for radio bearers mapped on RLC AM
- PDCP-related functions for below NSA function
  - secondary node addition
  - secondary node change
  - secondary node modification
  - reconfiguration with sync and key change
  - reconfiguration with sync but without key change

- PSCell change during inter- and intra-DU
- bearer-type change from spilt to Sequence Number (SN) terminated Master Cell Group (MCG) bearer
- bearer-type change from SN terminated MCG bearer to split bearer

- UL and DL COUNT Thresh hold
- PDCP discard timer
- DL and UL data rate limiting for ENDC spilt bearers using flow control algorithms

TS36.323 Technical Specification Group Radio Access Network
Evolved Universal Terrestrial Radio Access (E-UTRA)
Packet Data Convergence Protocol (PDCP) specification(Release 9) V9.0.0 (2009-12)

chapter_number	chapter_title	Feature Compliance details
-	Foreword	Information Only
1	Scope	Information Only
2	References	Information Only
3	Definitions and abbreviations	Information Only
3.1	Definitions	Information Only
3.2	Abbreviations	Information Only
4	General	Information Only
4.1	Introduction	Information Only
4.2	PDCP architecture	Information Only
4.2.1	PDCP structure	Compliant
4.2.2	PDCP entities	Compliant
4.3	Services	Information Only
4.3.1	Services provided to upper layers	Compliant
4.3.2	Services expected from lower layers	Compliant
4.4	Functions	Compliant
4.5	Data available for transmission	Not compliant
5	PDCP procedures	Information Only
5.1	PDCP Data Transfer Procedures	Information Only
5.1.1	UL Data Transfer Procedures	Compliant
5.1.2	DL Data Transfer Procedures	Information Only
5.1.2.1	Procedures for DRBs	Information Only
5.1.2.1.1	Void	Information Only
5.1.2.1.2	Procedures for DRBs mapped on RLC AM	Compliant
5.1.2.1.3	Procedures for DRBs mapped on RLC UM	Compliant
5.1.2.2	Procedures for SRBs	Compliant
5.2	Re-establishment procedure	Compliant
5.2.1	UL Data Transfer Procedures	Information Only
5.2.1.1	Procedures for DRBs mapped on RLC AM	Compliant
5.2.1.2	Procedures for DRBs mapped on RLC UM	Compliant
5.2.1.3	Procedures for SRBs	Compliant
5.2.2	DL Data Transfer Procedures	Information Only
5.2.2.1	Procedures for DRBs mapped on RLC AM	Compliant
5.2.2.2	Procedures for DRBs mapped on RLC UM	Compliant
5.2.2.3	Procedures for SRBs	Compliant

FIGURE 10–3 Example of a feature compliance matrix.

RF compliance testing

38.104 outlines the required RF tests to be performed to build a 3GPP compliant product. Obviously, these tests are mainly targeting RU systems.

Transmitter

RF transmit characteristics are analyzed using an RF signal analyzer capable of interpreting 5G signals, connected to the RF port of the RU.

Base station output power: adhere to local area (<24 dBm/port), medium range (38 dBm/port), or wide area output power targets.
Output power dynamics: Difference between maximum and minimum transmit power for an OFDM symbol for a 3GPP specified set reference conditions ranging between a single resource element (RE) and a full OFDM symbol. The upper limit of the total power dynamic range is the base station maximum carrier Effective Isotropic Radiated Power (EIRP) when transmitting on all resource blocks (RBs). The lower limit of the total power dynamic range is the average EIRP for single RB transmission in the same direction using the same beam. Measurement is done on PDSCH symbols (no Reference Symbol (RS) or Synchronization Signal Block). Transmitter testing includes transmission of two reference signals, one with a full OFDM symbol at 64 quadrature amplitude modulation (QAM) and the other with a single Resource Element (RE) at 64QAM.

In addition, the RE power control dynamic range defines the range of power control within a RE as we discussed earlier.

Transmit ON/OFF power: OFF power is defined as the mean filtered power over time around assigned channel frequency.
Transmitted signal quality: Frequency error (wide area: + / − 0.05 ppm; medium range: + / − 0.1 ppm; local area: + / − 0.1 ppm) as well as Modulation Quality/Error Vector Magnitude (EVM) [Quadrature Phase Shift Keying (QPSK): 17.5%; 16QAM: 12.5%; 64QAM: 8%; 256QAM: 3.5%].
Occupied bandwidth: Defined as B/W where transmit power is <0.5%—smaller as channel bandwidth.
Adjacent Channel Leakage Ratio (CACLR): Ratio of filtered mean power on assigned channel to filtered mean power on adjacent channel. 45 dB absolute limit. Cat-A wide area: −13 dBm/MHz; Cat-B wide area: −15 dBm/MHz; medium range: −25 dBm/MHz; local: −32 dBm/MHz.
Operating band unwanted emissions (OBUE): OBUE are defined from Δf_{OBUE} below the lowest && above the highest operating band frequency. Base station 1-C <100 MHz: 10 MHz
Transmitter spurious emissions: Over whole band from 9 KHz to 12.75 GHz excluding the frequencies covered by OBUE.
Transmitter intermodulation: Limits the generation of unwanted signals from interference from a colocated base station.

Receiver

Reference sensitivity level: Minimum mean power received at the antenna connector for which a throughput requirement (95%@QPSK, R = 1/3). Wide area: −95.7... − 101.7 dBm; medium range: −90.7...96.7 dBm; local area: −87.7... − 93.7 dBm depending on channel B/W.

In-band selectivity and blocking: Defines capability to operate in interference condition where interference signal is ~10 dB below wanted signal.

Out-of-band blocking: Defines capability to receive a wanted signal in presence of an unwanted interferer in outside bands.

Receiver spurious emissions: Defines unwanted emissions from the receiver, either generated internally or from interaction with a colocated transmitter.

Receiver intermodulation: Defines sensitivity from two interfering signals (say, f1, f2) where the interfering frequencies are set such that their third-order intermodulation products (2x f1−f1; 2x f2−f1) are such that they fall into receiver passband.

In-channel selectivity: Measures receiver capability to receive a wanted signal in its assigned frequency space (REs) with an interfering signal in the same channel but at different allocated frequency space (and at higher power spectral density).

Calibration

Analog subsystems never behave perfectly, and RF is an example of this. Hence, RF subsystems need to be calibrated to measure and enable compensation for inaccuracies. A few examples include:

Rx/Tx IQ mismatch: IQ imbalance or IQ mismatch is an artifact of a zero-IF RF demodulator. The demodulator multiplies the RF signal with an in-phase (0 degree) and a quadrature (90-degree offset) signal to generate I and Q baseband signals, but the 0 and 90 degrees are not necessarily perfect given that they are generated through an analog delay. Compensation can be done in the analog domain (adjust the delay to generate the 90-degree offset signal) or in the digital time or frequency domains. Measurement is done typically at boot time and tracked slowly over time given its semistatic nature.

Rx DC offset: The zero-IF demodulator multiplies the RF signal with a local oscillator (LO) frequency. Due to unintended coupling (capacitive coupling on the printed circuit board, radiated coupling between antennas, and so on) between the LO signal and other components (say, the receiver antenna), some of the LO signals is unintentionally received at the input of the demodulator. After multiplication with the intended LO frequency (so a multiplication of two signals of the same frequency), this signal comes out as a DC offset that can reduce the dynamic range of the system.

Tx LO leakage correction: Like Rx DC offset, transmit LO leakage means that the LO signal unintentionally makes its way to the RF transmit antenna and hence causes an unwanted transmission, potentially breaking RF transmitter specifications. Compensation for LO leakage can be done by generating a compensating signal in the baseband that is opposite to the LO leakage, thus canceling out the leakage.

Tx gain/phase error: As the transmitted signal passes through the digital-to-analog conversion and associated analog components before it is modulated to the RF frequency, gain and phase between I and Q signals will no longer be perfect as they were generated in the digital domain. The errors caused by this can show up as EVM errors in the transmitted signal, causing for higher bit error rates at the receiver.

Tx droop correction: During transmission, power amplifiers exhibit thermal effects which, specifically, in long RF pulses with low duty cycle can impact the PA gain/phase behavior.

Tools needed

These measurements are executed on a test bench such as shown in Fig. 10–4 for transmitter testing:

or in Fig. 10–5 for receiver testing:

Typical test tools include:

- RF spectrum analyzer and or 5G signal analyzer
- 5G/RF signal generator(s)

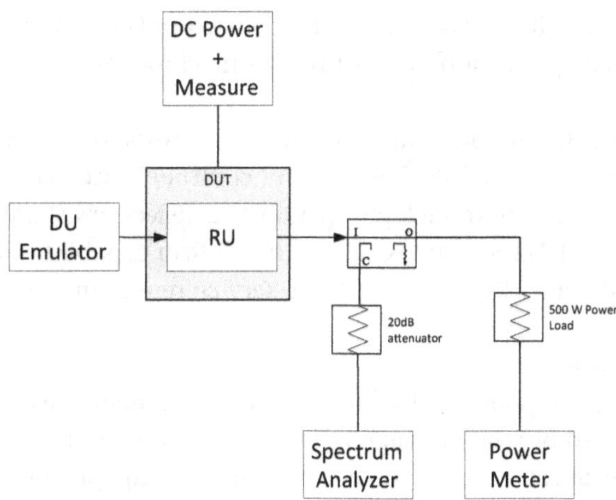

FIGURE 10–4 Transmitter RF testing.

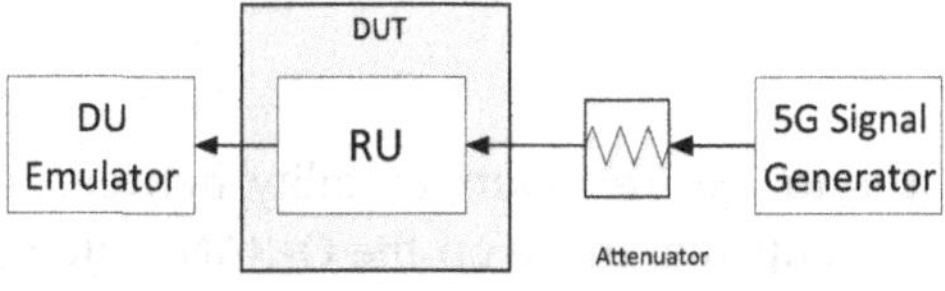

FIGURE 10–5 Receiver RF testing.

- RF power meter
- DU emulators and/or RU test software designed to generate/analyze signals in a predictable manner

Interoperability testing

Interoperability testing is done with a pool of test UEs, or with other vendor's core network, CU, DU, RU, or similar equipment. Typically, interoperability labs are both set up internally within the company, as well as at the customer.

Use-case scenarios

O-RAN defines several test use cases that are designed to confirm (likely) interoperability between components before entering plug fests. Examples include:

- V2X scenarios that are unique with regards to handovers that happen regularly and at high frequency;
- traffic steering across heterogeneous networks including Long-term Evolution (LTE), 5G/NR, and Wi-Fi;
- QoS-based resource optimization to ensure scheduler behavior is consistent with configured QoS slices;
- radio resource allocation for unmanned arial vehicle (UAV) application scenario to confirm radio resource allocation policies can be used in UAV applications;
- Quality of Experience optimization to ensure that machine learning or other algorithms are functional;
- massive multiple-input, multiple-output beamforming tests to validate optimization to balance loads, reduce intercell interference, and control electrostatic emissions;
- RAN sharing scenarios to confirm multiple operators can effectively share RAN resources; and
- RAN slice Service-Level Agreement assurance to confirm capabilities to create and manage customized networks to meet specific service requirements.

User Equipment test pool

In this test scenario, a large pool of UEs (mobile phones) is automated to perform voice/data transmissions in predefined or random patterns, targeting to confirm that there are no bugs in the combination of UEs (which can have unique behavior) or RAN equipment that is tested against.

The objective is that production-grade systems can be deployed in live networks without any (immediate) customer issues. Therefore the point is to use commercial grade rather than test equipment. This test is similar to "Safe For Network" testing required by some operators to ensure that a new device does not cause any harm to the shared network.

Plug fest

The scope of the plug fest is to ensure interoperability between equipment from different vendors, including all components that make up the O-RAN system (Fig. 10–6).

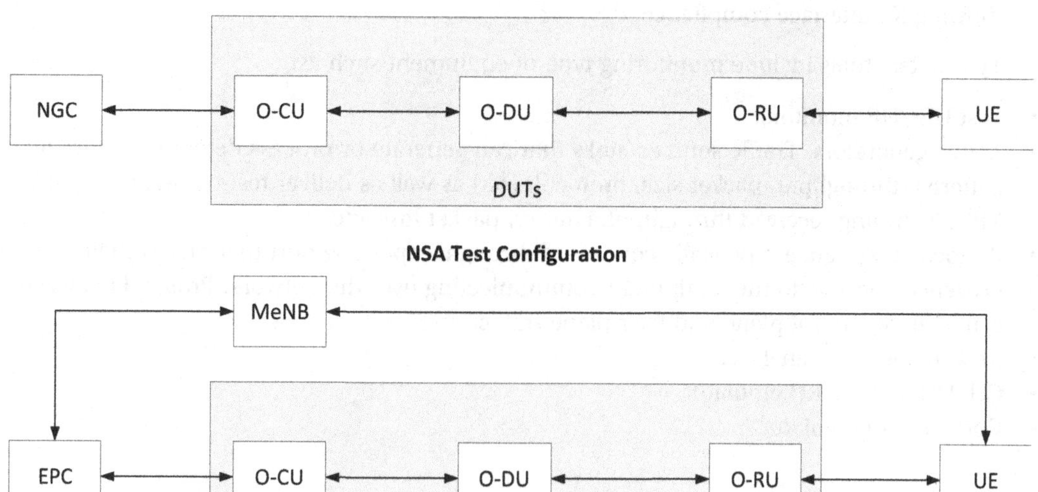

FIGURE 10–6 Scope of plug fest.

O-RAN is hosting plug fests for two main reasons. The first is to showcase development progress and give O-RAN member a platform to display engineering progress to date, specifically when targeting combined efforts from multiple companies. The second is closer to what the name indicates: to have an environment for interoperability testing for member companies. This validates published O-RAN standards (such as the fronthaul standard between DU and RU) and sets development milestones.

Plug fests are hosted in America, Europe, and Asia. The first plug fest took place in December 2019 with 35 member companies attending, with the second in the summer of 2020 and 55 companies comping together. Attendance is growing from here.

As example of what is achieved during these plug fests, we quote a few of them from the press release associated with the Indian 2020 Plug Fest as hosted by Bharti Airtel[2]:

- Altiostar and NEC showed interoperability between O-DU and O-RU including support for the C-, U-, S-, and M-plane protocols.
- VVDN, Xilinx, and Viavi demonstrated O-DU fronthaul using the Xilinx T1 telco accelerator card against Viavi test equipment. Testing included C-, U-, and S-plane protocols.
- Mavenir demonstrated coverage/capacity optimization and mobility robustness optimization, proving O1-interface compliance.
- STL and ASOCS demonstrated mobile load balancing and through that E2 interfacing.

- Altran successfully demonstrated ENDC X2 Setup and SgNB Addition procedures, showing X2 interface compliance.

  Typical test tools include monitoring type of equipment such as:

- Test UEs/UE monitors.
- Traffic generators. Traffic sources/sinks that can generate or process deterministic traffic patterns (throughput, packet size, protocol, etc.) as well as deliver test reports to capture KPIs, including received throughput, latency, packet loss, etc.
- Protocol analyzers are typically connected through a spoofing port that makes their existence invisible to the equipment communicating over the network. Protocol analyzers can analyze control plane and user plane traffic.
- RF and spectrum analyzer.
- CU, DU, and/or RU emulator.
- Core or core emulator.

Consumer application hardening

This type of testing centers around end-user application performance on the system under test. Rather than deploying artificial traffic patterns, end-user applications are exposed to the system under test to ensure that no unexpected behaviors are observed and to validate performance. Examples include the following:

- HTTP testing. Web browsing between a UE and a server that is colocated to the core network. This implies TCP traffic with a very asymmetrical traffic pattern and bursty traffic.
- Streaming TV, either using buffered or live (more latency sensitive) traffic sources and User Datagram Protocol (UDP) traffic.
- Gaming between multiple users on the wireless system under test or between a wireless user and a server that is colocated with the core network.

Performance testing

Performance testing can be done in two different ways: black box testing where the combination of the CU/DU/RU and UE becomes a combined system under test validated against preset performance goals, as well as low-level "white box" performance testing associated with identification of performance bottlenecks or validation of component-level performance as opposed to system-level performance against set values.

Black box testing

As the name indicates, this testing evaluates system performance looking "from the outside in" against easily understood metrics.

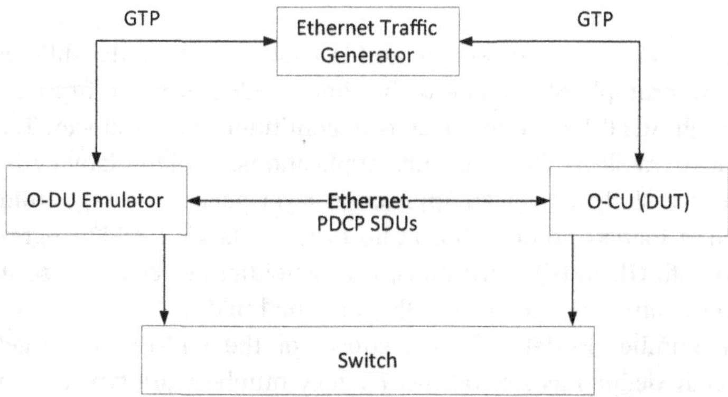

FIGURE 10–7 PDCP throughput test setup.

Throughput

End-to-end performance tests involve recreating a complete user workflow to get an accurate picture of performance. Performance needs to be measured over a wide range of parameters, creating a matrix of test scenarios that consider:

- packet size (small, IMIX, large packets)
- protocol configuration (PDCP cipher mode, RLC mode, scheduling priority, etc.)
- carrier configuration (carrier aggregation, dual connectivity, etc.)
- DUT configuration (maximum number of users, UE per slot, etc.)

Refer to RFC 2455, RFC 6349, and RFC 6815 for UDP and TCP performance measurement using different packets size and window sizes of TCP.

Given the abovementioned example of PDCP (CU) testing, consider a test setup like in Fig. 10–7.

PDCP performance is measured between the O-CU and UE/O-DU simulator. A standard Ethernet frame generator (either implemented as a software entity on a server or as dedicated test equipment such as from Spirent) is used as traffic generator, sourcing, and sinking GTP-U traffic for both the UL and DL streams.

The UE/O-DU simulator processes GTP-U traffic related to O-DU over F1-U interface as well as GTP-U traffic related to an emulated (in the case of ENDC testing) eLTE entity over the Xn Interface, as well as PDCP traffic related to the UE. Refer to RFC 2455, RFC 6349, and RFC 6815 for UDP and TCP performance measurement using different packets size and window sizes of TCP.

For this example, test configurations include a range of signaling/data radio bearer counts, packet sizes, cipher algorithmic configurations, and poll timer values.

During development phase, throughput testing can be done in isolation for L1 or L2 scenarios, for example, by so-called "L1 bypass" testing where the Physical Layer is bypassed through an emulated environment—for example, by communicating Transport Blocks over an encapsulated Ethernet frame with a random packet loss applied. Most (stand-alone) stack vendors implement this type of test environment to confirm their performance without being dependent on a (different!) Physical Layer stack implementation.

Latency

Latency testing is used to measure C-plane and U-plane latency under different traffic conditions and loads. An example of C-plane is the time needed to move from a battery efficient state (IDLE) through ACTIVE state to the start of continuous data transfer. Target for C-plane latency are in the 10 ms level for consumer applications. U-plane latency is defined as the time it takes to successfully deliver an application layer packet/message from the radio protocol layer 2/3 SDU ingress point to the radio protocol layer 2/3 SDU egress point via the radio interface in both UL and DL directions, where neither device nor base station reception is restricted by Discontinuous Reception (DRX), queue buildup, or other constraints. The target performance number is defined as a subset of the end-to-end targeted application latency and typically defined as several ms. Latency numbers are typically measured across channel conditions and associated Modulation Coding Scheme configurations.

Stability

Stability testing is a type of nonfunctional testing performed to measure efficiency and ability of a DUT to continuously function over a long period of time. The purpose of stability testing is to check if the DUT crashes or fails over normal use at any point of time by exercising its full range of use.

Stability testing is done to check the functionality of a DUT at and beyond normal operational capacity, often to a breakpoint. There is greater significance is on error handling, software reliability, robustness, and scalability of a product under heavy load rather than checking the system behavior under normal circumstances.

Stability test patterns are notoriously difficult to establish and defined with experience. Typically, a wide range of applications is operated over the DUT to identify issues that occur in mixed traffic scenarios. Examples include:

- throughput centric applications. Traffic generators emulating continued high-throughput use
- attach/detach scenarios;
- set up, tear down, and user of dual-/multicarrier configurations;
- nontypical use cases such as conference calls and emergency calls;
- combined traffic use cases where SMS, voice/video calls are combined with high-throughput HTTP traffic;
- Core Network Link recovery with Link Make/Link Break configuration;
- combined use of IPv4/IPv6; and
- exercises to the RRC state machines with idle/connected, handover, and DRX configurations.

Cell performance

Related to the RF and algorithmic performance of the RAN components, these tests confirm the performance of the deployed RAN combining RU and DU components. Examples include:

- Cell coverage and capacity. Overall range (distance) covered by the cell and capacity (in terms of user count and throughput) in various deployments scenarios. Cell coverage can include multicell deployments that incur intercell interference.

- Spectral efficiency. As a function of distance to the cell center.
- Mobility. Ensuring handover, Doppler, and other mobility-related performance aspects are covered.

White box testing

Aspects of performance analysis have been discussed previously in Chapter 9, Software Performance. Here we summarize how we take the "design-for-performance" mantra from system integration and test perspective.

An example of a development flow that includes performance targets is shown in Fig. 10–8.

The development flow includes these following phases:

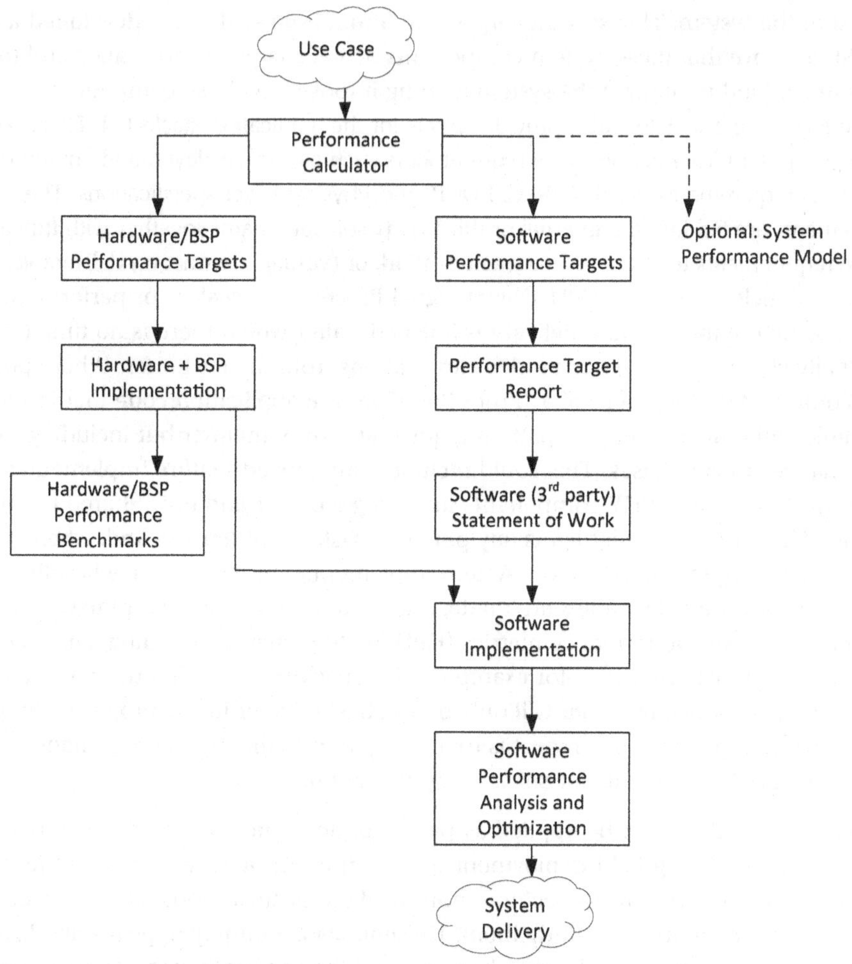

FIGURE 10–8 System lifecycle performance tracking.

- Performance exploration. Performance exploration is a static "paper analysis" phase that does not require final hardware or software to be available. During this phase, theoretical calculations on various system aspects as discussed chapter 3 are collected in a spreadsheet or a similar tool as we showed in Chapter 9, Software Performance. The spreadsheet allows us to plug in different use cases and predict various throughputs and other aspects. We can explore system corner cases (user counts, antenna counts, fronthaul latencies, and so on) and provide feedback to the requirements owner on "where the system breaks."
- This model can include power consumption exploration if chip vendors include low-level tools for thermals, which can feed into a board-level analysis.
- The static performance analysis delivers derivative requirements on hardware and BSP such as DMA (Direct Memory Access) engines, double data rate throughput, Ethernet bandwidth, and other components that can be individually benchmarked, independent of the rest of the system. This sets the targets for "Proof Points" that are developed as part of the BSP to prove that these system components achieve target performance and to characterize (and minimize) the system loading imposed by these components.
- The performance calculators also provide targets for the application stacks (L1, L2 stacks), translating system-level performance requirements such as carriers, layer, and antennas to lower level requirements on MAC/RLC, PDCP, and Physical Layer specifications. These requirements are fed into the internal or third-party software teams together with functional/feature requirements to define a Statement of Work or (virtual) contract on deliverables.
- Often, it is unclear what the CPU/Digital Signal Processing/accelerator performance should be before the software delivery is finished—after which there is no time left to potentially change hardware or modify expectations from the customer. Where possible, we recommend development of a "Proof Point" for the application code that implements the whole "Ethernet-to-sample" path in a quick-and-dirty manner but including the performance critical phases. This could include a minimized-feature implementation of PDCP, RLC, MAC, and PHY components for a single bearer and only Shared Channel support. The Proof Point highlights any potential system bottlenecks and, when padded with enough margin, can provide a system implementation performance baseline.
- When first software deliverables are made, automated regression test frameworks can be built to capture key performance metrics (CPU loading, memory loading, and so on) for a performance critical use case—for example, a loopback scenario that does not rely on actual wireless connectivity that will only be established later in the project (example in Fig. 10–9). This performance report is then compared to the initial performance projections to identify points for software optimization.

Code instrumentation can be applied as part of performance logs to be able to track performance numbers during field deployment or over time as bugs are fixed and features are added. This allows the system to become more resilient against "Heisenbugs" or errors that disappear once one attempts to study them. Heisenbugs commonly appear only during field deployment and are notoriously hard to fix given that they are both rare and often associated

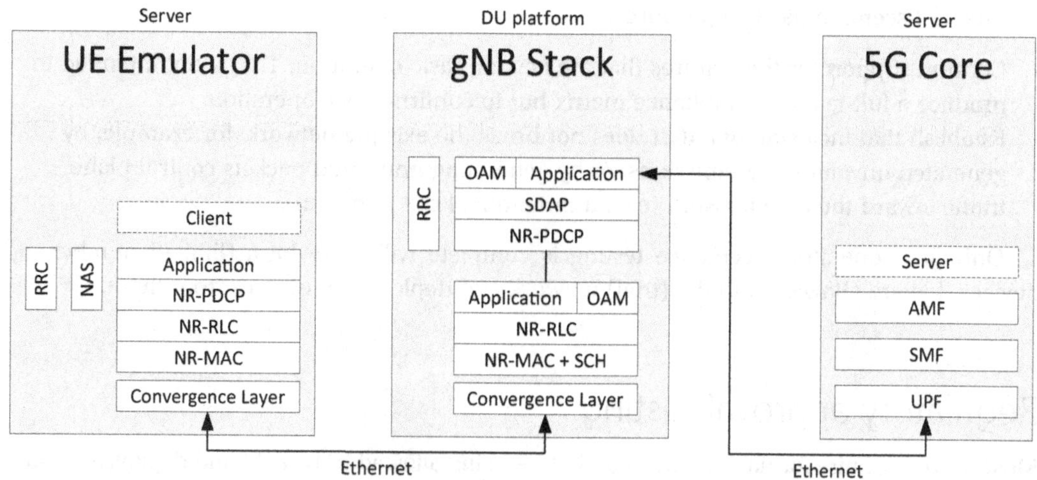

FIGURE 10–9 DU L2 stack unit test benchmarking platform.

with system timing that occurs in unenvisioned scenarios. These topics can often be traced back to unexpected performance issues.

Front-, mid-, and backhaul testing

Fronthaul networks have strict requirements set by the 802.1CM standard discussed elsewhere on packet delay (latency $<100\,\mu s$) and packet loss ratio (FLR of 10e-7) that can be independently validated with Ethernet test equipment.

Mid-haul (CU↔DU) and backhaul (CU↔core) requirements on the Ethernet network are similar and can be tested by standard test suites like defined in RFC2544 and Y.1564. This includes delay and throughput measurements. Typically, test strategies are hardware centric, with proprietary test equipment as a traffic source/sink.

In addition, (virtual) software-based testing can be implemented with open-source tools such as iPerf or proprietary tools. These tests give an end-to-end performance indication of the network infrastructure combined with the application stack running at the server and client sides.

Operator acceptance testing

Systems acceptance testing is done by the operator before new equipment is allowed into a live network. Besides validating that system and interoperability testing have been finished, acceptance testing performs a "sanity check" on the equipment. The acceptance test environment is typically a dedicated setup owner by the operator that replicates the operator network (e.g., core network) and includes typical client devices that are used by the end customer. In many ways, this test scenario is like UE Test Pool and Plug Fest testing but then in the context of operator-specific deployments.

This test scenario has two purposes:

- Confirm support for the features that are key for basic operation. This is not intended to produce a full-feature compliance matrix but to confirm basic operation.
- Establish that inclusion of a unit does not break the existing network, for example, by generated unwanted RF emissions, or by generating unwanted packets control plane traffic toward the core network (e.g., unintentional call initiation).

Only after operator acceptance testing is complete will there be a (limited number of) devices that are allowed onto the (trial) network and deployment evolves from there.

Regulatory approval testing

Most regions require formal authorization before being allowed to be sold and deployed, ensuring that the equipment does not cause harmful interference. For example, the United Stated requires Federal Communications Commission (FCC) authorization to the importer or manufacturer of (most) equipment that can emit RF energy—which is (rule of thumb) defined as any product that can oscillate at a frequency of 9 KHz (this includes CU, DU, and RU equipment, of course). The FCC website outlines the steps required for authorization as follows[3]:

- Determine the rules that apply to the product as per CFR47. Key documents include:
  - FCC CFR47, Chapter I, Subchapter A, Part 2—General rules and regulations.
  - FCC CFR47, Chapter I, Subchapter A, Part 15, Subpart A—Operation without an individual license of and intentional, unintentional, or incidental radiator. Also includes administrative requirements and conditions to device marketing. An intentional radiator is defined as a unit (e.g., smartphone or an RU) that requires to transmit energy as part of intended operation. An unintentional radiator is defined as a unit (e.g., DU or CU) that can potentially radiate energy as an unintended side effect.
  - FCC CFR47, Chapter I, Subchapter A, Part 15, Subpart B—Operation of an intentional or unintentional radiator not in accordance with the regulations in this part must be licensed.
  - FCC CFR47, Chapter I, Subchapter A, Part 15, Subpart C—Operation of an intentional or unintentional radiator not in accordance with the regulations in this part is prohibited unless specifically exempted.
- Determine the equipment authorization procedure. The FCC defines three ways to acquire authorization—where the first two have been streamlined into a single "Supplier Declaration of Conformity":
  - Verification of RF energy radiation by the manufacturer or (or a third-party laboratory) as done for equipment that either does not contain a radio or includes a preapproved/preintegrated radio (such as an 802.11 or 3G module). Compliant devices can be marketed without FCC approval.

- Declaration of Conformity as required for Part 18 (Industrial, Scientific and Medical—but includes fluorescent lighting, microwaves and similar) devices, requiring a laboratory to perform RF energy radiation testing.
- Certification as required for equipment that is most likely to interfere with other equipment and requires an explicit FCC ID to be labeled. O-RAN RUs fall into this category.

- Perform compliance testing. Formal compliance testing is done at an authorized third-party test laboratory but typically pretesting is done at development time. Regulatory testing includes development of an RF "test mode" application that radiates a predictable and repeatable pattern.
- Obtain formal approval for deployment by applying for an FCC Registration Number and Grantee Code from the FCC.
- Label conformance on the device and the reference manual. A self-approved device must be uniquely identifiable (e.g., by brand name and model) and include regulatory compliance information in the user manual. Certified equipment must include the associated FCC ID.
- Manufacture and/or distribute devices.
- Update compliance when modifications are done to the product.

References

1. Brooks, F. *The Mythical Man-Month;* Addison-Wesley, 1975. ISBN 0-201-00650-2.
2. Available at https://www.o-ran.org/blog/2020/10/24/second-global-o-ran-alliance-plugfest-demonstrates-the-accelerated-readiness-of-multi-vendor-o-ran-compliant-network-infrastructure.
3. Available at https://www.fcc.gov/engineering-technology/laboratory-division/general/equipment-authorization.

Differentiation by use case

Advantages of the combination of the protocol and timing flexibility of the 5G standard and the software flexibility coming from an Open Radio Access Network (O-RAN)-centric implementation permit O-RAN systems to be implemented in specific use-case vertical markets that will expand over time as the 3GPP standards evolve. We show a few examples of these use-case specific implementations in the following sections.

Ultra-reliable low-latency communication

Local industrial production domains (say, a robotic arm or the internals of an unmanned guided vehicle) are conventionally supported using (wired) Ethernet Time-Sensitive Networking (TSN) systems that are capable of latency in the 10 seconds of μs. Connectivity between different production domains within a factory is either provided by wired Ethernet or by industrial Wi-Fi solutions.

5G protocol extensions for ultra-reliable low-latency communication (URLLC) are defined for new services and applications from vertical industrial domains:

- smart factory and industrial automation
  - industrial control
  - robot control
  - machine to machine
- process control
- healthcare industry
  - remote diagnosis
  - emergency response
  - remote surgery
- entertainment industry
  - immersive entertainment
  - online gaming
- transport industry
  - driver assistance applications
  - enhanced safety
  - autonomous driving
  - traffic management
- manufacturing industry
  - motion control
  - remote control
  - Augmented Reality (AR)/Virtual Reality (VR) applications

Open Radio Access Network (O-RAN) Systems Architecture and Design. DOI: https://doi.org/10.1016/B978-0-323-91923-4.00002-1

- energy sector
  - smart grid
  - smart energy

Note how most of these examples are in geographical limited areas such as factory, harbor, airport, or campus. Typically, they are closed-loop control applications (e.g., robot collaboration in a factory) where latency constraints are obvious.

5G in industrial applications is enabled by several factors that are outlined here:

- 3GPP standards including URLLC
- deployment options such as edge compute and software-defined networking (SDN)
- spectral availability supporting private network deployment as we discussed in the chapter 1.

3GPP standards: ultra-reliable low-latency communication in 5G/NR

3GPP defines multiple deployment scenarios to guide requirements and protocol specifications in TS22.261, as summarized in (Table 11–1).

We identify the following key performance indicators (KPI) (3GPP 38.913):

- Latency. Defined as the (one-way) latency for delivery of an application layer packet/ message from the radio protocol layer 2/3 service data unit (SDU) ingress point to the radio protocol layer 2/3 SDU egress point via the radio interface in both uplink (UL) and downlink (DL) directions, where neither device nor base station reception is restricted by Discontinuous Reception or other energy-savin techniques. For URLLC, the target for user plane latency should be 0.5 ms for UL, and 0.5 ms for DL.
- Reliability. Reliability can be evaluated by the success probability of transmitting X bytes within a certain delay, which is the time it takes to deliver a small data packet from the radio protocol layer 2/3 SDU ingress point to the radio protocol layer 2/3 SDU egress point of the radio interface, at a certain channel quality (e.g., coverage-edge). A general URLLC reliability requirement for one transmission of a packet is 10^{-5} for 32 bytes with a user plane latency of 1 ms.

Table 11–1 Ultra-reliable low-latency communication use cases and requirements.

Scenario	End-to-end latency (ms)	Reliability (%)
Discrete automation (motion control)	1	99.9999
Electricity distribution—high voltage	5	99.9999
Remote control	5	99.999
Discrete automation	10	99.99
Intelligent transport systems—infrastructure backhaul	10	99.9999
Process automation—remote control	50	99.9999
Process automation—monitoring	50	99.9
Electricity distribution—medium voltage	25	99.9

- Availability/coverage. Max Coupling Loss (maxCL) in UL and DL between device and base station site [antenna connector(s)] for a data rate of 160 bps, where the data rate is observed at the egress/ingress point of the radio protocol stack in UL and DL. The target for coverage should be 164 dB. Link budget and/or link level analysis are used as the evaluation methodology.

Note that the time-critical scenario out of all of these is the discrete automation one which demands a 1-ms latency and 99.9999% reliability.

Building blocks to achieve these three KPIs are highlighted in the next chapters, which include:

- optimized numerology and integrated frame structure
- fast turnaround through mini-slots
- effficient control and data resource sharing
- grant-free UL transmission
- advanced channel coding schemes and interference management

Note that not all building blocks need to be used to be able to support industrial use cases. A mix and match of capabilities is expected to be standardized and developed to support specific market verticals. See the "Further Reading" section for a reference to the 3GPP study item that discusses these topics in more detail.

Orthogonal frequency division multiplexing numerologies

One area where 5G/NR is different from 4G and previous standards is in support of flexible numerology, where the subcarrier spacing (SCS) can be changed, mainly to support operation in different frequency bands. The 15-KHz SCS provides similar numerology to Long-Term Evolution (LTE), where at higher SCS, more symbols are placed in a single subframe (Table 11–2).

As the SCS increases, the slot time (minimum scheduling interval) decreases, allowing for lowering of system latency. In addition, NR defines the concept of a mini-slot (aka nonslot) of 1...13 symbols that allows for even lower scheduling latency. As such, low-latency scheduling can be achieved by picking a more aggressive numerology or by moving to mini-slot-based scheduling.

Table 11–2 Orthogonal frequency division multiplexing numerologies in 5G/NR.

Subcarrier spacing (KHz)	15	30	60	120	240
Symbol duration (μs)	66.7	33.3	16.7	8.33	4.17
Cyclic prefix	4.7	2.3	1.2	0.59	0.29
Minimum scheduling interval (symbols)	14	14	14	14	28
Minimum scheduling interval (ms)	1	0.5	0.25	0.125	0.125

Beyond changing numerology, 3GPP has other mechanisms to reduce latency. Think about Frequency Division Duplexing (FDD) (as opposed to Time Division Duplex) operation and flexible (on per-slot basis) scheduling of Rx/Tx frame split.

Slot/mini-slot structure

A key enabler for low-latency scheduling is the concept of mini-slots, especially when applied in <7 GHz spectrum. A mini-slot is a subset of the (typically 14) symbols within the scheduling interval, typically 2, 4, or 7 symbols. The example in Fig. 11–1 shows a 2-symbol mini-slot.

A (URLLC) mini-slot can be used to interrupt a "normal" slot [e.g., used for Enhanced Mobile Broadband (eMBB) transmission], analogous to frame preemption in (wired) TSN transmission. Preemption indication Downlink Control Information indicators are used to indicate preemption events in the next slot to other User Equipment (UE).

Uplink grant-free transmission

Traditional UL (UE to gNB) transmission in 3GPP systems is grant based. This means that the UE sends a scheduling request to which the gNB responds with a grant allocation—a time/frequency slot in which the UE is allowed to transmit user data. This handshake adds unwanted latency.

Two mechanisms are introduced to remove this latency: type 1 or semi-persistent scheduling (SPS, similar to LTE) where UL grants are regularly scheduled in an Radio Resource Control (re)configured pattern as well as type 2 scheduling where additional L1 signaling is used for fast modification of semipersistently allocated resources. This additional L1 signaling (de)activates the grant for a specific UE.

High reliability

The first step to achieve high reliability is to get a good link quality, as defined by signal-to-interference and noise ratio (SINR). Increasing the UE SINR is done through known radio techniques such as increased RF power, antenna diversity, adding redundancy, and so on.

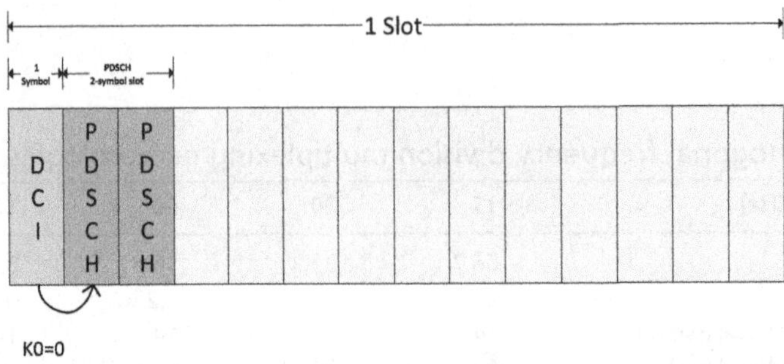

FIGURE 11–1 Mini-slot configuration.

- Antenna diversity at UE. Having multiple receive/transmit antennas in combination with rank 1 (single user, single stream) transmission increases the diversity order and SINR at the UE under fading channel conditions.
- High reliability PDCCH/PUCCH. Increased control channel signaling for URLLC is achieved through various methods. Transmission resources allocated to the PDCCH can be increased: lower the code rate, decrease the modulation order, or send redundant copies (repetition of scheduling information). Similarly, Hybrid Automatic Repeat Request (HARQ) ACK/NACK signaling can be enhanced—specifically NACK signaling where erroneous detection of a NACK can lead to degradation in spectral efficiency.
- HARQ enhancement. HARQ retransmission adds end-to-end latency, not only due to packet retransmission but also due to the signaling latency. The combined signaling and retransmission overhead, for example, in LTE is 8 slots. Given the relatively low retransmission rate, this may not impact the average packet latency by much, but it impacts the worst case latency and therefore system jitter. One potential mitigation is proactive retransmission which can be referred to as K retransmission. In this scheme, the UE gets resources for K transmissions. In case no ACK is received, the UE automatically retransmits the packet.

Interference management

By being a managed network, client-side SINR can be managed to techniques including Cooperative Multipoint (CoMP), frequency selective scheduling, and other techniques to reduce interference and increase received power levels. Many of such techniques can be implemented by a custom scheduling/self-organizing network algorithm that is optimized for industrial/local applications.

Deployment options including edge compute and software-defined networking

Industrial automation requires the use of private networks that are local to the premises (Fig. 11–2). Driving requirements for this include the following:

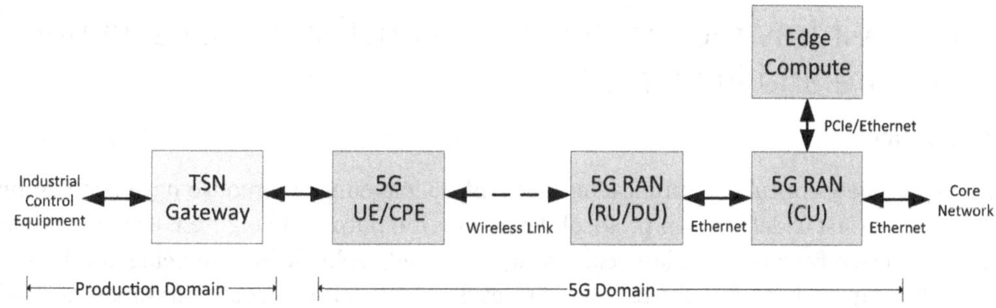

FIGURE 11–2 Production domain and 5G domain.

- Data locality. Production data is not allowed to go out of the industrial premises for data protection reasons. This removes the option of (remote) cloud processing and reliance on public networks.
- Latency. Public RAN latency is an order of magnitude higher than private network latency if public networks are optimized toward eMBB deployment and centralized compute. Even in an optimal (lightly loaded) environment, public network RAN latency alone is order of magnitude (10 ms).
- RAN performance. When relying on a public network, it is difficult to accommodate for guaranteed operational Quality of Service in every location. Operator owned small cells could be used to improve RAN performance but are intrusive on the IT infrastructure in the industrial environment.

Cost and (small) scale targets of industrial deployment are achieved by implementing SDN for control plane and core network functions. This removes the need to rely on high-end networking equipment that is designed for use by the mobile network operators.

The 5G RAN itself can be physically implemented/partitioned in different ways. For example:

- Integrated small cells where RF, Physical Layer, and 3GPP Medium Access Control (MAC)/Radio Link Control (RLC) layers are colocated into a single (small-cell) unit, connecting over sub-1 Gbps links to the centralized unit that takes care of 3GPP protocol termination, cell site routing, and connectivity to edge compute/core network.
- Distributed small cells where RF and lower Physical Layer are separated from the upper Physical layer and MAC/RLC functionality. In this case, connectivity between the two is provided by higher capacity Ethernet, typically dedicated 10/25GbE.

There are tradeoffs in cost, power, programmability/flexibility, and other factors between these two implementation options.

Specific URLLC protocol level features can be more easily implemented in an integrated small-cell environment (smaller slot times, low HARQ turnaround times) or in a distributed small cell (CoMP, interference management). The choice between implementation options will be use case dependent.

Vehicle-to-infrastructure (vehicle-to-anything) roadside unit architecture and implementation

Introduction

Even though safety-critical communication mechanisms targeting automotive have been around for many years (first widely known protocol defined for the purpose being IEEE 802.11p1[1]) interest has grown significantly in the last years. New, dedicated, technologies are being developed to give cars the capability to interact directly with each other and with road infrastructure. These interactions are known as of Cooperative Intelligent Transport Systems (C-ITS).

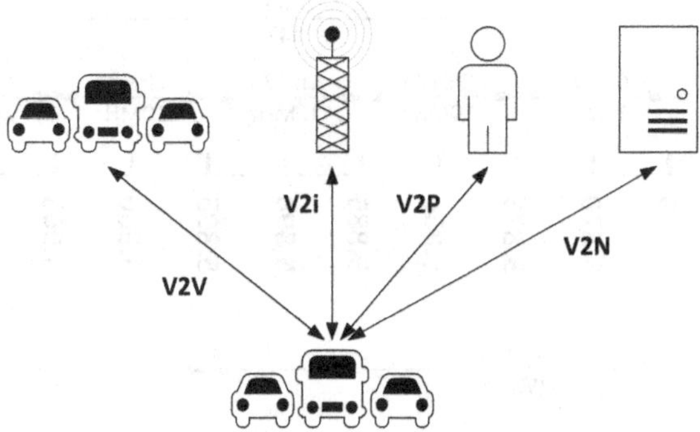

FIGURE 11–3 V2Anything communication.

C-ITS allows road users to share information that helps to make transportation safer and more efficient. However, to make C-ITS work, roadside infrastructure and vehicle communication need to be both standardized and more broadly implemented.

Known as vehicle-to-everything (Fig. 11–3), C-ITS covers multiple communication paths: vehicle-to-infrastructure (V2I), vehicle-to-vehicle (V2V), vehicle-to-network (V2N), and vehicle-to-pedestrian (V2P).

C-ITS supports two main application areas: safety-critical applications and additional services. This imposes challenges on appropriate prioritization (safety critical above anything else) and dynamic behavior with high relative speeds (vehicular) between transmitter and receiver(s), for both rural and urban scenarios, considering latency restrictions for safety-related applications.

Spectral aspects

Most global spectrum dedicated for ITS/vehicle-to-anything (V2X) communications is the 5.9 GHz band (5.850–5.925 GHz), even though there is little harmonization (but a lot of active discussion) of channel allocations within this band (Fig. 11–4).

One notable exemption is Japan which allocates 755.5–764.5 and 5770–5850 MHz spectrum to ITS.

RF requirements such as transmit power can be different across regions/standards. 3GPP stipulates an allowed V2X UE maximum output power of ~23 dBm with mandatory receiver diversity. However, regional differences can exist—for example, Korean deployments with 100 mW maximum transmit power or 2 W Effective Isotropic Radiated Power (EIRP).

Vehicle-to-anything standards and deployment timeline

Today, two main approaches for V2X communication are competing against each other: 802.11p (IEEE-based standard) and C-V2X (3GPP-based standard[2]). Both standards operate in the 5.9 GHz band described earlier. Within C-V2X, there is a 4G/LTE and a 5G version of the standard. Given relative slow adoption of technology, a 5G version for V2X communication is

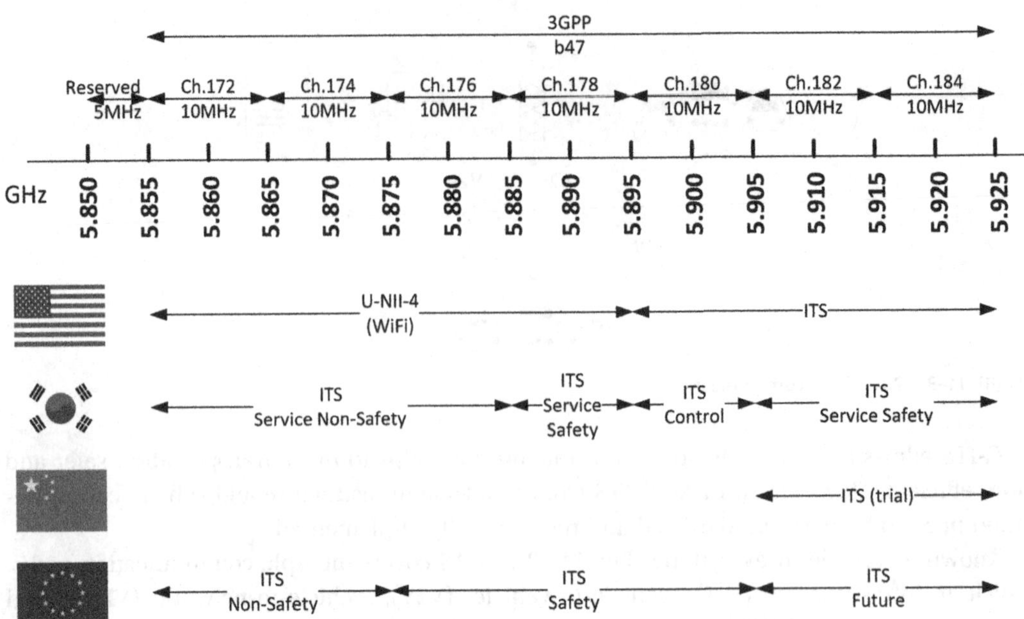

FIGURE 11–4 V2X spectrum allocation. *V2X*, vehicle-to-anything.

considered the most likely candidate of the two. This chapter summarizes 802.11p and C-V2X technologies before explaining the technology behind 5G sidelink communication.

802.11p

The 802.11p standard is related to 802.11a. It defines an orthogonal frequency division multiplexing waveform with 64 subcarriers with convolutional coding and a Carrier Sense Multiple Access with Collision Avoidance channel access method. Note that there are stricter spectrum mask and adjacent and nonadjacent channel rejection conditions.

4G C-V2X (PC5)

LTE C-V2X was introduced in 3GPP Rel-14 and Rel-15 (2017 and 2018) as an extension of 3GPP device-to-device Rel-12. It defines a "sidelink" direct communication between devices (e.g., V2V), also known as the PC5 interface. At the same time, the previously known device-to-infrastructure link (Uu) was adapted to support V2I communication. C-V2X refers to both the LTE C-V2X sidelink and the LTE C-V2X Uu interface. These different communication links (V2V vs V2I) are supported through C-V2X "modes." Mode 1 and mode 2 are standardized from Rel-12. Both are designed to support public safety application requirements, but unlike C-ITS applications, reliability and low latency are not required for such applications. Therefore the Rel-14 introduced modes 3 and 4 are most relevant for LTE C-V2X applications. Both modes 3 and 4 support direct vehicular communications but differ on how resources are allocated. In mode 3, vehicles are within the coverage of cellular network, resources are selected, allocated, and reserved by the base station (eNodeB). Mode 4 was

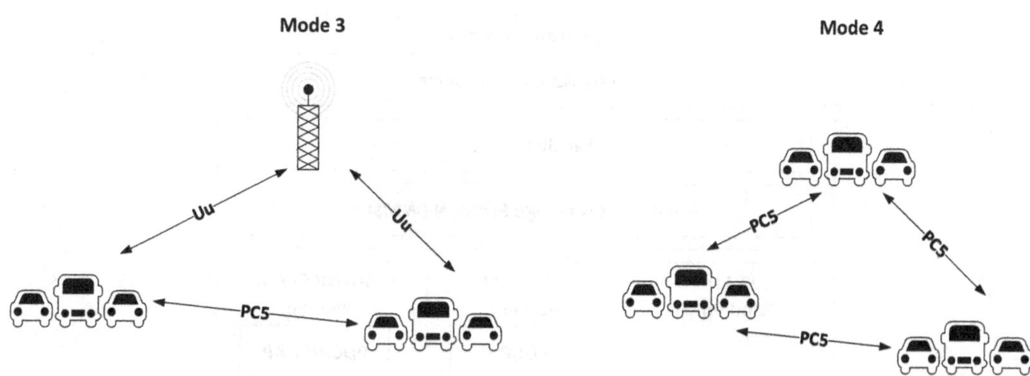

FIGURE 11–5 LTE C-V2X modes.

designed to work when vehicles are not under the coverage of cellular network. Resources in mode 4 are autonomously selected by vehicles. As it is considered as an ad hoc system, it is challenging to avoid collisions when selecting resources. Distributed scheduling algorithms and Distributed Congestion Control (DCC) algorithms are developed for this purpose.

Mode 4 is required for V2X since C-ITS safety-related applications must be established even outside network coverage. C-V2X is a synchronous communication method, which means that the system needs to get time aligned, either through Global Positioning System connectivity or by synchronizing to other vehicles/infrastructure (Fig. 11–5).

NR C-V2X (PC5)

3GPP's 5G standards have been finalized in Release 15 (Phase 1, September 2018) and Release 16 (Phase 2, Q4 2019). It is to be expected (given the deployment timeline as discussed below) that a 5G version of C-V2X will be more widely deployed than a 4G version given that the 4G version has limited legacy.

NR C-V2X is based on similar design principles as LTE C-V2X but supports flexible numerology and forward error correction using Polar and Low Density Parallel Codes (LDPC) as defined in the 5G standard.

Standards summary

The following table summarizes the different wireless protocol stacks:

	802.11p	**LTE C-V2X**	**5G C-V2X**
Synchronization	Asynchronous	Synchronous	Synchronous
Multiplexing	TDM	TDM + FDM	TDM + FDM
Channel coding	Convolutional	Turbo	LDPC/Polar
Retransmission	–	HARQ	HARQ
Waveform	OFDM	SC-FDM	OFDM
Scheduling	CSMA-CA	SPS + relative energy based	SPS
Channel width	10 MHz	10 or 20 MHz	Flexible

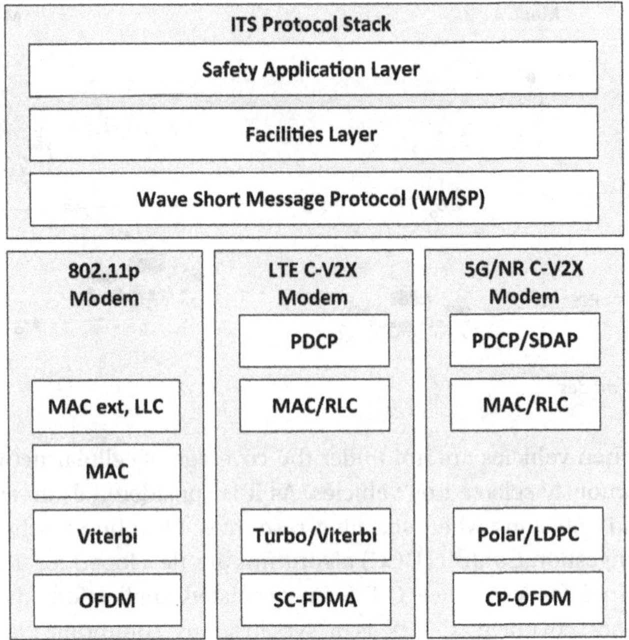

FIGURE 11–6 V2X protocol stacks. *V2X*, vehicle-to-anything.

The different wireless protocol stacks all communicate to the ITS protocol stack through the common Wave Short Message Protocol (WMSG) layer as is shown in Fig. 11–6.

Deployment timeline

Car makers are starting to introduce V2V communication in their vehicles. Toyota announced plans to begin broad deployment IEEE 802.11p vehicles from 2021. Cadillac (General Motors Group) is using IEEE 802.11p. Volkswagen enables V2X through IEEE 802.11p since 2019.

Ford announced plans to enable C-V2X in partnership with Qualcomm in October 2017, but this is not commercially deployed.

Overall, given the limited number of equipped vehicles, lack of harmonization, and limited roadside infrastructure, there is currently no single "winning" standard for V2X communication. Note that C-V2X and 802.11p can coexist if they are on adjacent/near-adjacent channels, allowing dual-mode roadside unit (RSU) to be built.

Roadside unit

The purpose of the RSU is to provide the infrastructure side of the V2I communication link. This includes both a communication function (V2I/V2X communications) and the intelligence/decision-making function to interpret received messages and generate the appropriate messages to be transmitted.

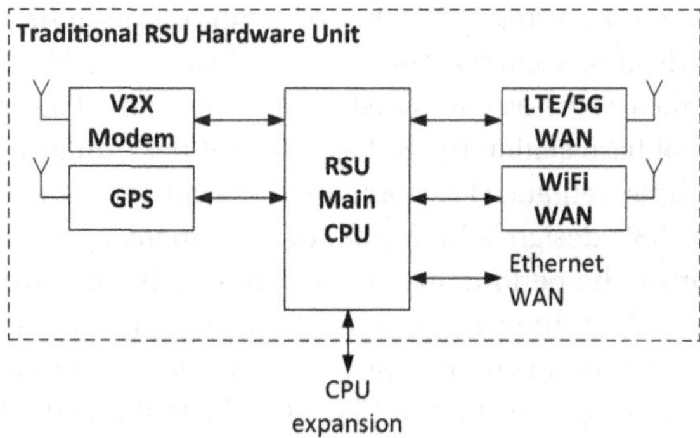

FIGURE 11–7 Roadside unit overview.

System architecture

A typical implementation of a traditional RSU hardware is shown in Fig. 11–7.

In this unit, the V2X modem itself is typically a modem, intended for vehicular use, that is repurposed for the infrastructure application. The main Central Processing Unit (CPU) provides an interface to the V2X modem as well as CPU horsepower to translate V2X messages to wide area network interfaces [Ethernet, LTE/5G wireless, or (less likely) Wi-Fi wireless] as well as to run application stacks needed to interpret and (more importantly) transmit safety-critical messages—the V2X application layer.

Challenges with this approach include (1) the V2X communication path and (2) scalability:

- V2X communication path challenge. As we explained earlier, typical RSU implementations reuse a vehicular V2X modem for V2I communication. Challenges with this implementation include the following:
  - Standards compliance. Whereas a client module is built with a specific use case, market, and lifetime in mind, an RSU may have to adhere to a wider range of standards (say, 802.11p and C-V2X both) as well as future standards enhancements that come with a lifetime of 10 + years in the field. These requirements can be readily supported with a more flexible (e.g., software defined and O-RAN based) implementation as opposed to a hardwired modem implementation.
  - RF performance. The V2X modem is implemented with an RF chain that is optimized for client (UE) use. Besides cost optimization targets that typically imply performance compromises, this also limits fundamental aspects such as antenna gain [antenna path count, multiple-input, multiple-output (MIMO)], RF output power (UE category is limited to 20–23 dBm), and so on. These limitations do not have to apply to an RSU that falls in the base station RF category.
  - Industrialization. Consumer- and industrial-grade components comply to different standards with regards to temperature range, operational lifetime, sensitivity to failure conditions, and so on.

  - Modem performance. In terms of number of communication messages/second, RF channels, RF bandwidth, sectorization topics, and so on, an RSU has widely different performance characteristics as compared to client systems. This is the same reason as why a conventional base station (gNB) has different performance requirements (and thus a different implementation) as compared to a conventional UE (mobile phone).
- Scalability challenge. RSU design is in its infancy, with many options for enhancements in performance to improve the performance/capabilities of the RSU with regards to traffic detection/recognition, algorithmic support, coordination, and so on. For example:
  - Inclusion of additional inputs to the RSU. Consider inclusion of RADAR/LIDAR; camera; wireless sensing (detection of LTE, 5G, Bluetooth, and other pedestrian-associated RF radiation); ultrasonic; and other sensors intended to enhance the scope of the RSU. This creates a scalability challenge where the main CPU of the RSU will end up becoming a bottleneck, both in terms of connectivity associated with all these sensors, as well as compute requirements to process all input data appropriately.
  - Associated with above is an architectural challenge in how to physically implement the RSU: should the RSU still be a single hardware unit that becomes more bulky/heavy as features evolve and is compromised in where it can be placed, or become distributed where each "sensor" can be placed in the physical location for it to perform optimally?

5G Reduced Capabilities (RedCap)

The 5G standard is designed with flexibility in mind, both in terms of hardware/software partitioning options for implementation as well as flexibility in use cases. The first standardized (Release 15) and well-published use cases, however, center around high-end applications including eMBB and URLLC. Both of these use cases are demanding on the hardware and software implementation complexity and therefore tend to be expensive.

As 3GPP releases evolve however, more use cases are being added, including V2X communication (sidelink communication channels) and, in Release 17, standardization of lower end use cases that allow for reduced hardware complexity implementations of the (client-side) modem itself. The reduced feature set is documented in 3GPP Study Item 38.8752[2] which defines a number of areas of complexity reduction targeting a simpler and thus lower cost modem implementation under the "Reduced Capabilities" (RedCap) name but also known as 5G-light, 5G-lite and similar names.

RedCap is important in embedded applications that are characterized by a requirement for long in-field deployment (and therefore do not want to rely on "outdated" standards such as LTE or WCDMA/GSM) and limited throughput/latency performance requirements. Consider various sensors, wireless cameras, and similar applications as called out in 38.875 (Fig. 11–8).

3GPP defines a number of simplifications to the UE deployment profile that allow for reduced cost hardware/software implementations that are documented as UE reduction

IoT Market Segmentation

High End, Low Speed: NB-IoT Picture, Audio, HF Monitoring Battery operated, Energy Harvesting	**High End, High Speed: LTE-M, 5G RedCap** Video, Voice, Realtime data. External power source, Rechargeable batteries
Low-cost, Low Speed (LPWAN): SigFox, LoRaWan, .. Monitoring, tracking Battery operated/Energy Harvesting	**Low Cost, High Speed:** Holy grail

FIGURE 11–8 IoT market segmentation. *IoT*, Internet of Things.

features with estimated hardware complexity (= cost) savings. We cover these at a high level here.

Reduced number of Rx/Tx antennas

The reduction is to limit the amount of transmit antennas to 1x Tx, both for FR1 and FR2 operation, and reduce the number of receive antennas to 1x Rx or 2x Rx. The number of MIMO branches is reduced correspondingly, leading to reduced throughput requirements. The associated cost reduction is established both in baseband as well as RF front-end complexity, where each Rx/Tx path has an associated cost for power amplifier, Low Noise Amplifier (LNA), modulator/demodulator, and associated components.

User Equipment bandwidth reduction

The study suggests a bandwidth reduction to 20 MHz (FR1) and 50 or 100 MHz (FR2) operation for RedCap implementations. The bandwidth reduction is assumed for both data and control channels and allows simplification of the analog chain as well as reduction in compute requirements for the digital modem.

Half-duplex Frequency Division Duplexing operation

Half-duplex operation, even in an FDD deployment scenario, prohibits simultaneous receive and transmit operation of the modem. This allows for the hardware to remove a duplexer in the RF chain (which can now be replaced by an Rx/Tx switch) but also opens up advantages for software-defined modem implementations that can share hardware across Rx and Tx operation.

Relaxed User Equipment processing time

3GPP standard 38.214 defined N1 and N2 values to indicate UE processing time associated with decoding PDSCH symbols (N1) and PDCCH indicators (N2). Extending the decoding time allows for a lower throughput LDPC/polar decoder, for example, when implemented as a soft implementation on ARM cores, GPUs and other more "general purpose" hardware.

Relaxed maximum number of multiple-input, multiple-output layers

RedCap assumes 1 or 2 MIMO layers to be implemented rather than the typical 2–4 layers (reference UE, FR1) or 2 layers (reference UE, FR2). This reduces the RF and baseband complexity both.

Relaxed maximum modulation order

Besides reducing the RF component cost (given that the performance requirements on the RF chain will be more limited), modulation order reduction—together with bandwidth reduction—limits the throughput and therefore the compute requirements associated with forward error correction and related components.

Combinations of abovementioned features

According to the 3GPP study item, combined savings can be up to ~70%, thus dramatically reducing modem cost and increasing competitiveness with other standards such as LTE and 802.11.

In addition to cost and time-to-market advantages, note that the 5G/NR RedCap standard is already being envisioned for feature enhancements in 3GPP R18 standards, including support for positioning, sidelink communication, and unlicensed spectrum operation. Hardware-centric modem implementations require next-generation device development for such features. A software-centric implementation does not and therefore can support a longer field deployment.

References

[1] D. Jiang, et al., IEEE 802.11 p: towards an international standard for wireless access in vehicular environments, IEEE Vehicular Technology Conference (VTC), p. 2036–2040, 2008.

[2] NR, Study on support of reduced capability NR devices, 3GPP TR 38.875, 2020.

Further reading

NR, Study on physical layer enhancements for NR ultra-reliable and low latency case (URLLC), 3GPP TR 38.824.

NR, Architecture enhancements for 5G System (5GS) to support Vehicle-to-Everything (V2X) services, 3GPP TS 23.287, 38.824: 2018; 23.287: 2018.

Index

Note: Page numbers followed by "*f*" and "*t*" refer to figures and tables, respectively.

G

S

Printed in the United States
by Baker & Taylor Publisher Services